Majesty in Monotony:

Everyday Things with a Cosmic Perspective

ISBN: 1539434109
ISBN-13: 9781539434108

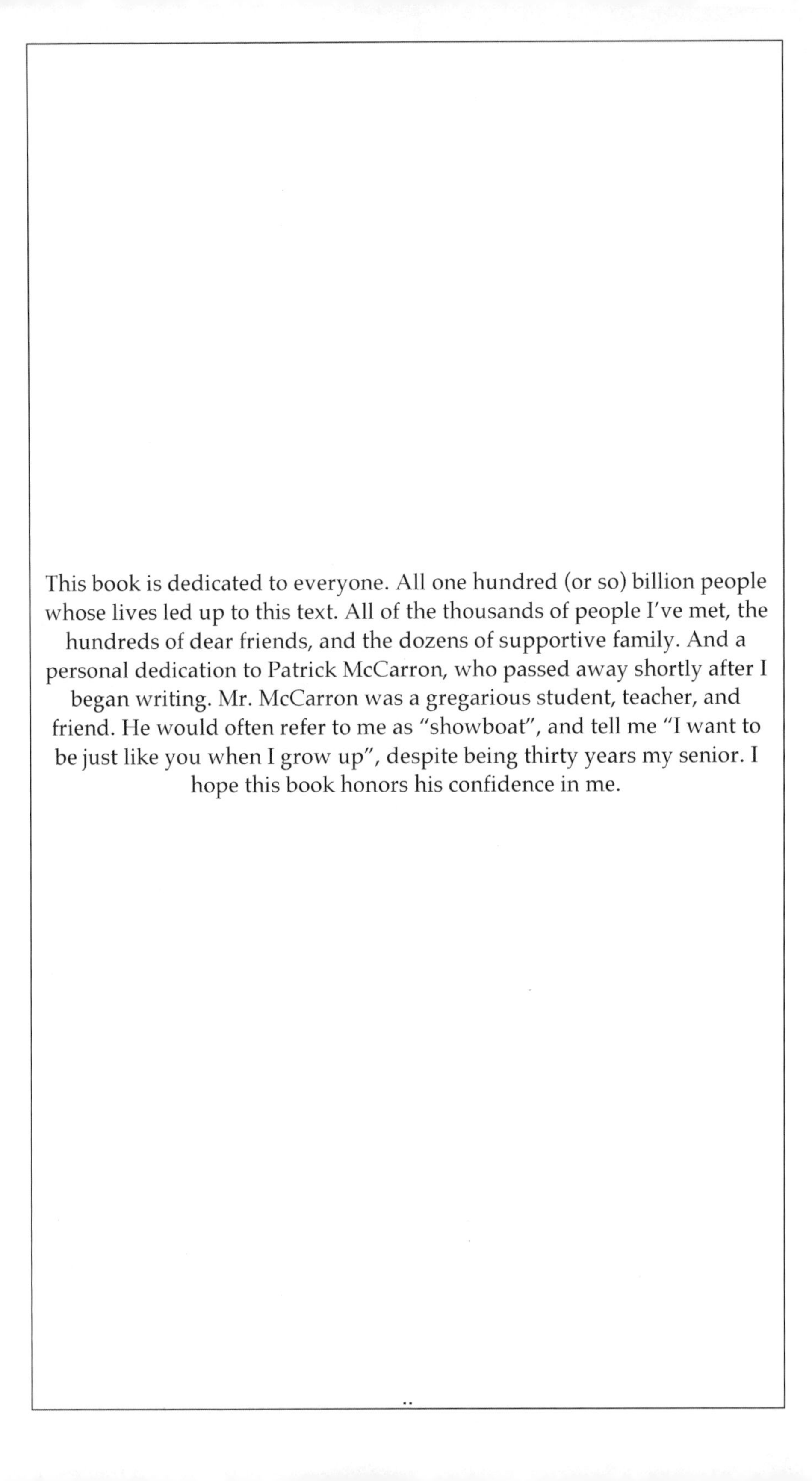

This book is dedicated to everyone. All one hundred (or so) billion people whose lives led up to this text. All of the thousands of people I've met, the hundreds of dear friends, and the dozens of supportive family. And a personal dedication to Patrick McCarron, who passed away shortly after I began writing. Mr. McCarron was a gregarious student, teacher, and friend. He would often refer to me as "showboat", and tell me "I want to be just like you when I grow up", despite being thirty years my senior. I hope this book honors his confidence in me.

Table of Contents

Preface

(Or, "How this book might be useful to you")

Bonjour! Annyeong-haseyo! Aloha! Jambo! Konnichiwa! Zdravstvuyte!

Greetings fellow Earthling, and thank you for taking the time to read this, my first full-length, action-packed, collection of cosmic thoughts and insights, better known as a book. We are going to embark on a journey through the immense eternity of the universe. Not just outer space, but all the complex and awe inspiring things that make it up. All of the cards are on the table, and by the end of it our goal will be to connect everything you can see, think, or know into one woven tapestry. Pretty good bargain for one book, don't you think?

An amazing thing, books. They allow people to communicate not just over vast distances, but also through time, giving people the power to develop ideas and innovations, to create influence and emotion. With a book you can see through the eyes of people who grew up in entirely foreign situations, or whom we may share intimate experiences, despite living centuries or continents away from one another. In many ways, a book is nothing more than a conversation, albeit a very one-sided, personal one. This is unique relative to other forms of writing and media. If read aloud, a single essay might become a strong, robust speech, a thunderous plea to shape the minds of our fellow humans, if even in a small way. A book, however, being so long and full of thought and experience, can't fit this niche as easily. Instead, one may read a book in much the same way that they hold a conversation with someone else. A book can be read anywhere that is sufficiently peaceful enough for thoughts to flow and ideas to sprout. You may prefer reading in the comfort of your home, your local park, your school, library, or wherever you are most comfortable. If you're at war, living in a dangerous territory, or isolated from all humanity, then a book might give you comfort and solace, like having a friend to talk to, who you can listen to and learn from, or argue with in your mind should you disagree. That is the power of not only books, but also your mind. It is a feat so profound that it is not done in any other place in the known universe. Among all the stars and the rocks, the plants and the beasts, the fact that we can have this conversation right now, possibly years or oceans apart, is a testament to the knowledge and intellect of the human race. With that said, I sincerely hope that you find our forthcoming conversation to be useful to you, if at least in some small way.

In my lifetime I have found it true that the best way to understand somebody and for people to talk to each other is to first be acquainted with the experiences they have both had and the lives they have lived. In other words, the most useful and meaningful conversations are held between people who have walked a mile in the other's shoes. It is important for the speaker to justify exactly why they are making different claims and coming to certain conclusions about the way the world works. So allow me to give you just a small bit of insight into myself, in order to better set the stage for our expedition into existence.

Like most people, my entire professional life has been spent balancing between being a student and a teacher. However, I do have more experience speaking in front of and communicating with people than most of my peers. For over a decade, I was a martial arts instructor, studying the Songahm style of Taekwondo, honing my mind and body in an effort to earn the trust and respect of my students, peers, and teachers. My academic history was turbulent early on, mostly due to apathy. However, I have always had a strong passion for knowledge about the natural world and the universe I live in. This passion eventually led me to a formal training in literature, mathematics, history, sciences, and music, not entirely uncommon in the part of the world I happened to be born in. But my education became unique as I entered adolescence, in that my courses had a strong emphasis on teaching a diverse, global mindset in every subject. We were taught to look at the world from every possible point of view, and most importantly to think critically about everything we heard, saw, or experienced. We had to not only embrace, but be able to justify radically different points of view, defending them regardless of whether we agreed or not. We were taught, in every way we live, to think for ourselves as individuals by carefully listening to everybody around us.

When I attended the University of Colorado Boulder, I developed into an astrophysicist and a philosopher. This has a tendency to confuse people, either because they don't know what those things are, or because they do but don't see any connection between the two fields of study. But in fact, they are direct relatives. Some of the oldest known people in history were celebrated as great thinkers and philosophers, and they were driven by an unusual and insatiable thirst for understanding. They were the precursors to the inquisitive minds of today, some of whom spend their days studying the physical laws, nature, and history of the entire universe. This is what has led to the astrophysicist of the twenty-first century, to people who use concepts and ideas developed by previous generations of researchers, dating back to the very first human, to understand what are ultimately the most profound questions ever posed by our species. People, including astrophysicists, who devote their lives to understanding the universe and our place in it may even be recognized as philosophers, as the pinnacle degree given to students of the world is to this day known as a doctorate of philosophy, or a PhD. Following graduation, I spent a summer at Harvard assisting with research on the origins of life, specifically the transition from chemistry to biology. This work was done with great influence from Nobel Prize laureate Jack Szostak, although we didn't interact personally.

People often ask me why I decided to study a topic that carries such gravitas, and has so much math. The answer that I give to save time is usually that "It just felt right", and indeed there are few things that I have found as gratifying as learning how the world works, and the fact that we can even remotely comprehend such ideas in the first place! However, I can recall the more concrete experience that gave me a passion for learning about outer space. Inside the Denver Museum of Nature and Science, within the Space Odyssey exhibit, lies the Gates Planetarium. The most astonishing thing I had ever witnessed up until that point was a movie that was shown there regularly, *Black Holes: The Other Side of Infinity*. For the first time, I was taken off of the Earth, flung thousands of light-years to the edge of the Milky Way. Seeing the immense eternity of stars and emptiness, when previously my entire world was nothing more than a state just a few hundred miles wide, was absolutely breathtaking. What's more, I came to realize that everything I had ever known, anything that anybody anywhere had ever known in all of recorded history, by my friends and family, and all of life on Earth, occurred in a piece of the universe so miniscule that it couldn't even be seen within this ocean of oblivion. Yet in no way did that eclipse the power and importance of all those things. Rather, seeing the full scope of all that there was made me realize that there is so much more to learn, all equally significant and influential to my being alive at all. Although I couldn't articulate it at the time, I knew with instinct and vigor that I must understand the world I lived in at all costs.

In order to make money during college, I had the opportunity to work at the planetarium we had on our own campus, beginning my very first year. It was quite serendipitous as well, as that same year they had conducted a new upgrade, with a digital system outclassing even the most modern IMAX theatres, as well as the ability to manually fly in real time through the universe. There my coworkers and I served as ambassadors to knowledge and curiosity, constantly meeting school groups, families, and businesses, all of whom shared the same thirst to learn that has plagued the likes of Plato and Einstein. I gave live presentations to groups large and small about things I had been studying for years, as well as new knowledge fresh from my classrooms and textbooks. If I wasn't giving a talk, I was serving as the navigator for one of my colleagues, flying at their command and taking audiences from low Earth Orbit to the edge of the observable universe. I have given talks about everything from planets to stars, quarks to quasars. I even used the planetarium to give a special talk to earn class credit, entitled *String Theory for Everybody: Finding the Fundamental Structure of Everything.* This was my final talk in college, and the final project I needed to complete in order to earn my undergraduate degree. I even had an official advisor, the very same person who acted as the science consultant for the black holes movie that inspired me in the first place.

My work wasn't just limited to education. Ironically, perhaps our greatest source of income came from doing music performances, using lasers and visual effects in sync with popular artist tracks. Although they were purely for entertainment, those music shows were some of the most crucial learning tools to help attain me a firm intuition for exactly where everything is in the universe. This is because in modern planetariums, operators have real time control to go anywhere throughout time and space. I have flown through the Orion nebula, dove through the Virgo Supercluster, zipped through DNA, soared through the rings of Saturn, and crashed into the Sun more times than I'd like to admit. All before lunch no less! Synchronize those visuals with lasers and pulse pounding music from the likes of Queen, Beyoncé, or Starset, and you have an experience that I would deem unrivaled in all human history, especially if they're live. Thanks to that, I've been to the very edge of the universe and back more times in a single hour than most people have been to a foreign country in their entire lifetime.

I have met thousands of different people throughout my life, from all around the world, and from all different walks of life. I have done this not just in person, but through the voices they left behind in their own writing, through cinematic depictions of their lives, or accounts from their contemporaries. Like all people of my generation, I have been fortunate to have access to the wide network of connections people who lived before me laid in place. I have learned about the ways in which people live, their ties to the environment and the creatures they share it with, and eventually peered into the fundamental pieces of the cosmos and understood how people came to learn about them. And what I've realized is that no matter where people come from, no matter what they've seen or done, the vast majority of them are simply trying to survive, and make sense of this strange and chaotic world we are born into, growing from a blank canvas, raised by the knowledge of their parents and surroundings, hoping to sustain themselves and pass on their knowledge to the next generation, all to just have some sort of influence. While there are plenty individuals who are boring, brash, and naïve, of all these people, I have found not one that I couldn't sit down and have lunch with.

But our connections go far deeper than just our own species. By far the most powerful impact of studying the universe is the context it gives. The school that I attended often likes to boast that its students are taught forty orders of magnitude. That is, in order to become an astrophysicist, one must understand the behavior of all things that comprise it, from atoms to galaxy clusters, and everything in between. In terms of actual sizes, that's one million billionth the height of the average person, up to the 46 billion light years we can see from Earth. That perspective has taught me that all the things you learn about in a classroom, all the stories you hear, the sensations you feel, the very thoughts you have, are all connected in some way or another. There is not a single thing that you can think, touch, or feel that isn't in some way attached to everything else in the cosmos. For we are comprised of material with a story as old as the universe itself, and by the very virtue that you are thinking and feeling at all is a testament to that magnificent fact. That idea will be the very topic of our conversation throughout this book.

When we were trained to read and write, our teachers would often have us consider not just what we were reading, but more importantly how we should go about reading different works. This can be a rigorous mental exercise, taking into account the various historical, cultural, and personal contexts and biases associated with each piece of work. Since this isn't a school project, and because we are going to be covering topics nearly fourteen billion years in development, I'm going to save you the trouble, and bestow some helpful concepts to keep in mind while reading this text.

The ideal reader of this text can be summarized with one characteristic, one that will make everything we discuss all the more remarkable. You must understand the concept of mortality. Not that you should obsess or in any way enjoy the idea, but that you are aware of your mortality. Even more than taxes, death is inevitable. We all face our time eventually, and everything that was once alive will eventually cease to be so. The majority of things that have been alive have gone extinct, but inanimate objects decay as well. Beyond the few decades we will live, some animals will return to the Earth after living for centuries, a planet which will be engulfed by the Sun in some five billion years, leaving behind its charred remains which will cool off and fade as the universe expands so much that all existence will become dark and empty. Regardless of what beliefs you have about the afterlife, nobody can deny the eventual decay of our physical bodies. But it is the very inevitability, I would argue, that gives meaning to life at all. The world is a vast and rich place, full of joy and sorrow, love and loss. It is so remarkably hostile in so much of the environment that life existing at all is astounding. You are alive now, and it's your job to make the most of what limited time you have. So do not fear the reaper. If you haven't made peace with this, then you might not yet be of adequate maturity to appreciate all the amazing things we are going to discuss on our journey.

As a heads up, if you happen to be a part of the 5% of people on the planet who don't use the metric system, I apologize. Partly because you're uncool, behind the times, and probably live in a stubborn country where everyone is afraid of change, and partly because I won't be using the same units you use on a daily basis. Because I'm a scientist, and metric units make more sense, and because 95% of people use them, it just makes sense democratically. So let's just rip this bandage off now shall we? A kilometer is about three fifths of a mile, a kilogram is just over two pounds, a joule is the amount of energy it takes to lift a quarter pounder one meter, and if you do that in one second, you've used one watt of power. Okay, let's move on.

The motif of this book is that everything in existence that you can think of, all the people you know, all of the things you've seen, everything within the borders of each living being within each country on this planet, in this cosmic disk floating on space and time, is connected. And because there are so many topics available, I have attempted to organize the text in such a way that one may cherry pick topics they find most interesting, and omit other sections. But of course you will benefit the most if you read this book in its entirety. We are going to be grappling with many profound ideas, and I will do my best to present multiple perspectives on each one. Also, as a philosopher and a scientist, I encourage you to take nothing I say for granted and to keep a healthy skepticism. If you disagree with me, that's fantastic, so long as you can validate your ideas and formulate consistent arguments against mine based on observation. Our goal is for us to reach a better overall understanding of how the world works, and knowing whether something is right or wrong can only help us attain further truth. We must be able to hold discussions with an open mind. With that said, I think it's time we got down to brass tax, and finally got this party started. We will begin with a brief tour of all existence.

Part I: What On Earth?

1 | Genesis of Today

"The nitrogen in our DNA, the calcium in our teeth, the iron in our blood, the carbon in our apple pies were made in the interiors of collapsing stars. We are made of star stuff."

— Carl Sagan, The Great Cosmic Communicator

For as long as there have been sentient collections atoms, made of congealed energy in the form of complex chemicals flowing through water, including but perhaps not limited to humans, they have asked the eternal question: what the hell is going on here? Why is the sky blue? Why am I hungry? Why did a lion eat one of my siblings and why did one of my parents fall asleep forever after getting cut by a tiny rock? Why is there evil? Why do stars shine and thunderstorms roar? Why do we love? Where did we come from? Who put bop in the bop, shoo bop, shoo bop? Is the cake a lie? Do we even exist, and if so how can we be sure? And of course, what is the meaning of life, and how is it possible that we are able to ask these questions at all?

Human beings are astounding creatures, not just for our curiosity, but our ability to communicate our ideas. Through collaboration, humans have been able to reshape the planet, and understand the inner workings of the world in a way unprecedented in the whole history of the universe. However, the trials and tribulations of everyday life make it all too easy to forget our humble beginnings. It's not at all uncommon, and perhaps reasonable at times, for people to think that the lives they live are absolutely normal, and that the world they see now has always been that way. Indeed our capacity for simplification is both a testament to our cognition, as well as our Achilles Heel.

But the fact of the matter is, everything you see before you has a history that spans far beyond even your oldest known ancestor. Those questions I just posed have been contemplated by dozens, thousands, perhaps even billions of individuals just like you and me, born into a confusing and chaotic world, with only the knowledge of their predecessors to light the way. Most organisms have a hard enough time just finding food to eat, a mate to pass on their genes with, and generally trying to not die, to try and find answers to these questions. But for those that had the good fortune to find peace and nourishment, if even briefly, it was an opportunity for them to develop knowledge that in turn allowed other people to build upon, eventually leading to today. We must understand the past in order to look clearly to the future, and that means not just understanding what happened, but why we even think those things are true. The story unveiled by our curiosity has shown that there is majesty to be found in every infinitesimally small corner of an infinitesimally vast space, and we find ourselves to be smack dab in the middle of that history, with the fragmented pages constantly being translated. And like any story, it is best told from the beginning.

Born from Infinity

In the beginning, there was nothing. Possibly. The first ten million, trillion, trillion, trillionth of a second is fogged in mystery. Then, there was everything.

All force and energy and all of time and space were infinitely close, making the universe infinitely hot and infinitely dense. Then, the universe began its lifelong process of expansion, and thus, cooling. Space expanded in four dimensions; up and down, left and right, forward and backward, past and future. One undecillionth of a second later, the universe grew at a rate that would prove unparalleled for the rest of its history, inflating by a million, trillion, trillion times in size within a trillion, trillion, trillionth of a second. Then, suddenly, the force driving the inflation slowed down, and the expansion followed suit. In every place there was light, with bosons, quarks, and leptons. Energy began to come together to form the most fundamental constituents of matter. In addition, every single one of these particles was formed with a twin, nearly identical to the other with the exception of having opposite charges. Just as they were produced by condensed energy, these particle pairs annihilated back into pure energy upon contact with one another.

When the universe was a thousand octillion times older, just a microsecond after expansion began, these particles also came together, with two of the six quarks grouping up in threes, forming positively charged protons with negatively charged antiprotons, as well as neutrons with antineutrons, both without charge. Energy also condensed separately to form six different leptons, most notably negatively charged electrons with positively charged positrons. However, there was a slight imbalance, an anisotropy between the particles and antiparticles, with there being one extra normal particle for every hundred million pairs that popped into existence. As a result, when the pairs annihilated with each other back into pure energy, all that was left were the few normal particles, drowned in vast amounts of light. About one minute after expansion began, protons and neutrons began to collide and fuse, forming deuterium. For around one thousand seconds, hydrogen and deuterium synthesized to form helium (with two protons), and trace amounts of lithium (with three protons) and beryllium (with four protons). At the end of this nucleosynthesis, the universe was left with three quarters hydrogen and slightly less than one-quarter helium, with all the other visible matter being lithium and beryllium, all these elements having various numbers of neutrons.

Following the first eventful minutes of the universe's beginning, everything would become relatively uneventful. At this point, the universe had a temperature of over a billion degrees. In a cooler setting, the negatively charged electrons would want to get closer to the bundles of positively charged protons fused together with neutrons. But in such an extremely energetic environment, electrons were moving far too quickly to stably surround these bundles. Not only that, but with everything in the universe being so cramped, light was constantly being absorbed and spat out by particles, making everything look like a thick hot soup. Luckily, space was constantly expanding, giving all these new particles room to breathe. This was a slow process, but eventually, when the universe was three hundred thousand years old and fifty billion times its age at the end of nucleosynthesis, everything was finally cool enough for the first full atoms to form. Moreover, with electrons comfortably surrounding atomic nuclei in equal amount to protons, light was free to travel throughout the universe for the first time, creating a background radiation that formed the outer edge of the visible cosmos at all locations. Most of this light though was only remnant from the early universe; inert atoms alone release almost no light. Thus ushered in a period of cosmic darkness. Fortunately again, there is a force of nature with reach over all space and time that ties every particle with mass together, gravity.

Since all of the matter and energy in space was almost perfectly uniform, it would normally be impossible for any structure to form in the early universe, as all the newly formed atoms would feel equal force from all the other particles in the universe. Thankfully, these miniscule particles have a strange property of never being in exactly one position, and so when the first pieces of matter were forming, small clumps were able to form in the universe that weren't able to keep up with expansion. Those tiny clumps of particles grew to form larger ones, giving the universe ever so slight ripples where there was more matter than average. These ripples became the seeds of larger structures that would begin to form as atoms came closer together. This process was the slowest yet seen in the universe, with thick clouds of inert gas collapsing together over hundreds of thousands of years. As these enumerable number of clouds collapsed, they began to heat as the atoms inside bounced off of each other more, and gravity's pull accelerated them faster and faster towards the cosmic seeds that broke uniformity. Eventually, the heat at the centers of these clouds grew to such a degree that the atoms again lost their electrons. Soon these free nuclei again zipped around in every direction, and despite being repelled from one another, over time were pulled by gravity, closer and closer to colliding. Hydrogen especially was the first to ionize in this way, having only one electron to hold on to. Then, when the cores of those clouds reached tens of millions of degrees, something amazing happened. Since these particles don't exist in a single position, but are instead distributed unevenly over space, eventually two protons got close enough that they jumped immediately next to one another. For the first time in hundreds of thousands of years, atoms fused together once again. Although it was not as efficient as the beginning of the universe, eventually this process created new helium, and enormous amounts of energy. Suddenly the clouds of gas ignite, and raw energy was shot out in every direction from their core, blowing away their outermost layers, leaving a glowing sphere of hot dense plasma. The first stars are born.

Throughout the cosmos, beacons of light begin to shine, pumping out high-energy light that stripped away electrons from nearby gas, filling pockets of space with ions of proton-neutron bundles and electrons once again. Most stars are born close together, and may even orbit around one another throughout their lives in pairs of two or more. These first stars come in a whole spectrum of sizes, most being small and red, with less hydrogen to fuel them. However, this small size comes with benefits. These tiny balls of plasma are small enough that they are able to use all of their fuel, with enough hydrogen to keep them shining for trillions of years, and when their fuel runs out they will quietly fade back into darkness. This is not the case for their larger cousins.

Some of them, being around ten times larger than the smallest stars, will be able to shine a much brighter yellow color. The cost of this though is a severe shortening of their lifetime, since they both burn brighter and are too large to access all of their fuel. But, their large mass allows for backup fuel to be accessed when their stores of hydrogen are used up. After several billions of years, these stars accumulate helium in their core, which is too difficult to fuse under the same conditions the star was in. The stars collapse, but as they do hydrogen surrounding the core actually begins fusing itself, saving the star from collapsing due to its own gravity. But the core continues to contract, raising the temperature to a hundred million degrees. Two brand new elements were born within these medium stars: carbon (with 6 protons), and small amounts of oxygen (with eight protons). Now, with both the core and a surrounding shell producing energy, these medium size stars grow anywhere from ten to a hundred times in size. Despite all the extra energy, the surface of these stars grow so much that they actually cool, turning the stars red. Eventually so much energy is released that gravity is no longer able to hold the surface plasma, and layer-by-layer, these stars will eject their bodies into space until their white-hot cores are laid bare. Just like the small red dwarfs, these white dwarfs will spend the remainder of their lives cooling and fading, keeping the carbon they fused locked away. The layers they ejected, though, can come together to form new stars.

Although most stars are born small, and some slightly larger, there are a select few that are able to accrue much larger amounts of material. The largest of these giants can outshine the dimmest red stars a billion times over with ferocious blue light. They are born over a hundred times larger and more massive than the smallest ones. But, as with all stars, the brighter they burn the shorter they live. Just as with the medium sized stars, as these giants run out of one kind of fuel, their core contracts, a fusion shell is formed, and the core begins to use heavier fuel, and each consecutive step lasts for shorter amounts of time.

In the largest stars, a vast number of new elements can be forged. For only a hundred million years hydrogen fuses to helium. In the next million years helium forms into carbon and oxygen. The following thousand years will be spent fusing carbon to neon, magnesium, and sodium. Next comes neon, colliding to form more oxygen and magnesium for just one year. Then, oxygen fuses and creates silicon, sulfur, and phosphorus. This step will last just a few months. Finally, after just hundreds of million of years and several layers, just a millionth the lifetime of a small star, silicon fuses to form calcium, titanium, chromium, copper, and most importantly, iron. At this point, the star has inflated tremendously, becoming so titanic that by comparison all other stars appear as nothing but specs. But this mighty scale won't last. The moment iron forms in the core; the star's death is just one day away.

Iron is the most stable element. Up until this point, elements that fuse relinquish tiny amounts of their mass and release it as pure energy, thus allowing stars to avoid gravitational collapse. If the nuclei were any larger, they could release energy by splitting apart, and possibly creating iron. For this reason, no star can sustain itself by fusing iron. The fusion of iron requires energy, vital for the star to avoid collapse. With the weight of trillions upon trillions of particles bearing down on the core, and nothing to hold them up, they collapse, instantly. The core shrinks to one ten-thousandth its original size as electrons are pushed into protons, transforming them into neutrons and releasing enumerable numbers of ghostly neutrino particles. The rest of the star tries to fall inward, but the outward forces from these particles, combined with the vacuum created during collapse, are too great. What follows is a catastrophic detonation, an eruption of apocalyptic scale, an explosion that belches more light and energy in a single moment than a normal star will release over its entire lifetime.

During these supernovae, elements forged on the periphery the star instantly fuse in large combinations not possible while inside the star. At long last, after over a billion years, the periodic table of elements is created, which would not be added to again for just less than thirteen billion more years. This process of core collapse can be further supplemented by another type of supernova, where white dwarfs sap away material from larger giant stars until the point where the entire dwarf undergoes fusion all at once, and explodes. What's left from these blasts are some of the most extreme bodies in the universe. The cores of most giant stars collapse to form stars made entirely of neutrons, some with intensely strong magnetic fields and precise beams of energy that shoot out from their poles. The largest stars have cores so massive that no force in all existence can hold them up, and they collapse to a single point to form black holes. These objects drift through space, collecting material or funneling it out, eventually forming the hearts of huge star communities, also known as galaxies. Galaxies too collide over the eons, growing bigger, feeding the black holes, and turning them into quasars with jets of energy that can be seen clear across the universe. Eventually though, these voracious monsters lose material to devour, as their central black holes gain the mass of millions of stars. At this point the universe begins to look like a web of a quintillion lights, connected by only thin streams of ionized hydrogen and dark matter.

This cycle of death and rebirth can repeat itself many times. Despite all the new elements being created, the vast majority of visible matter in the universe is still hydrogen and helium. But with new ingredients, unique and exotic structures can form. Molecular clouds accrue when heavy elements bind to both hydrogen and each other, making simple structures such as ammonia (NH_3), methane (CH_4), and dihydrogen monoxide, better known as water (H_2O). The remnants of dead stars can mix to form nebula so vast that it can take light tens to hundreds of years to cross them. It only takes a few generations before stars are no longer born alone or with other stars. Soon, while gas and dust collapse to form new stars just as before, pebbles of rock and ice begin to take shape around them. They can also be formed around or captured by nearby deceased remnants of stars. The first planets are born. One special star coalesced nine billion, two hundred and seventy seven million years after the big bang, was born at the edge of a swirling galaxy with bar in the middle. Here comes the Sun.

Birth of the Blue Marble

Stellar systems are born from gargantuan collapsing gas clouds, composed of countless particles moving in every conceivable direction. Most of these motions cancel out. Particles moving up hit ones moving down, gas swirling left runs into gas swirling right. But even though its individual pieces are moving every which way, the cloud as a whole will usually have an average rotation. As movements within cancel out, the cloud flattens until all of the particles are moving in the same direction, and begin to pile up into balls of material, perpetually falling around a central star in circles or ovals. Thus was the case with the Sun.

After several generations of stars, the system surrounding the Sun is filled with various elements weighing different amounts. The densest materials, rocks and metals, were near the center, and lighter materials such as gas and ice were farther away by the time the Sun formed. Over relatively short periods of time, these grains were brought together by their mutual gravity. Molecules into dust, dust to grains, into rocks, boulders, and mountains, every impact adding energy, and heat. Eventually, from the minute debris of stellar explosions, came a diverse cast of entire worlds, comprised of an eclectic mix of elements and molecules.

Among those closest in, formed a small but dense iron world, Mercury. Slightly farther out, there is more material that can collide to form a larger planet, but also some rebellious particles going against the flow of the solar system. One way or another, all these rocks coalesce into a planet that barely spins, and does so opposite to the rest of the planets, Venus. The next world out forms in a much similar way as the others, building from smaller pieces over time. But unlike all the other rocky planets, when the third world from the Sun was almost completely developed, it collided with an object more than half its size. In perhaps the only incidence of two fully formed worlds colliding in the history of the Sun's system, what's left in the aftermath would go on to become the largest of the Sun's rocky satellites, adorned with a companion of its own. This impact also tilted the planet just slightly; giving it balanced seasons through its orbit. This intrepid duo would become Mother Earth and her Moon. Farther still, a pint size twin of Earth also formed, making it the farthest rocky planet from the Sun. This is Mars. Just outside the orbit of Mars, there are many more grains of dust, but only enough to form mountains and one small dwarf world, Ceres. These clumps of rock might have enough gravity to cling on to wisps of gas, but the majority of their atmospheres will come from within, as gasses dissolved in their hot interiors escape through volcanoes.

Farther from the Sun though, dense bodies have much more access to hydrogen and light gas to hold onto, turning them into behemoths hundreds of times larger that the rocky planets. In order of distance from the Sun, the first of these gas giants is Jupiter, the greatest of all the planets in size. It would be able to fit a thousand Earths inside of it, and is composed of many different gasses that give it a striking striped surface. After Jupiter comes Saturn, with a much cleaner surface than the largest planet. Much better than the rest of the giants, Saturn collected material and ripped it apart with its gravity, giving it broad thin rings, and turning it into the crown jewel of the solar system. Moving outward, the solar system becomes frigid with ices of water, ammonia, and methane. The worlds that form here are forged in kind, and their blue hue reflects their icy build. These are Neptune and Uranus. But the Sun's influence spans far beyond the largest planets, and miniature worlds of ice collapse nearby one another, forming the Kuiper Belt. What's more, while the largest planets form a mostly flat disk around the Sun; the orbits of these dwarf planets are tilted, looking more erratic. They include Pluto, Makemake, Eris, and several others. And just as with the asteroid belt between Mars and Jupiter, comprised of heavy rock, a cloud of icy pebbles also formed around the Sun. Farthest from the Sun lays the realm of comets, the farthest of which may have formed into a spherical shell surrounding the solar system.

It takes hundreds of millions of years for the planets to clear out their orbits and fully form into whole worlds. The growing pains of this process involve explosive collisions, as rocks bombard the surfaces of the planets and smaller worlds get captured by larger ones, giving the planets moons and rings. As the planets settle into position, debris is flung in and out of the system, leading to more impacts. One of the most crucial of these events came around half a billion years after the birth of the Sun, when Neptune was kicked outside the orbit of Uranus, bringing Jupiter and Saturn closer to the Sun. The force from these planets shifting caused more debris to fall toward the rocky planets, including both comets and asteroids. This debris, in part, seeded the rocky worlds with the organic compounds inside them, with not just water, but also nitrogen from ammonia, and carbon from methane. Simultaneously, volcanoes spewed more of these gasses out of the rocky planets, and eventually they formed into thin atmospheres.

Unfortunately, holding onto an atmosphere is much more difficult than producing one. Mercury, being so small and close to the Sun's intense winds and scalding radiation, had no hope for an atmosphere. Mars, with its small size couldn't hold onto much of an atmosphere either, and its size also meant that what heat it did have got lost very soon after it formed, heat crucial for a magnetic field. As a result, Mars lost its water to the soil and air to solar wind. Venus, while it had enough mass to hold onto a thick atmosphere, also lacked a strong magnetic field due to it not rotating. It too lost its water, but held onto its carbon dioxide and sulfuric acid, insulating its heat so much that the surface temperature exceeded eight hundred degrees, hotter than even that of Mercury.

Only one world had a solid surface, a metallic iron core able to produce a strong magnetic shield from the Sun, and enough mass to hold onto an atmosphere, Earth. With these traits it was able to hold a surface of water, warm enough to melt from ice, and cool enough to condense from gas. This sprawling sea provided a haven for complex chemistry to take place, a tiny pocket of the universe safe from heavy radiation, bombardment, and the energy-sapping vacuum. For the first time, the cornucopia of elements forged in stars had the opportunity to interact, to dance randomly with one another, trying new bonds over and over to form molecules of all shapes and arrangements, with energy from the Sun and Earth rearranging them every moment, building structures made of thousands to millions of atoms. Chemistry continued to become more complex, nurtured on this small rocky womb. Then one day, spontaneously, those macromolecules split in two.

The Universe Gets a Mirror

There are many avenues in which life may have arisen. Some of the most basic building blocks may have been able to form in the interstellar medium, in the short time that water around stars was neither gas nor ice. This would have included basic sugars made of strings and rings of carbon, oxygen, and nitrogen. These pieces may have only been able to form in the presence of metals on Earth, which have a weak hold of their outer electrons and so may have catalyzed organic chemistry. Regardless, two factors were critical that could only have been nurtured in Earth's warm bodies of water. The first is a protocell, smaller than the head of a pin. This would act like a bubble, a barrier from the outside world where molecules could interact without interference from other random particles or chemicals that would have broken organic molecules apart. As long as these bubbles weren't broken, small molecules would be able to group together, and produce other more complex ones. Although its outer membrane would have protected this chemistry, it would have also needed to allow new material to be taken in and old material to be taken out. As more ingredients poured into those cells, molecules could be copied over and over, until eventually those bubbles split, each new bubble taking a little bit of material. In this way, the first strand of genetic information, ribonucleic acid (RNA), could form from only carbon, hydrogen, nitrogen, phosphorus, and oxygen. From this one molecule, which can both facilitate its own replication as well as the formation of new chemical machines like proteins, each composed of thousands of atoms, life would overtake every crevasse, crack, and cranny on Earth.

It took well over half a billion years for conditions on Earth to stabilize enough for life to be maintained. This time when the structure of the planets and solar system were stabilizing is called The Hadean Era. The outermost layer of the Earth cooled and hardened into over a dozen plates drifting above molten rock. Between these plates, mountains formed and lava was ejected, bringing visible land above the surface of the oceans. The earliest packets of genetic information could have been annihilated as quickly as they arose, as molten rock and noxious chemicals erupted from within the Earth as it was reaching equilibrium. However, having access to so many ingredients and energy meant that early cells could also synthesize more complex chemicals. These life forms would have been satiated on just the heat and fuel immediately available from the Earth's interior. Eventually ribonucleic acid would be replaced by deoxyribonucleic acid (DNA) as the primary means for storing genetic information, although the difference between the two being an extra oxygen atom and methane molecule within DNA. However, both would be crucial for making new proteins, which would be used to create new parts for the cells to survive. This continued into the era directly following the first half billion years, know as the Archean.

It took over a billion years after the Earth formed for cells to become abundant enough to visibly alter their environment. Tiny organisms developed in such numbers that colonies began to form on pebbles and rocks. Some intrepid cells found themselves outside the extreme environments where life was first conceived, diverging into bacterium that occupied the unconquered open sea. These early organisms developed the first means of locomotion, and may have even absorbed other organisms, using their innards to fuel more complex chemistry. When these cells split in two, their DNA was almost perfectly replicated, but not always. Sometimes a few pieces became altered or replaced, leading cells to differ slightly. These beings also mutated through random interactions with one another, transferring DNA and diversifying their population. One key mutation occurred around three hundred million years later, as bacteria drifted closer to the ocean surface. They absorbed sulfur and infrared light, becoming the first life forms to sustain themselves on energy from the Sun. Over the next seven hundred million years, bacteria would continue to diversify and migrate closer to the boundary between air and sea.

By the beginning of the Proterozoic eon, when Earth was two billion years old, cyanobacteria had evolved to utilize most of the light waves from the Sun. Instead of infrared, they absorbed more energetic visible light, and instead of absorbing sulfur from the Earth, they took in carbon from the air. In order to access this carbon, and turn it into energy rich sugars, they first had to separate them from the two oxygen atoms they were attached to. For the first time, carbon dioxide was transformed into oxygen gas, as these early photon feeders captured carbon and spat out the oxygen they didn't need. With consistent nourishment, their populations skyrocketed. So much oxygen was released that the very composition of Earth's atmosphere changed forever. This was no small feat, as oxygen is so reactive that before it could build up in the atmosphere, it first bonded with rocks, atmospheric methane, and other life forms that couldn't use it for anything. Free oxygen eroded the structures of countless organisms, killing off most life on Earth. This was further compounded by the depletion of methane, which reacted with oxygen to form water and carbon dioxide, which were not as effective at keeping the planet warm. To top it off, some of that oxygen bonded into triplets, forming an ozone layer that further shielded the Earth from high-energy light. More carbon dioxide led to more cyanobacteria, which led to more oxygen, until the whole Earth froze over. Together, these effects led to the first mass extinction event, as life nearly killed itself with runaway chemistry.

Luckily, even ice sheets as thick as mountains are still only paper-thin compared to the Earth, which was still broiling with internal heat. Tremendous eruptions released more greenhouse gasses, thawing the surface. Despite this delay, organisms that were lucky enough to be born with adaptations to their new environment survived the oxygen catastrophe, and passed on those traits to new cells. Aerobic bacteria developed, which could harness highly volatile oxygen and received far more energy than their foraging ancestors. So much energy, in fact, that they could sustain whole other organisms.

About three hundred thousand years after the great extinction, gargantuan cells absorbed both cyanobacteria and aerobic bacteria. Fueled by these new fuel sources, prokaryotes became eukaryotes. These highly advanced cells developed specialized regions for creating proteins, storing the genetic information that made them, and holding up their larger build. Some of these organisms began to band together in clumps, sharing resources. They also began to colonize land, after being trapped in the ocean for over two billion years. These radical organisms perhaps could have fed on the remains of cyanobacteria built up near shores, developing into the first fungus. Some six hundred million years after fungus, small proto-plants found themselves symbiotically trading water for nutrients with fungi on land.

While early fungi and plants began to adapt to their new environment, clusters of cells still at sea began to grow as large as small stones. Some of these colonies, made of millions of cells, grew into simple sea sponges eating nutrients that drifted by. Others, rather than staying put, began to actively hunt for food. They went from having one layer of cells, to many, and those layers all needed to eat. Clusters of cells formed into stream lined tubes. In order for these worms to function effectively, their cells adapted to perform specialized functions of their own. Some transformed into protective coats, some specialized in absorbing nutrients and delivering them to fellow cells. Others held on to cells specially made for reproduction. This innovative adaptation provided a huge advantage over other life forms; because instead of one organism making identical copies, two separate organisms could share half of their genes with one another using specialized sex cells. This meant that offspring would be more diverse, thus more likely to be born with random variations that would allow species to adapt faster to a changing environment. Finally, some cells became lines of communication, taking chemical information from each other and their environment to observe the world. Organisms became highly receptive to chemicals, light, and motion. For the first time life could adapt moment by moment in response to its surroundings to actively promote survival, the very first glimmer of sentience.

After two billion more years of genetic trial and error, the Earth was already four billion years old, and life was finally prepared to transform the inert landscape and dominate the planet. The end of the Proterozoic ushered in the period known as the Cambrian, when life exploded into huge diversity. Organisms that fed on microbes developed into organisms that devoured others their size. They were insanely complex, made of trillions of cells, all collaborating for the singular purpose of survival. Cells developed new proteins and enzymes that produced new structures. Animals began to be born with hard outer shields, and innovative hunting techniques. The oceans became filled with exotic species; armored arthropods, gelatinous jellyfish, and early fish without bones. At this point, eukaryotes might pass on their genes asexually via copying, or sexually using male and female counterparts of one species. Males became designed with small, highly mobile haploid cells containing half of their DNA. Females developed with less mobile haploid cells that also contained half of their DNA, and these eggs would develop into new whole organisms when combined with the male haploid. To reflect form, the function of the sexes began to develop. Females often required more energy for offspring, and so became less physically aggressive, but both sexes began to compete for mates by demonstrating their health and fitness. Some even developed with both male and female cells, able to reproduce even when mates were scarce.

Fast-forward fifty-seven million years, to the beginning of the Ordovician period. By this point fish have developed jaws for devouring prey. Soft-bodied mollusks with hard shells and long appendages, the first slug and squid-like organisms, roamed the ocean easily consuming smaller organisms. These first animals created a complex food web as species specialized to different environments and available food sources. Organisms began to fill specific niches, keeping others from using the same resources and terrain. Predators drove prey to adapt new means of protection and defense, such as armor, agility, and camouflage, or else died out. In order to feed, predators adapted new techniques to swiftly catch prey, such as heightened senses, stronger weapons, and toxins. Of course, traits were not exclusive between the two; often the same adaptation could serve multiple tasks for both predator and prey. In addition, every organism fed on one another, as eventually prey and apex predators alike became consumed by microscopic infections resembling the first living beings. As cells within tissues continued to specialize, organisms also developed cells designed to eliminate foreign invasion. After twenty-four million years of growth, everything came to a halt as life again faced a mass extinction, likely caused by several factors, including volcanic eruptions that turned the oceans toxic.

Only those able to quickly adapt to their environment through the generations emerged to usher in the Silurian period. For the next twenty-four million years, soft fish developed the first internal skeletons, allowing them to keep their bodies structured, grow larger, and protect newly refined internal organs. Among these new skeletal structures included the first vertebrae, spinal chords to protect the nerves that communicated throughout their bodies. Hard-shelled arthropods and fish alike began to develop bundles of nerves in the front of their bodies, as well as varied amounts of hormones, showing unique individual behaviors within a species. By communicating and making decisions based on each other, these organisms would have laid the foundation for society and culture. The first horseshoe crabs, a hugely successful species, also developed during this time. In addition, with their exoskeleton for protection and water retention, arthropods became the first animals to join the plants and fungi on land, temporarily being able to breathe without oxygen in the ocean. But they wouldn't be the only animals ashore for long.

Following the Silurian period was the Devonian. During that time arthropods fully adapted to life on land, leading to the first insects. Slimy fish also took advantage of these new uncharted lands, their descendants eventually growing into the first amphibians. These slimy creatures were still closely tied to the water, needing to lay their eggs in the moisture where they were born. But once they matured, they had tremendous advantage over the other animals, being able to escape aquatic predators, while simultaneously feeding on abundant levels of insects. By this point, plants had diversified into many ferns, shrubs, and algae, slowly creeping over the dead landscape. Sixty million years after the beginning of the Devonian period, Earth again saw a massive extinction at sea. The lucky few who made it ashore were able to avoid this catastrophe, and continued to diversify. Animals began to appear with four limbs, allowing them to travel more efficiently on land. Arthropods, on the other hand, began to take flight, out competing all other land animals as the dominant species, with access to three dimensions of movement once again.

Thanks to their billion-year head start, plants were able to fill the landscape with lush greenery. For the next sixty million years, small land plants grew to towering trees, again enriching the air with oxygen. This time however, animals were able to utilize this extra fuel, and too grew enormous. Bush sized spiders, centipedes, and dragonflies were all abound throughout the land. Some amphibians continued to spread throughout land as well. Eventually they evolved into the first reptiles, able to live their whole lives ashore thanks to their hard-shelled eggs and scaly skin. Unlike insects and amphibians, they also didn't need to go through any type of metamorphosis, being almost fully developed right out of the egg. At this time the Earth's continents aligned such that all the landmass was unified, making the supercontinent Pangaea. These diverse fauna lived, bred, and died on land, their carbon piling up for millions of years into coal and oil. This was the carboniferous period.

While initially facing great danger from the arthropods that once ruled the land, reptiles eventually found open space to diversify. Over the next forty-nine million years, vertebrates began to develop adaptations that allowed them to dominate the landscape. Among them included the first herbivores that fed on native plants. Despite requiring more energy for digestion, these animals were able to take advantage of plant immobility, and survive off of a larger and consistent source of food. As always, predators developed to take advantage of the nutrition already cultured in other animals. One type of predator included synapsids, strange animals with great fins on their backs, which allowed them to become the first animals able to regulate their own body temperature. Over time this adaptation would lead to the first mammals. The Permian was punctuated with the largest mass extinction in the entirety of Earth's history, as nine out of ten animals in the ocean became extinct, followed by seven out of ten animals on land. Once again it was Earth, releasing its pent up fury in the form of lava and lethal gas, which likely heralded this great dying. But at this point, life was no stranger to adversity.

Fauna that survived trekked on for the next 50 million years, during the Triassic period. Pangaea began to split apart, and the landscape was treacherous. With all the land mass together, seasons fluctuated from scalding hot to frigid cold. The temperature of the Earth overall, however, was too warm for ice caps. During this time reptiles also began to flock back to the sea, refilling niches lost in the last extinction. Lizards became large while the developing mammals shrunk in size. The first dinosaurs emerged, accompanied by several other reptiles. It wasn't long until a vast number of species again succumbed to a mass extinction at the end of the Triassic. The animals that emerged from this event would become immortalized as the very symbols of the ancient world. With new opportunities on land, dinosaurs took their place as the dominant life forms.

During the Jurassic period, lasting fifty-five million years, the great continent split into many. As a result of this isolation, organisms saw vast diversification. With their small size, mammals developed fully into ectotherms, able to produce their own heat, aided by the development of fur to help them stay warm. Dinosaurs evolved to become the apex animals on the planet, reaching scales of size never before achieved by any animal. Sauropods towered several stories over other animals with their enormous tails and long necks. Herbivores like stegosaurus developed plated armor for combat and defense. Pterosaurs with several meter wingspans ruled the skies. Over this timespan, plants began to develop primitive flowers in the form of cones as their means of reproduction. By the end of the Jurassic, plants fully developed the first true pedaled flowers, and highly intelligent feathered birds began to split from the reptiles. Insects coevolved with them, resulting in many pollinators, including the first bees.

The great success of the dinosaurs continued for eighty million years after the end of the Jurassic. During this time, known as the Cretaceous, animals had developed vastly complex and powerful organic features. Giant sharks and mosasaurs ruled the seas. Birds took flight, developing powerful talons and beaks, and too became capable of internal heat regulation. The land was home to giant crocodiles, velociraptors, triceratops, and the royal bird-lizard, tyrannosaurus rex. These creatures were remarkably successful, ruling the planet for a total one hundred and thirty-five million years. But even after their long reign, they had no way to prevent the calamity that would dethrone them. One faithful day, a mountain sized asteroid collided with the Earth, striking the Mexican Gulf and releasing land shattering devastation, throwing debris across continents. Aided by volcanic eruptions, the Earth would have been clad in thick dark smoke. With less access to light, plant populations fell, breaking the food chain all the way to the apex predators, killing off three quarters of living species, including the giant reptiles. With resources being limited, only small animals survived, including the shrew-like mammals, which were able to avoid much of the devastation.

The era of mammals is known as the Cenozoic. With the ruling animals gone, these creatures quickly replaced the niches they left behind, radiating across the land in many new varieties. In the sixty five million years following the asteroid impact, the decedents of those meek and furry creatures would spread to the skies and oceans as well. Building upon the autonomic nerve bundles first formed nearly two and a half billion years ago, these creatures distinguished themselves with large brains, resulting in highly complex group behavior. Aiding in nourishing their young, mammals evolved nutrient rich milk, and with the exception of just a handful of species, gave birth to live young rather than laying eggs. Bats, bears, whales, warthogs, honey badgers, horses, hedgehogs, mice, meerkats, felines, canines, pachyderms, pandas, kangaroos, camels, all of those and more descended from these tiny creatures. Another crucial newcomer wasn't an animal, but a plant. Hard to digest grasses sprouted, and would develop into grains, rice, and corn. Ice caps developed and the poles grew for the first time since the Carboniferous as well.

One branch of the mammalian family adapted to live in trees. These squirrel-like insectivores branched off into the first monkeys. Around thirty-five to forty million years post-dinosaurs, these monkeys split as some migrated to Central and South America. Those left behind on the African continent populated forested areas. Eighteen million years after that, many new species had branched out from the monkeys, including great apes that lost their tales. While many animals at the time had been able to develop social networks and fashion tools, selective pressures would force these creatures to develop ever more expanded neural networks, leading to greater intelligence. Apes also developed an extremely unique trait that would separately arise in bats and elephant shrews. Female members of the species would release their eggs in a regular cycle lasting about twenty-eight days, the same amount of time it takes the Moon to orbit around the Earth. These eggs attach themselves to the endometrium, a thin outer layer on the uterus that gives the egg access to nutrients through the female's blood. In most mammals, if the egg isn't fertilized this lining is reabsorbed into the body, but eggs in primates bury themselves much deeper into the uterine wall, requiring a thicker layer.

Instead of being reabsorbed, this lining is removed from the body. As a result of this process, it might have been possible for both the mother and embryos to get ample nutrition while allowing the embryo to burrow farther into the womb, or simply to reduce the odds of abnormal offspring. As these primates became more intelligent, they required more resources to raise their young, making it even more crucial that offspring were developed in optimum conditions. Two groups of these primates became isolated, and grew into completely separate species. Four million years later, the African climate shifted, replacing forests with plains. One of these isolated species began to travel along the flat African savannah.

In order to increase mobility for hunting and fleeing, they adapted an upright posture, walking solely on two feet. This was Australopithecus, and they too diverged into new genera. One million, two hundred thousand years after they left the trees, the genus Homo emerged from the genus Australopithecus. The first of these was Homo Habilis, who was able to live off the land and master stone tools. Over a million years pass, during which time selective pressures force the genus to adopt more advanced social structure and communication, lest they die out. One of the decedents, Homo Ergaster, learned to harness rapid oxidation, or fire. Fire provided them with protection from predators, but more importantly it allowed them to cook, sanitizing food and unlocking nutrients otherwise unavailable. This accelerated their cognitive development even further, resulting in more advanced communication, thus better hunting and defensive capabilities. These apes lost their thick fur, and instead used the protective coats of other animals. As with any other successful animal, they spread over the continent and beyond. Several species within the genus homo emerged over the next one million, three hundred thousand years. Just a few of these included Homo Heidelbergensis, Homo Neanderthalensis, Homo Erectus, and the fateful Homo Sapien. These hominids coexisted for over a hundred thousand years, living in small communities, hunting and gathering. With their strange and dexterous hands and thumbs, they created art, musical instruments, tools, and most importantly, they learned and built upon knowledge gained from their parents. Generation by generation, they became the first animal to preserve knowledge after death. At long last, after thirteen billion, eight hundred and twenty thousand years of cosmic expansion, nearly four billion years of genetic shuffling, and after sixty five million years of primate development, beings finally arose that could look out at the world and asked, "why?"

Over time, Homo Sapiens would outcompete and eliminate their cousins, and one hundred and ninety thousand years after they emerged, they were the last humans to remain. While this certainly could have been due to war and genocide, equally likely our relatives simply could not adapt quickly enough to environmental changes, most notably the eruption of the supervolcano Toba, which occurred around the same time the last of our relatives went extinct. By this point, people had migrated through and out of Africa. Some groups trekked through the southern tip of the Red Sea, from the Southern Middle East, through the Indian subcontinent, past Malaysia, and finally ending up in Australia. Others went through the middle of the Red Sea to fill Asia and the base of India. Others headed northward, into the Fertile Crescent and to the edge of Europe. Groups in Asia were able to journey across the northernmost frigid ices of Russia, onto the Alaskan tundra. From there humans radiated Southward all the way to The Drake Passage below Chile, where the Atlantic meets the Pacific, thus populating all of America. Those that migrated away from the equator were exposed to less sunlight, and so lost melanin in their skin, and if they migrated back they regained it. Then almost simultaneously around the globe, people began to take edible plants, and regrow them. This was the birth of agriculture. They cultivated the land, and became the first animal to have a consistent surplus of food.

When people had a surplus of food, they were able to do things impossible for any other animal. After that people settled the land, and built perimeters to protect it and their families. The first professions were developed as people could focus on tasks besides eating, supporting development even further. Populations exploded. Villages became cities, cities became kingdoms, and kingdoms became empires. People grew more and more connected over the millennia and throughout the world, spreading ideas, resources, and technologies. When resources ran low people waged wars leading to many genocides, slavery, and arbitrarily shifting borders. Many religions came and went as humans tried to grapple with the awe inspiring world they found themselves in, creating centers of culture worshiping weather, kings, and carpenters. They created huge monuments to their deities and glamorous epics about their origins, including one odd tale with a garden snake. Empires stretching as far as the eye could see rose and fell. Over a hundred billion individuals lived and died, leaving their experiences to live on in the next generation. Eventually they were able to harness the very elements christened in the stars to mold the Earth, tame the decedents of their million year old ancestors, and fling themselves onto the moon ever briefly, all with nothing more than their highly adaptive brains. In a mere ten thousand years they created flying machines, bacon cheeseburgers, and the Internet, where one can study everything humans have ever learned and even share their own experiences. And then I was born. Then I wrote this book. And now you're reading that book. And all the wisdom gained between then and the first human will be our topic of discourse. I hope you enjoy.

2 | Who Are You?

"When I discover who I am, I'll be free."
— Ralph Ellison, Author of Invisible Man

"Have no fear of perfection; you'll never reach it. Nothing in life is to be feared; it is only to be understood."
— Marie Curie, Wonder Woman of Science

How do you define yourself, and what makes you unique as a thing in the universe? Well, more than likely you consider yourself to be a human being. So then are you a person? Have you always been the same person, from childhood to adulthood? What if you moved and people started calling you an alien? What if other people already call you an alien, even though you both grew up in the same place? Can other people take your humanity away? What if you commit a crime, do you relinquish some of your personhood? Are people who rape, molest, or slaughter less than human? Most people would almost unanimously say yes. But does the severity of the crime matter, or is all crime equally detestable? If you became a drug dealer, would it be just as bad as if you became a terrorist? What if you were fighting for the freedom of an oppressed people, or the independence of your country from some imperial power? Very quickly it becomes apparent that being a "person" is more complicated than just the actions you make, the things you do, or even how other's perceive you. Perhaps we can come to a satisfactory answer by looking more objectively at ourselves.

Were you always a person, even when you were just an egg in your mother's womb, even though an egg can have no more than half of your chromosomes? Are your skin cells people because they have your DNA? What if you were born in a laboratory, made of corpses, or of oil and wires instead of blood and organs? Are you only a person if you have a developed brain? If so, human infants would be less human than chimpanzees, dolphins, elephants, pigs, cats, dogs, octopuses and parrots. Or is the potential for humanity more important? This would suggest that a human infant is still more of a person than any animal. But what if you are an adult with a cognitive disability? Cats and dogs are some of the few animals that can have neurological disorders just like people do, and they are certainly more intelligent than many animals. It would seem that our brains are crucial elements in determining what makes us human. Are you a person if you feel pain? Again, this wouldn't be enough to make humans distinct from many of the animals we eat for dinner. Perhaps so-called humanity comes in degrees. You might be more of a person as an adult than when you were a baby, and less of a person when you pass away. But if that's the case, then any random star, plant, or cow could potentially become human, since they're all made of the same thing, and may even all be integrated into a person before lunch. How about after a traumatic experience, like the loss of a relative, a psychedelic hallucination, or after coming home from war, are you the same person after as you were before?

These questions are some of the most difficult that anyone can ask themself. They may be uncomfortable, and perhaps even terrifying. But there is also a certain fulfillment to be gained by answering them for yourself. By at least thinking about these quandaries, we can gain a richer and fuller understanding of not just ourselves, but also the world around us. What's more, our ability to even ask these questions in the first place should be seen as a testament to our ability to think critically, which is by far one of the most human things we could ever do. As it turns out, you are far more complex and beautiful than what a simple glance in the mirror may suggest. So, in order to try and find some answers, let's take a moment to discuss a few of the many ways modern thinkers have tried to grapple with these daunting questions, to find out just what we are made of. We'll begin with a modern model of what scientists have discovered over the last few centuries, and discuss the actual experiments that led to those discoveries more deeply in Chapter 6.

What are we made of?

Look at your hand, what do you see? On the surface, there is skin, full of marks and creases that might reflect the things you've experienced. But beneath the surface, lies a microscopic world of individual living organisms and structures, each specially designed with the sole purpose of keeping you alive. Blood flows through your arteries, taking oxygen from your lungs and nutrients from your digestive system to each cell. They take these building blocks and use them to produce all of the enzymes and proteins that build your bones and muscles, and provide them with the energy to do so. You also have your own individual army, dozens of cells made with the job of eliminating foreign invaders that might try to steal your resources in order to multiply. There are trillions of these cells that make up your body, and when they die, new cells made of the food we eat replace them, and the old cells are passed through our waste. Inside every one of us is a multitude. What one normally considers a single entity is really an aggregate of several trillions of individual organisms vying for survival. You are many.

But what are these cells made of? Well, they too are made up of chemicals, thousands of different kinds, contorted into many different shapes. These chemicals are comprised of many different elements, the very same ones produced by stars billions of years ago. The countless different structures that make up a cell are nothing more than various combinations of these elements, stacked together like puzzle pieces. For example, the protective outer layer that makes up your cell's walls are actually made up of millions of duplicates of a single molecule. Each one of these molecules looks somewhat like a string of atoms, and each one has two distinct ends. One end is polarized, which means it has parts that have slightly more negative charge, and parts that are slightly more positive. Since water is also polarized, other polar molecules (like sugar) will dissolve easily into it. Molecules that aren't polarized (like oil) will stick to each other and not dissolve in water. The phospholipids that make up the cell wall have one end that is polar, and one that is not. This makes the cell wall into a bilayer, with two phospholipids next to each other, with their polar sides facing outward toward water, and the non-polar sides facing each other. This layer provides the cell protection from the outside world, and allows a cornucopia of far more complex molecules to form and develop inside the cell. So what are these molecules made of? Just like all of the parts of the body, and every living organism for that matter, these molecules are simply different combinations of the same building blocks, known as atoms. The majority of which include hydrogen, oxygen, carbon, and nitrogen. Incidentally, the order I just listed these elements goes from the most common to least common, not just in terms of the abundance of elements in the human body, but similarly the abundance of elements in the universe. However, if you recall, helium is by far the second most abundant element in the universe. But for reasons we will discuss shortly, helium is essentially useless to living organisms.

In school many students learn about these elements in their chemistry classes through the periodic table of elements. When people first learn about all the different elements, they often think about what those elements would look like if they were to hold them in their hands. When I think about carbon, I usually think of black coal, gravelly and easy to crumble, which is very different from something like gold, which is glimmering. On the large scale, it is easy to think of elements being completely unique substances. After all, they look and behave in completely different ways. But in reality, the only difference between the calcium in your bones and the iron in your blood is the number of protons and neutrons each of those elements has. By definition, helium is just a hydrogen atom with another proton inside in the nucleus. However, because they have the same type of charge, two protons stuck together all alone are highly unstable. This is where the neutrons come into play. An element is defined by how many protons it has in its nucleus, but the same element can have a different number of neutrons. Scientists call these isotopes. Lastly, atoms have negatively charged electrons surrounding the nucleus. Just like with neutrons, the same element can have a different number of electrons surrounding it. These varieties are known as ions.

Let's try and put this all together with an example. Carbon is the element that has six protons inside its nucleus. The most stable isotope of carbon has six neutrons inside of it, which is why we call it Carbon-12. If the atom were electrically neutral, it would have six electrons surrounding it, to balance out the charge from the six protons. If you were to add an electron it would have more electrons than protons, becoming negatively ionized. If instead you were to take one away, then it would have more protons than electrons, making it positively ionized. These subtle differences in charge result in not just all of the chemical reactions that keep you alive, but also every interaction between atoms that occur in the universe. Wow! But we aren't done yet.

Atoms are ludicrously tiny. If one were to count the average number of atoms inside of a grapefruit, about the size of both of your fists together, there would be as many atoms inside of that grapefruit as the number of blueberries that could fit inside of the Earth (if one could do such a thing). But this is just the atom. Remember, an atom has two main parts, the nucleus in the center, with the protons and neutrons, and the electrons surrounding the nucleus. The distance between the electrons and the nucleus is astounding. In a simple hydrogen atom, with one proton surrounded by one electron, the size of the atom is one hundred thousand times greater than the nucleus alone. The electrons are so ridiculously far away, that if the atom were the size of a football stadium, the nucleus would be the size of a marble in the middle. The craziest thing though, is that nearly all of the mass of the atom is packed into that tiny nucleus! Matter at such small scales behaves very strangely compared to objects we can hold and touch. When one delves into the smallest scales of the universe, they find that the most fundamental constituents of matter have nearly exact properties, from how they behave alone to how they interact with other particles. That is to say, they are quantized. To understand what this means, and how molecules form from atomic interactions, let's look at the atom more carefully.

Unlike the planets, which could orbit around the Sun from any distance, electrons can only surround a nucleus at precise locations, called orbitals. In many ways, electron orbitals are like stairs, except those stairs can be different heights. An atom by itself will have electrons orbit as closely as possible to the nucleus. This closest position is known as the ground state, and it is the distance from the nucleus where the energy that the electron has from moving is half the energy that it would take for the electron to fall into the proton due their opposite charges. In order to go from one orbital to a higher one an electron needs energy. It has to be the exact right amount of energy, otherwise it will stay in place, kind of. We'll get to that in just a second. Electrons will jump to a higher energy level when they absorb energy; this energy comes in the form of photons, which are little packets of light. Furthermore, since all things in the universe want to minimize energy, after being energized the electron will go back to its ground state, and in doing so release a photon exactly like the one it absorbed. Since every atom has different energies that its electrons can orbit (think different sized stairs), that means each atom and molecule have their own unique photons that they absorb and emit. This gives each element and molecule a sort of "fingerprint", a unique spectrum of exact colors that it absorbs and emits. This is how scientists are able to learn what everything in the universe is made of, from the atmospheres of planets to the contents of stars and galaxies, and even rocks on the Moon and Mars. Alright, now that we've generally covered how individual electrons behave around atoms, we can move on to what happens when atoms have multiple electrons, which will tell us how molecules form.

Perhaps the strangest concept revealed by quantum mechanics is that at small scales all things are described by probability. While quanta may exhibit particle-like properties, like a soccer ball, they also simultaneously behave like waves. But these waves aren't like waves in the ocean; instead they are waves of probability. These waves describe where things like electrons are likely to be, but in truth they have some chance of being anywhere. These probability waves describe where electrons are most likely to be, but not where they are exactly. This is probably (get it?) the most unusual and uncomfortable aspect of quantum mechanics, to student and researcher alike. This idea was also the main source of contention for Albert Einstein against quantum mechanics. When the probabilistic nature of the quantum world began to reveal itself, he famously claimed, "God doesn't throw dice", to which Niels Bohr—one of the forefathers of quantum mechanics—would reply, "Stop telling God what to do."

Indeed, probability is inherent to the nature of the universe, and it has been experimentally confirmed time and time again. However, the universe is not completely random. There are certain outcomes that are more likely than others, and through fiendishly complex mathematics, one can calculate such probabilities. The ground state of a hydrogen atom, for example, is the location where the electron is most likely to be, and when one goes to calculate what this might look like, they would find the electron to be distributed around the nucleus in a spherical shell. This seems to be reasonable, as the electric force only depends on charge and the distance between charged objects, and every point on the surface of a sphere is by definition the same distance away from the center. But things quickly get more complex as more protons and electrons are added. This leads us to the very first concept taught in my own quantum mechanics class, that of spin.

~By the way, if you have trouble visualizing electron orbitals, you are in good company. People far more clever than myself have spent decades refining measurements and rerunning calculations before coming up with the model we use today. If you are interested, there are numerous free resources that have different visualizations for what quantum atoms look like, but for the purpose of understanding what our bodies are made of, the previous information should suffice. ~

Both atoms and molecules are subject to the quantum concept of spin. If one uses the analogy of a planet orbiting a star, then the electron spin is similar to how a planet spins like a top while also orbiting the star. In other words, it's the difference between a day and a year. But it's dangerous to take this comparison too seriously, exactly because quantum mechanics and gravity don't behave the same way. Instead, one might imagine a single hydrogen atom as a dense positive sphere of charge surrounded by a larger hollow sphere of negative charge. But, the outer shell is also rolling in free space, at an exact and constant amount. Not only is it rolling at the same rate, but also in only one direction. That is to say, if the electron were a ball on the ground, it could roll toward or away from you, to the left or the right, or it could spin like a top clockwise or counterclockwise, but never a combination of these.

This aspect of spin results in all the ways electrons fill their electron shells. Take the first shell, which looks like a sphere and can hold two electrons, each spinning in opposite directions (chemists know this as the *s* orbital). This means hydrogen and helium look very similar, with helium essentially like one hollow ball inside of another, both rolling in opposite directions. The next shell can hold eight electrons. Like the first shell, it has one 's' orbital, which is referred to as its own subshell. However, the second electron shell also has what is known as a *'p'* subshell, which has three distinct orbitals. Now, recall that every orbital can hold two electrons, each with opposite spin. So, the second electron shell holds eight electrons, four orbitals, and two distinct types of subshells, distinct because they have different shapes. The best analogy for the 'p' orbitals I have seen are balloons tied together, which is brilliant because like electrons, two balloons can't occupy the same space, and the average balloon resembles the shape of a 'p' orbital. Lastly, the reason there are six electrons in this orbital is because electrons there are spinning in all three dimensions of space: up to down, left to right, side to side, and the reverse of all these. Lastly, electrons will tend to spin in one of these three directions around an atom before they will spin opposite of each other. It's easier to be unique than contrary. Once an atom has ten total electrons, both the first and the second shell are full, and any additional electrons will follow the same pattern, first creating spherical clouds around the nucleus, and then dumbbell shaped clouds around the same distance from the nucleus. This continues until the nucleus has twenty-one protons, and then electrons begin to fill the *d* orbitals, and the mathematics begins to look especially ridiculous. The elements that use those *d* orbitals are the metals, such as the famous shiny rocks gold and silver. Metals are unique in that the electrons farthest from the nucleus are very loosely bound, which is why they are often so good at conducting heat and electricity. But seeing as the vast majority of living things aren't made of metals, at least for now, we won't need to discuss them. Instead, essentially all organic molecules don't use elements beyond the first two electron shells, with a few exceptions. This makes discussing them much easier (if one can even call quantum chemistry easy).

~ As an aside, the reason orbitals are referred to as s, p, d, and f has nothing to do with what atoms actually look like. Instead, it is actually based on the light spectrum different types of elements produce, an anachronism created before the advent of quantum mechanics, which is why they don't make any sense. The key thing to remember is that s orbitals look like spheres, p orbitals look like dumbbells, and both f and d orbitals look like the electrons decided to have a rave! ~

Finally, we can discuss how molecules form. Basically, atoms will bond with each other for one of two reasons. First, they might have opposite charges. This should immediately strike you as odd, because atoms alone are electrically neutral, since they have the same number of electrons as protons. This leads us to the second, and probably most important, reason atoms bond. That is, they are more stable when they have filled electron shells. Allow me to elucidate. The way electrons surround the nucleus is based on how they spin. Any object that is spinning also has some angular momentum. Furthermore, because electrons are electrically charged, the fact that they spin means that they also produce a magnetic field, which in turn can affect how molecules move and interact with other molecules. Taken together, this means that if an atom doesn't have its outermost shell filled, there are going to be unpaired electrons whose spin isn't cancelled out by another electron, making it unstable. These atoms will teeter about like a poorly thrown football or a top that's about to fall down. This makes these atoms more erratic if their electron shells aren't filled, and more calm if they are filled.

Since all things in nature want to minimize their energy, atoms will actually be more likely to gain or lose electrons—and thus become charged—if it means filling up their outermost layers (also known as valence shells). This is the reason noble gasses such as helium, neon, and argon don't react with anything; they already have full valence shells. This behavior is how table salt forms. If you happen to have a periodic table handy, you'll notice that sodium, an alkali metal, is far on the left, which means that it only has one electron in its valence shell. Sodium will readily just get rid of that electron if given the opportunity, resulting in a positive charge. On the other side of the table, you will find chlorine, a halogen. Chlorine will do the opposite of sodium, readily grabbing onto a free electron, even though this gives it a negative charge. When these ions come together due to their opposite charge, they can form an ionic bond. In this case, they form sodium chloride, possibly the most delicious of all molecules. Now, when it comes to alkali metals and halogens, such as sodium and chloride respectfully, when you put them together it is often easier for one to give up an electron to the other, but that isn't common.

Much more frequently atoms share electrons in order to reach stability. Molecules that share electrons form what are known as covalent bonds, because they share valence electrons. They include all of your favorites, from water to essentially every molecule in every cell in every organ of your body. Let's begin with water, the elixir of life. The human body is comprised of around sixty percent water. If you were able to isolate a single cell, be it bacteria or eukaryote, you would find that it was essentially a microscopic sack of water with come other stuff inside. That's why all living organisms need to drink water to survive. The very first living organisms developed inside of water, and they weren't able to survive on land until they evolved tools for retaining water. Plants have thick cell walls for instance. This is why fish evolved into amphibians, which start their lives in water and can't survive without a large source of moisture or humidity. It's why amphibians evolved into reptiles, which were able to lay eggs on land and had dry hard skin that allowed them to survive dry climates. Finally, the need to retain water is also why arthropods were able to make it to land before fish and develop into insects and arachnids; they already had thick outer shells. So, why is water so essential to life that astronomers look specifically for planets with liquid water to find extraterrestrial life, and what types of elements are the best building blocks for life? It all goes back to the bonds, specifically the special properties of oxygen and carbon.

Oxygen is an element with a voracious appetite for electrons. It's so reactive that if there were much more of it in the atmosphere, your hands would catch on fire whenever you rubbed them together too quickly. Oxygen fuels fire, causes metal to rust, and caused the first mass extinction on the planet, all because it just wants to fill up its electron orbitals. It only needs two electrons to do this, which is why water is one part oxygen and two parts hydrogen. But since oxygen has far more protons, the shared electrons are pulled closer to the oxygen. In addition, recall that electrons will prefer to orient themselves in different directions. Since oxygen has two empty spots for electrons, when it bonds with hydrogen, those atoms fill in those adjacent spots. As another way to think about this, imagine that you were an oxygen atom. In your outermost shell, you have an electron cloud to your left, to your right, above you, and behind you, but nothing in front of you or below you. Due to this side effect of quantum spin and oxygen's strong electronegativity, water becomes a polarized molecule, with a negative charge by oxygen and a positive charge by the hydrogen. This is key. Thanks to this lopsided charge, water is deemed by many to be the universal solvent and this is why we use water to clean our belongings and ourselves. It is able to act a medium between chemical reactions. Sometimes oxygen just outright kicks off one of the hydrogen that it's bonded to, transforming it from water into the negatively charged base hydroxide. These free hydrogen quickly bond to other water molecules, turning those into positively charged and acidic hydronium. This is technically happening constantly in all liquid water, and it's thanks to this property that the very chemistry of life is possible.

Carbon is slightly different than any of the other elements we've discussed thus far. It doesn't have a full valence shell, but it isn't predisposed to gaining or losing electrons. Carbon has a total of six electrons, which means it has four in its outer shell ready for bonding. This gives carbon immense diversity in the way it can bond with other atoms. It can easily form bonds with multiple different atoms at once, donating multiple electrons to single atoms if need be. Carbon atoms can bond with each other easily as well, forming long strings or very strong rings. Furthermore, carbon is the smallest atom with four valence electrons, meaning that it is more abundant than any other elements with the same property, and also meaning that the nucleus is able to hold onto its electrons more strongly, giving carbon compounds amazing stability. Thanks to this versatility, basically everything that is alive, was alive, or was made by a living thing, has carbon as its main building block. The element is so intimately tied to life that the very definition of an organic molecule is one that contains carbon in the center, surrounded by all sorts of other things.

What are amino acids that make up proteins and muscles made of? Those are just carbon chains with nitrogen and two oxygen attached to it, one of which swiped two electrons, with hydrogen bonding to everything else. What about carbohydrates, the source of energy for living organisms, like glucose, sucrose, and lactose sugars? Those are nothing more than chains or rings with equal parts carbon and oxygen, again with every other bond being filled with hydrogen. How about lipids, such as the fat that stores our energy, or that build our cell walls, or makes up our vitamins? Again, they are just complex arrangements of carbon, oxygen and hydrogen. It's what's in your food, what makes up your clothes, your hair, your nerves, your veins, your hormones, and your DNA. Carbon is what's left behind millions of years after plants die and what living plants siphon out of the air to grow. It is the very infrastructure for which life was able to crawl out of the sea and overtake every nook and cranny on the surface of the planet. Carbon, carbon, and more carbon, facilitated by water with some nitrogen and phosphorus tossed in to add a little bit of spice

So now we have a better understanding of what connects us to all the vast numbers of species throughout the planet. But there is a greater connection that spans far beyond our home. In order to fully understand what makes us unique, we must understand the ingredients that make up the cosmos.

What is the universe made of?

What's inside of an atom? It's sort of an ironic question, considering the ancient Greek origin of the very word, *atomos*, means indivisible. But as we have seen, the different behaviors of each element are a direct consequence of the fact that they are comprised of various quantities of three different ingredients. However, there are pieces of matter confined even within these incredibly tiny objects as well. They are so difficult to pry out, that people could not observe them in action without the aid of city sized tubes with the power to smash those particles together. But with some international collaboration, state of the art technology, and a whole lot of quantum mechanics, teams made of thousands of researchers have developed what is perhaps the most successfully verified theory in all of science, the standard model of particle physics. It will take us to the very forefront of human understanding into what makes the universe tick.

The standard model lays out the foundation for how to build the universe, with just one key exception, which we will discuss shortly. It lays down the building blocks for not just matter, but all the reactions between their building blocks, and they have been given by far some of the most futuristic names of any natural phenomena. First we will set out the nomenclature, which alone can be daunting, and then follow up with the physics of how they make up the world around us. Let's begin with the particles that make up all the tangible stuff there is, such as you, planets, and stars. These particles are known as fermions, and they in turn come in two types, called quarks and leptons. The names of the quarks are up, down, strange, charm, top, and bottom. While all of these are fascinating in their own right, the ones of most interest to us are the up and down quarks, because they define protons and neutrons. At any given time, protons are made up of two up quarks and one down quark, whereas neutrons are reversed, made of two down quarks and one up quark. The second types of fermions are called leptons. The different leptons include the famous electron, in addition to muons, taus, and neutrinos. Neutrinos are ghostly particles that barely interact with any other form of matter, and there is one distinct type for every lepton. That is, there is an electron neutrino, a muon neutrino, and a tau neutrino.

The second fundamental pieces, besides fermions, are known as the gauge bosons. These particles can be thought of as mediators, or particles that carry the forces of nature. There are four fundamental forces of nature, some of which should sound familiar. They are gravity, electromagnetism, the strong and the weak force. The gauge bosons are exchanged between the matter particles, which allow matter to interact, like people tossing a ball back and forth between each other. There are four known gauge bosons; they are the photon, the gluon, the W, and the Z. Lastly, there is also the Higgs Boson, which is thought to give some of the fundamental particles mass. This is crucial, because particles without mass will only travel at one speed—the maximum speed anything can move in the universe—the speed of light. In essence, interacting with the Higgs weighs down some of the particles so that they aren't aimlessly zipping around through the universe.

Alright, review time. First, we have four bosons, which carry forces. Second, there are eight total particles of matter, six of which are quarks, and six of which are leptons. Of these, three of them make up the matter we are all made of. Three, the neutrinos, almost don't interact with anything, and are nearly massless. The rest are actually more massive than the particles you and I are made of, which means they take large amounts of energy to produce, and only exist for short amounts of time before decaying. The final key aspect to remember about this whole picture is that for every particle we've discussed so far, there is also a so-called anti-particle to match. The only difference between the "normal" matter that we are made of and anti-matter is that it has opposite charge. The best example of this is the electron, whose antiparticle is the positron. They are exactly the same in terms of mass and spin, but the positron has a positive electric charge while the electron has a negative one. If a particle happens to not have a charge, then it can be (but isn't always) its own anti-particle. Lastly, matter and ant-matter annihilate when they come into contact with one another. They can't touch each other and still exist, because if they do they will turn into pure energy, which I think is one of the most awesome aspects of particle physics. That brings us to a crucial concept, which we will need before delving deeper into the subatomic realm.

There is one idea in physics that is known almost universally, whether by plucky science enthusiasts, news reporters, or time-honored researchers. It was put forth by the first physicist to ever become a worldwide celebrity, and one of the only scientists to become synonymous with intelligence itself. His name was Albert Einstein, and this idea is best expressed as an equation:

$$E = mc^2$$

Einstein taught us that mass and energy are really one and the same, and the amount of energy that an object has (at least while it's not moving) is proportional to the speed of light (given by the letter c, which is always the same number) multiplied by itself. This has profound implications for how one can look at the world around them. It means that a cup of coffee is actually heavier after you heat it up, since objects that are hot have more internal energy than objects that are cold. It means that everything we can see, including you and me, is at its core just energy. This is also why matter and anti-matter turn into pure energy upon contact. Critically, this also means that the reverse is true; if there is enough energy, a particle will form along with its anti-particle twin. This equivalence is also why the Sun and all the stars shine. Recall the composition of helium, two protons and two neutrons. While this is only a summary, essentially a helium nucleus is formed from four separate protons, two of which turn into neutrons. If one had a tiny scale to measure the helium nucleus, they would find that it was actually less massive than the sum of the four hydrogen originally used to make it. The extra mass was released as energy in the form of light.

Incidentally, this also brings up a seeming paradox when one considers massless particles such as light. After all, if photons carry energy, then why don't they have mass? It's very subtle, but it has to do with how there is energy in movement. The famous version of Einstein's equation, as it turns out, is actually just an approximation for particles that aren't moving. The full version of the equation actually needs to take into account the momentum that particle has (given by the letter p):

$$E^2 = (mc^2)^2 + (pc)^2$$

You'll notice here that if a particle has no mass, then m is zero, and the energy is just a product of the momentum and the constant speed of light. In addition, if the particle isn't moving, then p is zero, and presto! You get the original, famous version of the equation that everyone knows, and this is why it is often referred to as the *rest energy* of a particle. This is the answer often given to undergraduate students, and frankly, I kind of loathe it. This sort of explanation always seemed lackluster to me, partially because it evokes equations and mathematics, which can seem intimidating if you don't speak the language, but more importantly because it skirts around the issue by bringing momentum into the picture, and momentum is taught to first year physics students as being defined BY mass! While a professional physicist is taught how classical physics developed before 1900 is too just an approximation, and barely even applicable to the subatomic world, this has always been a serious point of contention for me. Instead, I personally find it easier to work from the ground up with particle physics.

One way of understanding what makes all of the fundamental particles in the universe different is by understanding how they interact with one another, and what forces apply to each. Let's compare the electron to the photon to see how this works. I mentioned before that the particle responsible for giving particles mass is the Higgs boson, but this isn't precisely the case. Beyond quantum mechanics, where basic matter can be thought of as particles and waves, actual tangible things, there is something known as quantum field theory. A full description of this would require several textbooks worth of information, but in essence, quantum field theory states that all the fundamental pieces of the universe are in fact fields. They permeate the entire universe, and there are separate fields for every fundamental particle, and the things we observe as particles are actually just excitations in those fields. For example, this model describes there being an electron field, an up quark field, and so on. The common analogy used is to think of a calm body of water as being a quantum field. When the water is jostled, ripples form along its surface, and in some places the water will be higher than before, and these excited locations are equivalent to the particles of nature we observe. Some of these fields can interact, and cause ripples in each other. While this is indeed highly abstract, it is helpful for understanding how the Higgs mechanism gives particles mass. Take for example the electron and photon. The electron has a tiny bit of mass because as it moves, it interacts with the Higgs field that is everywhere in the universe, and this causes them to "slow down" in a sense. This interaction gives electrons mass, and thus keeps them from traveling at light speed. Massless particles, such as photons, do not interact with the Higgs field whatsoever, and so always travel at light speed. This same logic can be applied to all the fundamental pieces of nature, and it will allow us to understand how each of the four fundamental forces ties the universe together.

We'll start off with the strong force, also known as the strong nuclear force, or if you're a physicist who gets their sole pleasure from confusing non-physicists, the color force. This is the force of nature that binds the quarks together to form protons and neutrons, and in turn binds protons and neutrons together in the nucleus of atoms. The gauge boson that is the force carrier for the strong force is the gluon (think "glue"). It has an incredibly short range, which makes sense seeing as it only works on the scale of a marble compared to a sports stadium. So gluons are being shared by quarks, jumping between them to and fro, and this keeps them bound together. But why do gluons only work on quarks, and not leptons? The answer comes from another horribly named property of matter, known as color charge. It's poorly named because it has nothing to do with any sort of color whatsoever, and while it seems like a conspiracy thought up by mad scientists, it's important to remember that they sort of had to make it up as they were going along. They were discovering properties of matter never before seen by human eyes, and they just had to come up with a way to describe them that they thought best fit what they were observing, so I suppose we should cut them some slack. Anyway, you may think of color charge in a similar way to the more familiar electric charge. Just like how something can have either a positive or a negative charge, quarks can have either a "red", "blue", or "green" charge, and this means that anti-quarks have charges of "anti-red", "anti-blue", and "anti-green", which just begs confusion. But this is why color charge is referred to using this nomenclature. For you see, quarks can't exist on their own. If you were to try and pry a quark out of a proton, the energy you would need to do so would be enough to produce a quark—antiquark pair. Furthermore, quarks can only exist together if all of their charges cancel out. The reasoning comes from mixing different types of light. When you observe all colors of light at the same time, you get white light. Similarly, in order for multiple quarks bound together to be stable, their colors need to cancel out, making them "white". When quarks form stable particles, like protons and neutrons, they form what are known as hadrons. Hadrons with three quarks are known as baryons, and when you have a quark—antiquark pair (also bound by gluons), then you have what's known as a meson. Apologies for all of the jargon, but at least the names of the particles are fun to say, and now you know what they are!

As an aside, the idea that hadrons need to be "white" in order to be stable really rubs me the wrong way, although I'm sure that I'm just overthinking it. The charges could have been called any number of things after all. They could have been called bacon, lettuce, and tomato. Or perhaps Harry, Hermione, and Ron. Or Id, Ego, and Superego. They could have been named Navarro, Frenk, and White. If it needed to be simple they could have called it x, y, and z charges, although that might cause some false correlations with the dimensions of space. Rock paper scissors would have worked just as well (or toads snakes and slugs if you're in Japan). Heck, they could have just been up, right, and front for the quarks and left, down, and back for the antiquarks (although this would admittedly be biased toward people who are right handed an have eyes in the front of their bodies). But instead they chose something like "anti-red", seriously? That's just lazy. But I digress, back to physics.

The way protons and neutrons are held together is that they share quark and anti-quark pairs between each other. The meson that binds protons and neutrons together is called a pion, which is made of a down quark and an anti-down quark. The reason these particles are produced is because particles and anti-particles are always popping in and out of existence due to the high energy environment inside protons and neutrons. This is actually where most of the mass in the nucleus comes from. Sure, the three quarks have mass from the Higgs mechanism, but that mass is only around one percent the mass of a proton. The rest of that mass actually comes from the energy within that small space, from the energy of the particles zipping around the gluon field. So, your mass really is intangible energy (wow!), and that's enough energy for matter—antimatter pairs to form, one example of this being the pion. The thing is, these pairs will hit each other and turn right back into pure energy almost as soon as they appear, which is why they only exist for short amounts of time. Protons and neutrons are stuck together because pions spontaneously pop into existence in the high-energy environment of the nucleon, and fly into another nucleon. The anti-quark annihilates with the down quark of the second nucleon and is replaced by the down quark produced in the first one. This also explains why nucleons need to be so close together in order to bind, because they are kept together by particles with anti-matter. But without this force, stars wouldn't shine, plants wouldn't grow, and the whole universe would be nothing but a soup of quark-gluon plasma. Next lets move on to what is probably the least known fundamental force of nature.

The weak force, or weak nuclear force, governs how subatomic particles can transform into one another. You are probably most familiar with it as the force that makes atoms radioactive. It turns up quarks into down quarks or visa versa, therefore turning protons into neutrons or back again. It is with this force that the normally anti-social neutrinos shine. As a rule of thumb, anytime there is an interaction involving a neutrino, it is because of the weak force. There are actually two gauge bosons that mediate the weak force, fittingly with the most cryptic names, W and Z. These abbreviations actually make some semblance of sense, the "W" stands for weak, and "Z" comes from the idea that this boson was the last one needed to explain subatomic interactions. The Z boson mediates the transfer of momentum and energy between neutrinos and other matter, and has no electric charge. However, it is crucial to remember that this almost never happens. The reason I referred to them as "ghostly" earlier is because a single neutrino could travel through a light-year of lead as if it were empty space. In fact, thanks to the other gauge boson for the weak force, some six hundred and fifty trillion neutrinos are constantly steaming through your body every second from the Sun. They are so difficult to detect that scientists need to build hundred thousand gallon tanks of water just to observe them.

There are technically two kinds of W bosons as well, if this whole thing wasn't convoluted enough. But the reasoning is sound, because this boson is the key player that causes radioactivity. The two W bosons, each with different electric charge, mediate the transformations from protons to neutrons via electrons and positrons, and visa versa. A neutron might decay for several reasons. It might be because an atom is unstable. This is the case for most elements heavier than iron, and unstable isotopes of elements (like a carbon with four or eight neutrons instead of six). They are radioactive because they have so many protons packed into their nuclei, that the electric repulsion between those protons makes them unstable. If the nucleus is quite large, it will simply vomit out a helium nucleus, but other times its neutrons may decay into protons, changing one element into a new one. This occurs as the neutron releases a negatively charged electron and an electron antineutrino. This is rather remarkable, to think that protons, electrons, and neutrons are connected in such a way. Not only can neutrons turn into protons, but protons can turn into neutrons as well, instead by releasing a positron and a regular neutrino. This was the crucial missing piece we needed to explain how the Sun and other stars fuse hydrogen into helium.

Finally, the weak force plays a critical role in the phenomena that make both planets and life possible, supernova explosions. Within the cores of supermassive stars, atoms are put under so much pressure that the only reason they don't collapse into black holes is because the weight of all that matter is balanced by the outward force of light from fusion, and because fermions cannot occupy the same place at the same time. But when fusion no longer produces enough energy, the octillions of tons of star matter bear down on the core without restraint. If the electrons are like balloons around the nuclei, then this weight is enough to make them pop. The core collapses all at once, like a stadium instantly being crushed to the size of a marble. All the mass of our entire solar system goes from being the size of the Earth, to the size of New York, and a neutron star is born. The electrons crash into the protons, they neutralize each other—exactly opposite of a neutron decaying into a proton—and electron neutrinos fly out in all directions from the newly formed neutrons, so many that it stops the rest of the star from falling into the core, ejecting all that material out into the cosmos, thus allowing planets, and people, to exist. That's how neutrinos, thanks to the W bosons, play their role in making life possible.

However, none of this would mean anything if the neutrinos couldn't actually push on other particles, and this is where the Z boson comes into play. While the two W bosons can change the electric charge of particles through neutrinos, the Z boson allows those neutrinos to actually change the way other particles move through space. On their own, they will decay into matter-antimatter pairs, but they are also the only means neutrinos can affect the other fermions. We already discussed how electrons spin, and how an electron can only spin an exact amount and in only one direction in one dimension. What I have yet to mention is that bosons also spin, but twice as much as fermions, and this is also what distinguishes the two. In addition, an electron's spin isn't set in stone. As a matter of fact, an electron moving in a straight line through space will constantly switch between spins. This is done through the Z boson, which can flip the spin from going one way to the other. This is the key to the famous Higgs mechanism. Just as some particles have the color charge and electric charge from the other two forces, the weak force also has an associated "weak hypercharge". Any particle that has this charge interacts with Z bosons in such a way, constantly changing the properties of that particle. They do this through the Higgs field, which is sort of like an infinite pool of weak hypercharge. Because theses particles are constantly absorbing and emitting Z bosons, and changing their momenta, they become confined and gain mass. It should be noted that professional particle physicists who have been studying these phenomena for decades may rephrase and refine these definitions, but this is sufficient for a basic understanding.

Now that we've covered the atomic forces, we can zoom out of the nucleus, and discuss the force that binds molecules together, electromagnetism. As anyone who has ever seen a battery knows, electricity involves positives and opposite negatives. Same charges repel, and opposite charges attract. Similarly, magnets also have two opposites, namely north and south. The reason they are paired together is because they are actually one and the same! Electricity creates magnetism, just as magnetism can drive current. If you were to place a compass over an active wire, you would notice the compass move. Similarly, you could move a magnet through a loop of wire, and notice that the wire would suddenly have current flowing through it.

As it turns out, the reason they are the same is thanks to Einstein's special theory of relativity. While the details and consequences of special relativity are complex, anyone can understand the basic idea just by sitting in a car or train. If you're driving and moving at a constant speed, not slowing down or speeding up, and then pass by somebody else, they will look like they're moving backwards. However, if you were the one standing on the sidewalk, obviously you would see that the car was in fact moving. Now let's use an example that isn't as obvious. Imagine that you are an astronaut, floating free in the infinite expanse of space. There is no air to slow you down, and your jet pack is all out of fuel. Then all of a sudden another astronaut flies past you. How do you know who was moving? The answer is that you can't, so long as neither of you change how fast you're moving. The situation would look exactly the same regardless of who was standing still and who was actually moving. The answer would only be different depending on who's perspective, or reference frame, you are talking about. This is why the electric field made by a charged object creates a magnetic field only when it's moving, because it's all a matter of perspective. What one object sees as a magnetic field another object sees as an electric field, and visa versa. This is important, because it also gives insight into the properties of the boson that mediates this force, which we call light.

Everybody is familiar with light. It falls into our eyes and lets us experience the world around us. We cannot go to distant planets or galaxies now, but we can learn about them by studying the light they emit. It's so critical for our daily lives that one of the most common human fears is darkness, which is nothing more than the absence of visible light. But we don't just need light to see with our eyes. Rather, it is the near universal way all things, living or not, interact with one another. First consider what matter is. Whether it is a rock, the air, water, or humans, all the things we see and interact with every day are held together by molecular bonds at the subatomic level, nothing more than atoms filling their valence shells in countless unique patterns, like so many puzzle pieces. The electromagnetic force holds those bonds together, which is subsequently maintained through the exchange of photons. The reason that you aren't pulled to the center of the Earth due to its gravity is because the outermost electrons of the ground repel the outer electrons of your body. How are skyscrapers built and what allows trees to grow several stories tall? Again, rigid molecular bonds. What's the difference between a lump of coal and a diamond? How the carbon atoms are bonded together by sharing electrons. Now think about the other senses you use to understand the world. All the things you feel are just chemical reactions, atoms coming apart and back together, technically at random. Sight works when packages of light (photons) land on nerves in your eye and send electrical signals to your brain. You smell when molecules land on receptors in your nose, which sends electrical signals to your brain. You hear when vibrations traveling through a quintillion molecules of air hit your eardrum and vibrate sensors in your ear, sending electrical signals to your brain. You taste when molecules in your food bind to receptors on your tongue, which send electrical signals to your brain. And of course, you touch by actually placing your molecules on the molecules of other objects, which excites the nerves on your skin, and well, you get the idea. Our whole existence is tied to electricity, and in a certain sense, to light itself.

But humans aren't just limited to these classic five senses. You and I have a whole set of electrochemical tools for reacting to our environment, just a few of which include balance, pain, and my favorite, temperature. Some of us even have a sense of right and wrong! It's crucial to note that the light we use for these various senses is very rarely the same light we use to see. As a matter of fact, light comes in an entire spectrum, where the light we see consists of only a small sliver. This spectrum varies with light of different energies and wavelengths, which you can think of as how close the peaks of the wave are to each other. Light waves are sort of like springs in this way; the more tightly they are packed, the more energy they have, and the more they are stretched out, the less energy they have. Most commonly, the light we see is broken into a rainbow, with the least energetic light being red, followed by orange, yellow, green, blue, and finally violet. The visible part of the spectrum is more or less in the middle, with lower and higher energy light going on in either direction. It has incredibly small wavelengths, just less than a micrometer. One beautiful piece of evidence for the small size of the light we see can be captured on soap bubbles, which can be seen under a microscope to have complex rainbows of light adorning their surfaces.

Light more energetic than the light we see begins with ultraviolet (or *above* violet), which is the light that causes sunburns and can be seen by some birds and butterflies. More energetic than U-V light are X-rays, which doctors use to look at your bones, and are energetic enough to damage human organs. Lastly, the most energetic light come in the form of gamma rays, which may cause mutations, but are far more likely to kill you than give you super strength. Light with less energy than the light we see starts with infrared (or *below* red). Infrared light is probably the most integral for our survival, because it is the light we interpret as heat. Infrared light has the same energy it takes to wobble small three or four element molecules, including water, carbon dioxide, and methane, which is what makes them such effective greenhouse gasses. Objects that are around room temperature also emit infrared light due to the random motion of their atoms, which means that you and I are actually glowing, just in the infrared. That's how IR cameras and snakes can see in the dark. Less energetic than IR are microwaves, which we use to heat up our food since they too excite water molecules. On the opposite end of the electromagnetic spectrum from gamma rays are the radio waves, which we use to communicate and listen to music on long road trips. There aren't defined boundaries between each class of light, but it is useful to be able to distinguish between them.

So what exactly is light anyway? Like all subatomic particles, they can be thought of as both waves and quantized particles, which again we call photons or electromagnetic radiation. The waves of light are made of two perpendicular pieces, a wave of electric field, and a wave of magnetic field, which create each other as they move through space. But as we have already discussed, these two waves are really one and the same. This is sort of insightful, because it allows us another way to visualize the more foreign gluons and weak bosons. It might be useful, although perhaps not completely accurate, to think of a gluon as a wave of color charge and the W and Z bosons as waves of weak hypercharge, just as photons are waves of electric fields.

That concludes our review of the standard model. Let's try and put it all together to see what can we take away from all of these discoveries. First we have the six quarks. They have color charge, electric charge, and weak hypercharge, and thus interact through the strong, electromagnetic, and weak forces respectfully. There are up quarks as well as charm and top quarks, which are essentially just more energetic versions of the up quark. There are also down quarks, which also have more energetic counterparts in the form of strange and bottom quarks. Then we have the six leptons. Three of these have electric charge and weak hypercharge, while three (the neutrinos) only have weak hypercharge. Again, the muon and tau are basically the same as the electron, just with more energy. Together these eight particles make up all matter in the visible universe.

Interestingly, the number of forces a particle is subject to also illuminates a trend amongst fundamental particles. One can see a direct correlation between the masses of these particles and the number of forces they are subject to. Specifically, neutrinos are the least reactive, and the least massive. Next are the electrons, which are slightly more massive, and react through two forces. Finally, quarks are the most massive fermions, interacting through a whopping three separate forces. This would suggest an answer to the question of what mass is. We know that matter is energy, and so perhaps greater mass is based on nothing more than how confined that energy is by other particles. Particles could subsequently be thought of as extremely bunched up waves. This is an oversimplification however, as while gluons are massless, they do interact with each other, and the weak bosons have more mass than any of the fermions. But hey, no one said particle physics would be simple.

Now wait just a minute. We've spent so much time discussing all of these forces, but we've neglected to cover the very first thing you might think when you hear the word force; gravity. Here's the problem: gravity isn't part of the standard model. It's the most egregious hole in all of contemporary physics The force that tethers planets together, keeps your feet planted on the ground, the same force that ultimately kills stars, doesn't exist in the depth of the subatomic. Their distances are so miniscule, that the same force that causes whole galaxies to collide essentially disappears. Some jest that if they were to base their entire worldview by watching quantum particles, they wouldn't have the slightest clue that gravity even existed. More remarkably, using our current understanding of gravity, which has thus far allowed us to send giant cameras to every other planet in the solar system and twelve humans to the moon, all of existence shatters at the scale of quanta. In the depths of a black hole and the birth of the universe, there is a gigantic blank page inside every student's textbook. To understand why, we must again consult Der Physiker himself, Einstein.

According to Einstein's general theory of relativity—so called because the special theory didn't account for when the astronauts use their jet packs to change their speed—describes the fabric of the universe as a whole. Quite literally, the most common analogy is a bed sheet fabric. In essence, the universe consists of three dimensions of space, and one of time. This of course is obvious. You only have board meetings at a certain time and location, and nobody makes a date without knowing when and where. You need to know how high or low, how far left or right, how much forward and back, and for how long. It's so obvious that it's put far back in our minds and we actually forget about it. Now, when you move through space from one place to another, eventually you're going to need to start moving and then eventually stop moving. When you start or stop moving, or change your pace in any way, you feel something resisting you.

When you crash your car, get decked in the schnoz, or ride an elevator, you feel that resistance. As a matter of fact, if you were riding a rocket powered elevator in space, without the weight of the Earth below, as long as you were speeding up, you would have no way of knowing that you were in a vacuum. Acceleration and gravity are one and the same thing.

Now think of it this way, if you were skydiving, then you replace the firm ground you normally stand on for the air that blankets our planet. But you still feel a force pushing you upward against gravity. Both forces are molecular, and they will push up against your weight falling toward the center of the planet. However, the ground is solid, while the air is merely a sparse fluid, and you are a condensed sack of salty water much thicker than air. But the Earth doesn't care. No matter where you are, you are still falling down. You just don't usually notice because fat sacks of salt filled water are still less dense than rocks. Here's Einstein's genius insight: If gravity is the force that makes the planets constantly fall around the Sun, then they must be falling through something, and we know that we live in four dimensions. This, coupled with his insight that the speed of light is exactly the same no matter how fast you are going, led Einstein to claim that space and time themselves are a real, changing, tangible thing, and that thing can warp. In order for the planets to be falling in something, that something has to have geometry, shape. And what shape allows things to fall in circles the way planets do? The answer is an indent, like a bowling ball on a bed sheet. If there are no other particles around to hit the ball, it could move along the fabric for billions of years, just the way planets moons and galaxies all do. That is general relativity in a nutshell. Matter and energy warp the four-dimensional fabric of spacetime, and more massive objects—in other words more confined energy—warp that fabric even more. Energy tells space how to bend, and space tells energy how to move.

Special relativity ties into this picture rather well. When one travels very fast, near light speed, special relativity describes that space is contracted and time is dilated. In other words, if you saw someone in a car driving half the speed of light, they would look squished, and your time would be passing more quickly than theirs. If they went around the world in their car enough times, they could end up years into your future before their dinnertime. Space scrunches and time slows for objects that are moving relative to those standing still, because nothing in the universe can move faster than a massless particle. For objects like photons, time doesn't exist, and they experience the universe all at once. It's sort of like a boat speeding through a lake; moving objects drag spacetime along with them as they move. But the problem is, there is so little energy bundled up in the size of a quark, that these warps are not apparent.

The second problem is that gravity is by far the weakest force. The weakness of gravity isn't hard to prove to yourself either. Just pick something up, anything. Any object you could hold in your hands is being pulled to the ground by the weight of an entire planet that itself weighs sixty thousand billion, billion people, with nothing more than your hands. You out-lift the entire planet on a daily basis; you just couldn't lift two at once. If you were to quantify the strength off all the forces, gravity would be a million billion, billion, BILLION, times weaker than the weakest subatomic force, which itself is still a million times weaker than the strongest force that binds atoms and quarks together. As such, scientists are still at work daily attempting to unify gravity with the realm of quantum fields. They do have a name already picked out for the boson they think might carry the gravitational force; it's called the graviton. There are several proposals and theories to unify these two pillars of physics, including string theory, but at the present they are all awaiting experimental verification. The person or group that figures out the correct theory of everything will become the most acclaimed physicists since Einstein.

Pat yourself on the back, we've just covered all of modern physics and he fundamental laws that govern the expanses of the cosmos. Now we get to cover the strange anomalies that elude our explanation. Specifically, there are two pieces of the universe that rebel against our best efforts to understand nature. To add to the shame, they comprise some ninety-five percent of all the matter and energy in the universe. As such, I get full reign to tell you whatever I want about them, so long as it coincides with our observations. They are of course dark matter and dark energy, similarly named but wholly separate phenomena.

Dark matter is the excess curvature in space that cannot be accounted for by any of the matter that emits some sort of electromagnetic radiation. We can observe it by measuring how light bends around galaxies, bending caused by a lot of matter packed in the same place, just like galactic bowling balls on a cosmic mattress. It just so happens that those bowling balls are made of glowing plasma, and we can see other bowling balls behind the ones blocking our vision because the warps they create act like lenses on a pair of eyeglasses. We can also track how fast stars orbit through their galaxies, and we see that galaxies are spinning so quickly that those stars should fly out into the empty void of space. The only way galaxies can spin so fast and stay together is if there is more stuff that we aren't seeing. So what is it? It would make sense for it to be neutrinos, because they don't emit any light. They also have such feeble mass that one would expect them to orbit on the outermost fringes of galaxies for the same reason supermassive black holes are at the center, and models predict the majority of dark matter is on the fringes of galaxies. We also know that neutrinos are constantly being emitted by the Sun and other stars, and so presumably they must collect somewhere, if they feel even a little big of gravity. However, we can calculate the amount of neutrinos that should be emitted by stars and released after the big bang, and it doesn't account for all of it. We also have evidence that galaxies form by merging and then subsequently devouring smaller galaxies. Our own Milky Way galaxy has several satellite galaxies and clusters. Perhaps black holes remnant from the early universe could be orbiting around galaxies as well. They could be captured, which may allow them to orbit rather than sink to the centers of galaxies. In any case, more data is needed.

Dark energy comprises three quarters of the universe's inventory of matter and energy. It drives the very expansion of the cosmos. As we gaze farther and farther away from Earth we observe galaxies progressively flying farther and farther away from us. You read correctly, the universe is expanding, and that expansion is speeding up, which is crazy! That's as if you tossed a ball into the air, and instead of slowing down, stopping, and coming back to you, it just kept flying upward, faster and faster. It's unfathomable, but that's what we see happening, and we can verify it multiple different ways. But it's not so much that the universe has an edge and is being stretched out. Rather, it's more like the amount of space in the universe is somehow growing, in all places at once. In places with plenty of matter, the force of gravity keeps objects together, but on the scale of galaxy clusters and superclusters, everything is getting shoved apart, as the amount of empty space between them growing. Well, not exactly empty space.

As it turns out, some hints into the nature of this energy counteracting the gravity of all the matter in the universe come from the subatomic world. Recall the hydrogen atom, with the dense sphere of gluon quark plasma like a marble, and a shell of an electron zipping around it like a stadium around the marble. What's in between the two, filling up all of that vast space? One might assume that it is empty, but quantum mechanics tells a different story. In those gaps lie electric fields, mediated by photons, both of which carry energy. There isn't a single point in space that is truly a "vacuum"; there is always some energy in empty space. There is even a quantum principle that supports this. The famous Heisenberg Uncertainty Principle—which states that it is impossible to know both the location and the motion of a particle with equal certainty—also allows for energy to pop into existence for certain periods of time. The more energy it is, the less time it can exist, and the less energy, the more time it can last. This was first demonstrated mathematically, but it can also be visualized somewhat from the idea of quantum fields. If there are these so called particle fields pervading the universe, like lakes that produce fermions and bosons when they get agitated, and if they can interact with one another, then why wouldn't they be constantly jostling, maybe to the point that matter-antimatter particles actually form? If this model is correct, then it doesn't seem to be all that surprising that we observe there to be energy inside of a "vacuum".

The second law of thermodynamics fits into this picture perfectly as well. It states that the universe tends toward disorder, or entropy, as energy is evenly distributed throughout a confined space. The universe is expanding, so it would seem reasonable that the confined energy in matter is gradually being dispersed into a diffuse vacuum. So there isn't more empty space coming from nowhere, it's all just the same energy being spread out evenly though the universe. When the universe began it was hot and confined, with all the forces being unified, and as time past that congealed energy just diffused into "empty" space. This however, still doesn't fully explain why the vacuum in the universe is expanding. The problem is again a conflict between quantum theory and observation. The expansion we measure of dark energy dictates that within a cubic meter of a vacuum, the energy is equivalent to about a thousand quarks. However, quantum field theories suggest that there is more energy in a cubic centimeter of empty space than all the energy in the universe, which is a problem. Perhaps as matter fields dissipate their energy, they get lost inside of the vacuum itself. It may very well be that there is a sort of "ether" in the universe, it just so happens that tracking photons like Michelson and Morley is the wrong way to check. In any case, more data is needed.

~A reminder, if you're dying to know how on Earth I can be so confident in saying all of these things, we will go over some actual experiments and observations in Chapter 6, but for now just have fun and allow your wonder to wander~

That's it! That's the checklist of all the things our universe, ergo humanity, are made of. And it only took us a few thousand years to figure it out too! Not bad for some hairless apes. There are still a few holes, like what charge and energy are, and how they affect space, but all in all not too shabby. Yet even with centuries of astronomical discovery, we have hardly gotten any closer to answering the question of what it means to be a human-like person-type character. But, with this background, we might be able to lend credence to the many other ideas that surround what makes humans the number one organism to date. Speaking of which, how do we tell the difference between a protozoa and a person?

Are you your Genes?

All living organisms on the planet Earth have two things in common. For one they both have an outer layer made of some sort of carbon-based membrane, hence why the most basic units of life are called cells. What lies inside each of these cells is a long chain of genetic information, which acts as the blueprints for each and every protein manufactured by the cells to help them survive. The cells of animals, plants, fungi, bacteria, and viruses all have the same genetic information within them. The only difference is how much and in what order.

However, you aren't simply your cells. You actually get a new body every score or so. After fifteen years, your outer layer of skin, both inside and out, will have served their duty, passed away, and been replaced by fresh young recruits. Your skeleton, muscles, liver, essentially everything except your brain, get replaced several times over the course of a lifetime. But they all have something in common. Each and every cell in your body has the same package of genetic instructions, with tiny pieces used to create every protein and enzyme that makes every piece of your body. These instructions are made up of only a few ingredients, each piece having about two dozen atoms of carbon, oxygen, hydrogen, nitrogen, and phosphorus. These units stack together, creating a spiral staircase like structure, except with only one rail. The various groups of stairs on those staircases encode the genetic information that leads to the physical traits of all organisms on Earth. Within the cells of the most complex creatures, there is a second membrane inside of the first, and inside these enormous shells chains of these instructions are stored, where they form around partners, creating a complete spiral staircase. This is the famous double helix, which commands the cell to create all the wondrous complexities of nature, including all of humanity. This life-giving molecule is of course the one and only Deoxyribonucleic Acid, or DNA.

Inside of our less complex ancestors, there are single strands of a molecule slightly smaller than DNA, and likely far older. With one additional oxygen, one less carbon, and two less hydrogen, long strands of this one-railed staircase are known simply as ribonucleic acid, or RNA. Unlike DNA, which get wound up with itself and around charged proteins into chromosomes, RNA coils up in a wider variety of ways, such that it can decode both itself and DNA. Actually, many of the cell's organs, or organelles, are comprised of structures made from amino acids that are brought together by bundles of RNA. All of the things inside of a cell are exactly identical to the things we need to eat. Lipids make protective membranes, amino acids build together to form proteins that do the various work of providing structure, transport, and energy production from breaking apart carbohydrates. You and your cells need these nutrients to replicate so that you can grow and heal, and in order to make all of the different chemicals and structures you need to live. The way different cells read your DNA dictate their function and shape, leading to every one of you organ systems, from your digestive system to your lymphatic system. The only difference between a human, a banana, and a bacterium, is the number of genes they have, the order they are in, and the proteins that they create. Over the last four billion years, the development of every creature on Earth from one ancestral single celled species is thanks entirely to the growth and development of these nucleic acids.

The instructions for living organisms are molecules of deoxyribose sugar rings with two other molecules bonded to it. On one side, the sugar is bonded with a phosphorous atom surrounded by four oxygen atoms, and on the other side, it is bonded with one of four unique rings of nitrogen and carbon. All DNA and RNA have the same backbone of phosphate, but they differ in the type of sugar and the types of nitrogenous bases they use. The elder molecule, RNA, uses ribose as its sugar, a five-sided molecule with a variety of atoms bonded to each point. DNA on the other hand uses deoxyribose as its sugar, which has one less oxygen in its assortment (hence the "*de*-oxy"). This may affects how malleable the molecules are. With another voracious oxygen, RNA may be able to contort itself more easily than DNA.

If the binding of the book of life was the phosphates, and the pages were the sugar, then the words would be one of the five nitrogenous bases, each one having one letter. Inside both nucleic acids are the bases Adenine, Cytosine, and Guanine, with the fourth bases being different for RNA and DNA. They are uracil and thymine, respectfully. Again, the similarities far outweigh the differences. The uracil inside RNA has a lone hydrogen atom connected to one of the carbon atoms that make up its shape. In its place, thymine has an additional carbon, which itself has three hydrogen bonded to it (methane). Unlike the sugars, the nitrogenous bases are hexagonal, at least in general. Cytosine, Uracil, and Thymine are purely six sided shapes, of course with additional molecules on the edges to fill every outer electron shell. Guanine and Adenine are actually slightly larger. They essentially look like hexagons attached to pentagons, and this is a good way to remember how your DNA is structured. The famous double helix is formed with exactly precise electrical bonds between the nitrogenous bases of two DNA strands. But they don't share electrons per se. Instead the charges of each base are such that only certain bases will be attracted to them. In DNA, these base pairs are between one large and one small nucleobase. Guanine always pairs with cytosine, and adenine always pairs with thymine or uracil. Put together a few trillion of these tiny nucleobases, and voila, homo sapien.

There are so many of these molecules bonded end to end, that if you lined up all of them from a single one of your cells, that strand of DNA would be taller than you. If you lined up every single one from all the tens of trillions of cells in the human body, they would stretch to the Sun not once, but SIX-HUNDRED TIMES. There isn't a single person on the planet with the exact same combination of DNA as you, but we share 355 genes—around 7% of our DNA—with every other living thing that scurries, soars, or swims on this planet. Different animals may use the same protein in different ways as well. One protein, keratin, is the same molecule that makes up both the hair on our heads, as well as the horns of a rhino, which is why neither one can be used as medicine. These genes also code for similar functions. Our outermost skin, the foremost protection against invasion from lethal pathogens, is actually dead. The living cells are down beneath the outer layer, and when they die they collect into a protective shield that sheds its outer layers constantly, taking harmful microbes with them. The same function can be seen with the bark of trees. But the similarities don't have to be quite so abstract. One can see similar genes at play everywhere in nature. The same genes code for bilateral symmetry, as well as the eyes, ears, and even the skeletons of animals everywhere. We can chart the order in which different species developed on Earth as well by observing which features they have in common with all other living organisms. These common features outline our evolutionary history, which is why if you were to ask a biologist what you are, they would say: A Mammalian amniotic tetrapodal sarcopterygian osteichthyan gnathostome vertebrate cranial chordate. Each one of those words describes a common trait we share with other species.

It's almost miraculous how DNA goes from being a code made of four bases, to the gargantuan proteins that make up organisms. How does it work? In eukaryotes like us, it all starts in the nucleus, where a globular enzyme called RNA polymerase is constantly trailing long strands of DNA. Every gene on a DNA strand has terminal ends to denote where each gene begins and ends. When the polymerase finds the beginning of a gene to be expressed, it bonds with the double helix, unwinds it, and allows polymers of RNA to form, using the DNA as a template. The newly made genetic strand gets molecular caps placed on each end to prevent interaction with any other nucleotides. It then begins its way out of the nucleus. Here's something interesting; some of the genes in the new RNA are extraneous, and those pieces are actually cut out of the RNA by spicing enzymes. Once outside of the cell's nucleus and inside of the cytoplasm, the messenger RNA finds its way to a ribosome, essentially a clump of genetic information. Here, nucleobases on the strand are translated into functional proteins. This is done with three bases at a time. Then a separate RNA, acting as another enzyme, transfers amino acids one at a time based on every three base matches. As it goes through the messenger, a new amino acid is added to the previous, until eventually a protein is made, and once it travels through other organelles for polishing, it is ready to go to work for the cell or in the body. One by one, proteins are built from only twenty amino acids. Their electric charge warps them into unique folds, bends, and structures. Proteins can merge with others, creating superstructures with specialty functions of their own. That, in extreme brevity, is all it takes to go from a chemical string to a living organism!

Now in order for that organism to survive, its cells need to be able to multiply. In order for the average cell to divide, it must undergo mitosis, duplicating all of its internal structures in order to create a clone of itself, including their genetic material. Astoundingly, the process of cellular division is remarkably precise. However, this tight efficiency makes it difficult for single cells to adapt to their environment. If clones are perfectly identical, then even a slight environmental change or a widespread pandemic could wipe out large populations. To compensate, single cells and other asexual organisms multiply with large numbers, increasing the likelihood that random genetic differences will occur in the instance something catastrophic happens. This is likely why life was so simplistic for billions of years before the first multicellular organisms emerged. Eventually, organisms began swapping DNA, sharing half of their genetic information to create more diverse offspring. Thus organisms began to develop with two sexes. This is why only sexually dimorphic animals could develop higher levels of complexity, greater genetic diversity, and faster adaptability to their environments. Interestingly, all genetic information is transferred from parent to offspring randomly, and as such might be seen as a natural consequence of the probabilistic nature of tiny particles. In that sense, evolution is an organic product of quantum mechanics.

Biological divergence is also why there are still organisms that are nearly identical to their ancestors that date back to the first life forms. Crocodiles date back tens of millions of years, horseshoe crabs hundreds of millions. Even archeans, the modern decedents of the very first living things, all retain their attributes from when they first diverged from a more basic organism. Dinosaurs may have gone extinct, but their decedents still survive as countless birds and reptiles. Those species that haven't changed have become so specialized to their environments that few other animals can compete with them. Thus, simple organisms remain simple, while life as a whole grows more complex. DNA is also intimately tied to the means of which cells acquire and move energy as well. Plants and animals have special organelles that produce high-energy molecules for them to function. Animal cells have mitochondria, while plant cells have mitochondria and chloroplasts. The thing is, neither of those structures came from the cells themselves. They actually have their own separate DNA. It seems as though it would be more accurate to think of them as the remnants of old prokaryotes that simply weren't digested by primitive cells.

However, complex life might not just owe its existence to bacteria. Viruses, the enigmas of the microscopic world, and the epitome of pestilence, may actually play a crucial role in our survival as well. Viruses are the smallest individual units that carry genetic information, but they can only replicate by invading other organisms, inserting their RNA into a cell and hijacking its resources to make copies of themselves. This is far more insidious than bacteria, which merely devour resources needed by other cells. Viruses actually erode away a host until it explodes with thousands of new drones ready to spread an infection. Since they carry genetic information, but cannot self-replicate, many biologists debate whether they can really be classified as living things. My favorite explanation for these strange bundles of genes is that they may provide the key to understanding the origins of life itself.

When most of us think about viruses, we generally think about things such as the flu, the common cold, or a global pandemic. But despite their bad reputation, viruses may actually play several key roles in the development of complex organisms, and may still help animals survive today. We know that viruses have been around nearly as long as life itself. They live along side the archaea, some of the strangest and most ancient organisms still alive today. The vast majority of these organisms are prokaryotic, like bacteria, but they are also extremophiles. Archaea live in some of the most ridiculously lethal places on Earth, places that would normally be too acidic, too salty, too hot, or too dark for any other form of life to survive there. These extreme environments are similar to those that would have been far more abundant on the Earth after it first formed, and these organisms have the simple genetic information to support this idea. But they aren't alone. Even in those hellish landscapes, there are still viruses parasitizing larger organisms. This leads to an interesting hypothesis. It may be that viruses and the first complex cells diverged from a common ancestor, one becoming more complex, and the other staying simplistic, and simply evolving to "eat' its larger cousins. This idea is supported by a paper titled *Viruses in Extreme Environments,* published by Springer Science and Business Media, which goes so far as to cite viruses as part of the biosphere. Viruses may have very well been the first predators. It seems to make sense that a smaller organism would have developed before a larger one. Furthermore, viruses may not be able to replicate without a host or respond to their environment, but they are subject to evolution. This is why flu shots change every year, and how viruses are able to affect bacteria, plants, fungus, as well as animals even as they themselves adapt to the environment. As such, it seems apparent that viruses are a form of life; just one that has lost the ability to self-replicate in order to breed through a more efficient avenue.

Now, it is clear that viruses provide important population control on other potentially harmful pathogens, in addition to being dangerous to us. But it turns out their means of reproduction may actually spur adaptations in larger organisms. Research conducted on prairie voles suggests that viruses may actually be able to modify the genomes of cells without turning them into rampant virus factories. The experiment actually involved the mating psychology of these animals. They were deliberately infected with a manufactured virus carrying the gene for vasopressin, a neurotransmitter that has been observed to play a role in making animals monogamous, and indeed the voles displayed consistently less philandering behavior with the injection. This is exciting research for making use of retroviruses to artificially transmit genes into animals is still young, with potential implications for medicine and agriculture. Being able to add genes into another creature's DNA also leads to the question of whether viruses may have added to the genome of all organisms through history, leading to the creation of new and complex species, and giving evolution an occasional push forward.

The idea that all matter is energy becomes more fascinating in the context of this biology. Even though many cellular structures and organic molecules are often portrayed as shapeless blobs, it is important to remember that these are just approximations. Even DNA is often portrayed as nothing more that a solid object, but as we know in reality is made up of miniscule nuclear dots surrounded by huge clouds of electrons, and those clouds give the illusion of solidity. Cells are even more monstrous compared to those building blocks. For scale, a chromosome of bunched up DNA can be around a million times larger than an atomic nucleus. An animal cell is about a hundred thousand times larger than an atom, or around ten billion times the size of a nucleus. The width of a bacterium is about a tenth that of a plant or animal cell, and most viruses are a tenth the size of a bacterium. These figures mean that a million cells could fit onto a single grain of sand, but ten thousand viruses could fit on the average animal cell. There are of course exceptions. Mimivirus, Megavirus, and Pandoravirus are nearly as large as some bacteria. They show signs of being nearly large enough to self-replicate, but still need a host cell to multiply. Yet, somehow, these organisms are just run on particle physics. It is unclear how this can be, but it certainly applies to all living things, including us.

The relation of living organisms can be charted along a tree of life. It appears as a tree because as organisms become isolated, they evolve in response to their new environmental conditions, and split apart. This is the same as your own family tree, only charted over millions of years. If two groups become different to the point where they can no longer produce fertile offspring together, then they are considered to be new species. In that sense, species branch off from one another, and continue to diverge until things such as plants and animals become so dissimilar that they don't even remotely resemble one another. In that tree, species aren't allowed to jump between branches. Time only goes forward, and so descendants of new species are forever separated. However, as you go back in time, those branches come back together. Go forward in time, and environmental changes will make branches keep separating from one another. This is why humans didn't evolve from chimpanzees. That doesn't make sense because we are alive at the same point in history. We are cousins, who had a common ancestor that diverged into separate species. Go back far enough, and one sees how all mammals and birds developed at separate times from reptiles, who diverged from amphibians, who diverged from fish, from worms, to sponges, which adapted from the complex single cells that diverged into plants and fungi, all the way back to the first single celled packages of RNA.

Even so, it's still no wonder why people get uncomfortable when discussing evolution. After all, looking at chimpanzees is quite surreal, and mildly terrifying. They are so close to being human, and they are definitely thinking about something. The same is likely true for whales, elephants, gorillas, and some birds like parrots or ravens. But chimps are just so close. They have fingernails exactly like we have. They play in water and hunt with spears for things like termites and bush babies too. They regularly solve complex problems in order to survive. They might even replace us if we don't get our act together, wandering into laboratories and driving our cars. Those animals all recognize themselves and the faces of individual people. That's probably why people hate to admit how animal we are, because we often see ourselves as superior through communication. But it's more astounding to know how similar we truly are, because it means some nature has grown inherently intelligent throughout the ages, and we also have more friends to make. Plus none of them have rock and roll or sitcoms, so we would definitely take at least a few centuries to replace. Come to think of it, *Planet of the Apes* was slightly less fictional than most apocalypse movies. Also, no other organism can read. So if you make fun of an animal, never do it to their face, they can understand that mischief. Learn how to eloquently articulate yourself, and write it down. They'll never know, and you can vent some steam.

However, also remember why our closest living relatives are still so far from being human. After all, we were far from the first hominids to develop. We weren't even the first ones to figure out stone tools, fire, or cooking. All of our closest cousins from the last four million years have gone the way of the dodo. We out-competed four million years of natural selection, effectively burning down the nearest branches of our family tree in order to build a fancy tree house. If any of the other humans had comparable intelligence to an adult human, they got wiped out. The only ones left behind when we were through were the chimpanzees, which are barely as smart as preteens. But our similarities are undeniable. Even though we appear hairless, we do technically have as many hairs per square inch as great apes, as well as the same blood types. The reason humans have proven the more adaptable organism is because of a mere five percent of our DNA, which encodes for the development of highly robust and flexible neurons, as well as the organs to produce complex spoken language.

And this, at long, long last, finally leads us to the one thing that irrefutably, undeniably, distinguishes what it means to be a human being. We may be made of cosmic dust, and we may even have the same genes as every other living thing. However, only the human animal has the genes that code for those atoms to arrange themselves into a complex social brain.

Are you your brain?

All things inherently human originate in the mind. Or, to put it more accurately, the entire network of nerves throughout the body serve the function of letting us interact with the world around us. The nervous system, from fingertip to forebrain, is like quantum biochemical circuitry. It can even short circuit at times, and it too runs on electromagnetism. Without these connections, humans are little more than a lump of some trillions of cells unable to communicate with one another, barely more complex than any plant. The energy utilized by this vast network is enough to power an incandescent light bulb, and an entire fifth of your food is needed to power it. It is the most complex object in the known universe, trillions of connections crammed down to the size of two clenched fists. Welcome to mission control.

It's difficult to express all of the ways our brains shape us as human beings. The brain is such an enigma that humans have only been able to study it since the early twentieth century. For nearly all of human history, nobody knew what it was that gave us consciousness, language, and motor control. Now we understand that every emotion you have ever felt, thought you've ever had, and thing you've ever done, all take place between your ears, and some of it even gets memorized. After searching though the cosmos, we can see that what really defines a person truly lies here. It is so powerful that our thoughts can actually change our physiology. The placebo effect is the quintessential example of this. People can be given pills that a professional claims will do something, while really only being fake sugar tablets, and people will still feel as though they were given a real drug. They may even request a drug they know is fake just because it makes them feel like they are taking some sort of medication! Thinking happy or sad thoughts can alter your body, just as smiling or sulking can affect your mood. Seeing or hearing other people tell you something can affect your performance at certain tasks, and just telling yourself something enough times can make you believe it's true, regardless of reality. The brain allows athletes, engineers, and great leaders alike to perform tasks that might seem god-like to others. Some even claim that they can heal their wounds at a faster pace and perform other supernatural abilities with nothing more than sheer willpower, also a product of the brain. In many regards, the brain is existence incarnate.

As it is the source of our humanity, it is no surprise how difficult the brain is to study. You can't take it out without murdering someone, nor can you observe its function after it dies. The first researchers into the neural landscape were little more than clever counselors, who eventually developed the field of psychology. It wasn't until a better understanding of electromagnetism led to the development of magnetic resonance, that researchers were able to see the brain in action for the first time. But even with new advances, neuroscience is still in its infancy. As of yet we cannot get finely detailed images of a working brain, nor observe someone's brain without sitting them down with expensive and bulky equipment. But even with these limitations, we are still able to learn much about the organ that makes us human.

Some of the oldest means of learning how the brain affects us have come from observing people who have had parts of their brains damaged. One of the oldest and most famous case studies comes from Phineas Gage, a railroad foreman who had a metal rod shot though his skull in 1848. The rod went through the front of his brain, yet he miraculously survived for more than a decade after the accident. Common reports cite that he became far more irritable and less of a gentleman after the accident, but these stories were likely embellished, as there are no first hand accounts of this being the case. I can't think of anyone who would be happy after getting a hole in their face, although he did become a celebrity afterwards. Nevertheless, it does beg the question of how lost pieces of brain affect psychologically, and what parts matter most. The twenty-first century has given even more fascinating glimpses into the human brain. One heroic tale comes from a little girl named Cameron Mott, who had half of her brain surgically removed in 2007. Cameron grew up with a rare disease known as Rasmussen's Encephalitis, which led to her having heavy repeated seizures that may have lasted several minutes at a time, almost completely debilitating her. Miraculously, after her surgery, the half of her brain she has left has been able to compensate for its other half, and she now dreams of becoming a dancer.

While Cameron and Phineas demonstrate the resilience of the human brain, there are still examples of brain trauma that occurred without ever cracking the skull. A striking example of this is seen in Clive Wearing. Once a musician and conductor, Clive lost his ability to retain and create new memories after being infected by a herpes simplex. Ever since, he has become unable to remember anything for more that a minute at a time, forgetting conversations immediately after they occur. He has only been able to recognize a handful of people, most importantly his dear wife Deborah, who he excitedly gets to see everyday as if they had spent years apart. She has stayed with her husband, retelling him memories and helping him increase the amount of time he can recall events. Thankfully, he never lost the ability to play beautiful melodies on his piano. More interestingly, even without any memories of his past, his personality was still able to change, going from highly frustrated in his youth to more mellow in his age.

So what is this strange substance between our ears? Many people are likely familiar with the pink walnut shape with wrinkles and the texture some describe as being the consistency of mushroom. Of course, like all parts of the body, the brain is made of cells. Specifically, they are nerve cells, and they come in all shapes and sizes, serving various different functions depending on their location throughout the body. Nerves that sense acceleration, temperature, and touch are lined throughout the body, from the skin on the edge of our being to and around our organs. Many of these nerves are extraordinarily long, such as the nerves that run the length of our legs. They convene at the spinal chord, which transmits signals to and fro with the brain, and can signal to cause the body to move. All of these signals are of course facilitated by the large mass of nerve cells that compose the brain, which has two distinct cell types; neurons and glia. These cells are so tightly packed that a hundred thousand could fit into a single grain of sand, culminating into over eighty-six billion neurons and eighty-five billion glial cells. Together, these bodies create a complex network of connections able to store anywhere from ten to a hundred terabytes of data, more than ten times the data held on the Internet at the dawn of the new millennium.

The pieces of the brain can be thought of as rooms added to a billion year old house. Like our DNA, each additional piece of the brain also corresponds to our evolutionary history. Beginning at the neck is the brain stem, which is attached to the spinal chord. It contains structures such as the famous medulla oblongata, and regulates all the vital functions we don't think about everyday. Like for all the basic animals, the most fundamental brain functions include keeping the heart beating, our blood pressure regulated, the lungs breathing for oxygen, and our intestines digesting to obtain nutrients. Moving upward from the stem is the cerebellum. This part of the brain is also found amongst many animals, as it regulates movement, posture, and balance. When one commits something to muscle memory, they can thank their split-second reflexes on the cerebellum, tucked neatly under and behind the most prominent brain structure. This is the cerebrum, what most people normally think of when they think of the thing that thinks about thinking things. The first cerebrum likely developed during the Cambrian era, when large animals first began to spring up throughout the world's oceans. All of what we would call "thinking" is done here, but at different capacities for different organisms. Fish, amphibians, and reptiles all have basic cerebra. However, brain circuitry wouldn't become advanced enough for complex thinking until birds and mammals first evolved. In all of their shapes and sizes, cerebra give animals memory, and in the most intelligent creatures also grants reasoning, emotion, speech, and maybe even self-awareness. It processes senses as well, and can even mix them up in some cases, causing synesthesia in the case of humans, allowing them to hear smells and taste sounds. Within the center of the cerebrum is another old and crucial area known as the Limbic System, where emotion and fear are processed. This is the area that gives animals the famed fight or flight response to danger.

For neurons, as with people, the source of intelligence is communication. All of these billions of cells send signals through their bodies within a millisecond to transmit information that keeps animals alive and alert. Neurons are classified by their functions, specifically what ways they send signals to other neurons. The nerves that live in your sensory organs, like the skin, send signals to only one other neuron at a time, hence are called unipolar. They essentially just send signals straight to the central nervous system. The eyes have unique bipolar neurons especially for interacting with the three main wavelengths of light we see; red, blue, and green. Finally, there are the superstar multipolar neurons. In many ways they resemble an octopus with one long leg, and potentially dozens of shorter ones. The smaller extensions are called dendrites. They are nearest to the nucleus of the cell and act as their "ears", receiving messages from other nerves. The longest extension is known as the axon, and it branches off as well to send signals to other nerve cells. They are used to send signals from the brain to the body, but perhaps most crucially they are the reason for that weird thing where we know we exist. All neurons send signals through the same mean: synapses, which lay at the ends between axons and dendrites. Furthermore, these synapses might communicate with each other one of two ways depending on how far apart they are. They might send an electrical signal to converse immediately, or they might exchange chemicals known as neurotransmitters, for a more specific and relaxed conversation.

If you have your periodic table nearby (as all good thinkers should), you'll notice that both sodium and potassium are alkali metals. As we've discussed, this means they are most stable without that pesky outer valence electron of theirs, and losing that electron gives them positive charge. This is particularly fascinating because in order for current to flow, it is often easier to set up opposite charges and divide them—just like with a battery—and when you want current to flow, you allow charges to move from a place of similar charge to a place of opposite charge. The nerve cells accomplish this through the means of large negatively charged proteins and the sodium-potassium pump. This remarkable feature that these cells have uses energy to polarize the cell and give it a net negative charge. Specifically, each pump uses Adenosine Triphosphate, known as the cellular currency for energy, to kick out three sodium ions and take in two potassium ions. So thanks to the proteins, even though both ions have a positive charge, the nerve cell accumulates an overall negative charge. Interestingly, in their most common isotopes, potassium is about seventeen nucleons heavier than sodium, which may mean that it's easier to move. When the cell is stimulated, other valves open to let sodium flow back in, causing a current to flow from cell to cell. These gateways might open for several specialized reasons. Mechanical gates may open when the neuron is physically moved or when other gates open. Ion channels will only open when the voltage within the cell reaches a critical value of 55 millivolts. In other words, when enough sodium floods into the cell from other stimulation, the charge within the cell goes down, until a certain point when more gates open and truly allow a signal to pass. Otherwise, it will just be a false alarm. Lastly, there are ligand-gated channels, which only open when specific chemicals attach to them, stimulating the neuron.

When a neuron is stimulated, the part of the neuron between the cell body and the axon—called the axon hillock—generates the action potential that moves current, which then travels down the axon toward the axon terminals via a change in polarity. Mechanical sodium ion channels open first, enabling sodium to rush into the axon causing depolarization. Then potassium channels open and sodium channels close, causing potassium ions to rush out, establishing repolarization, which prevents the action potential from moving backwards. Then when the action potential reaches the axon terminals, calcium is released into the terminals, causing vesicle bubbles filled with neurotransmitters to fuse with the membrane and release chemicals into the space between axon terminals and dendrites. The neurotransmitters bind to receptors on the dendrites, inciting a signal to travel to the cell body, starting the process all over again. Then your arms move, your hands close, your mouth salivates, and you open the refrigerator to cook lunch.

While neurons get all of the publicity, there are also unsung heroes of the brain, the glial cells. As the name implies, when they were first observed, they were thought to be a sort of "glue", or scaffolding that held the brain together. It truth, they serve all of the necessary maintenance that allows the neurons to do their jobs, which is why they make up half of all our brain material. They regulate ion flow, assist in the transmission of electrical and chemical signals, and connect neurons to the blood stream, circulating nutrients. They also insulate the neurons, allowing signals to travel more quickly, like the rubber lining of electrical wires, thus insulating and cushioning the nervous system. They even provide immune support to the brain, protecting them from pathogens and destroying dead neurons. Together with neurons, these cells make up more connections than stars in the observable cosmos, masterfully melding the laws of chemistry, biology, and quantum physics. Thus leads to the wondrous complexity of what it means to be a human being. The complexities of the mind have led to some of the most profound questions about human existence, and now we are closer than ever to finding answers to them.

For instance, do you have a soul? It was once thought that the human spirit was altogether independent of the soft pudding that our bodies are made of. How can it be that something that can contemplate the wonders of the universe could be as brittle as the human body? Amazingly, it turns out that our fragile being is the stuff of stars, arranged so compactly and with so many connections that it leads to intelligence. But this still begs the question of how inert matter becomes sentient. Not only that, but what makes humans different, and how can we see this through our brains? What genetic or neurological roll of the dice leads people become great leaders or maniacal dictators? We can see that when certain parts of the brain are underdeveloped or lack communication, people lose their ability to do things that most of us find innately human, including feeling empathy for others. That's the only difference between a saint, a sociopath, and a serial killer, what's between the ears. And what of super geniuses? Great minds of math, science, and art, or even autistic savants who lack social skills but are prodigies of nuanced skills, what makes them special?

This of course leads back to the age-old dichotomy of nature versus nurture. What human acts can be traced to predisposed genetics, and what actions are caused by our life experiences? Are we all clockwork oranges—just mechanical organisms—or do we have free will? How is it that Nazis can be taught that tossing babies into the air for fire practice is ethical? How could Huck Finn grow up in a society where befriending a slave was a one-way ticket to hell? How is it that masses of human beings, just like you and me, can commit such heinous crimes and live proudly with those actions? Clearly both genetics and upbringing are crucial, but it's not altogether clear whether they are always independent of one another. The brain has miraculous plasticity, being able to adapt to losing half of its mass and being able to change old behaviors. People can fight their genetics through willpower and dedication, or at least come to understand their own predispositions and thoughts. It may even be the case that modern medicine will do away with all genetic defects that keep people from living happy lives. Clearly, there are no simple answers to these questions. Our mind is an expansive web of unique thoughts and connections, and every single cell contributes to that chaos in some way. But it is that same chaos that has enabled humans to conquer the planet Earth in the cosmological blink of an eye. The world we see every day is far more than meets the eye. After discussing all this, try looking at your hand once again, and imagine all of the wondrous complexity going on every moment just beneath your fingertips, from your skin cells all the way down to your quarks. Think about everything that goes into your brain moving the muscles in your hand and interpreting the light coming off of it, and that you are able to wonder about all of it at all. But don't stare too long. Somebody might think you're on drugs.

So, who are you? You are a collection of all the past experiences you can remember, some of which may be from the experiences of someone else, as long as you assume they are telling the truth. You can have these memories thanks to your brain and several trillion specialized cells working in unity, all manufactured by some assortment of proteins coded in your DNA, which developed over some three and a half billion laps around the Sun, from some tiny particles crammed together due to gravity, first forged in the uniform energy of the universe. You are an enormous citadel made of living specks, and a speck sitting on a gargantuan drifting pebble floating in the great ocean of spacetime.

So, you are you. Probably probability. At least possibly. Let's move on to something a little less abstract.

3 | Where Are You?

"You develop an instant global consciousness, a people orientation, an intense dissatisfaction with the state of the world, and a compulsion to do something about it. From out there on the moon, international politics look so petty. You want to grab a politician by the scruff of the neck and drag him a quarter of a million miles out and say, 'Look at that, you son of a bitch'. "

— Edgar D. Mitchell, Apollo 14 Astronaut

"The first day or so we all pointed to our countries. The third or fourth day we were pointing to our continents. By the fifth day, we were aware of only one Earth."

— Sultan bin Salman bin Abdul-Aziz Al Saud, STS-51-G Astronaut

You might have heard that space is big. This is true, but mostly just because we are so small in comparison. Maybe you've been told that the universe is old. Perhaps it is, but only because we only live for a few decades at a time. Really it's just a matter of perspective. Whenever I hear space enthusiasts, documentaries, or non-astronomers talk about the scales of the universe, they always describe them as, "unfathomable", "beyond our comprehension", or something of the sort. Then I went to college and actually took classes about these things, learned how to formulate astrophysical calculations, and suddenly all of the inconceivable lengths of time and vast distances became commonplace. When ideas and people who think about those ideas surround you on a daily basis, anything can become normal. When you're an astrophysics student, nobody bats an eye when you discuss the angular diameter of a nebula large enough to engulf eight of our solar systems. Nobody loses their breathe when you have to calculate the moment of inertia of an entire planet about the axis of one of its mountains. There are no second thoughts about describing the fate of the universe based on various parameters of visible matter, dark matter, and dark energy. We bang our heads against the wall, sure, after all those things aren't trivial to calculate or describe. But we still do it, and there isn't anything wondrous or unfathomable about it. It's just something you have to do, just like any other field of study. Of course, one can do all the math and take all the measurements they please, but that still doesn't make it easy to visualize what the universe actually looks like, because astronomers look from the Earth out into the universe, and most haven't seen it from the other way around. But I have, and I've shown it to thousands of people as well as a part-time job. As a matter of fact, that was how I paid for the majority of my rent in college. Nothing is glamorous once you've done it a thousand times, and that includes visualizing the entirety of the universe. Anyone can do it, you just need a little bit of confidence and neuroplasticity, and I'll show you how.

Our Place in Space

One of the most delightful aspects of working in a digital planetarium is that you have a literal universe of topics you can discuss with the public. Whether your forte is planetary science, astronomic culture throughout history, solar physics, or astrobiology, there is no shortage of interesting topics that in some fashion shape the way we live. But perhaps the most breathtaking talk a presenter can give their audience is the quintessential tour of the universe, where the audience is taken from low Earth orbit to the edge of our visible horizon in just under fifteen minutes. This ego shattering experience is given as the de facto short presentation for all demographics. It highlights the power of modern planetariums and inspires awe in everyone, from school groups to small businesses renting out the theatre for whatever reason. Never before in all of human history has it been possible for people to visualize so accurately their place in the cosmos.

Of course, everyone has different methods for presenting the universe tour. After showing my audience footage from a real life rocket launch loud enough to rattle their seats, I often like to begin the journey orbiting a satellite in low Earth orbit, like the International Space Station or the Hubble Telescope. Probably the most difficult aspect of studying space is visualizing the size of objects relative to one another, and accounting for the near-unimaginable scope of the cosmos. For this reason, it is often extremely useful to begin with objects that have sizes we can visualize, and then think of other objects relative to those. These analogs are crucial for perspective. Anyone can hold a globe in their hand or look at a flat image of Earth on a map, but neither one of those can capture the overwhelming immensity of our world compared to its inhabitants.

The Space Station is around the size of an American football field, and the Hubble Telescope around the size of a school bus, which may sound very different, but on the scale of a planet they are nearly identical. From their orbit, which is only around 370 kilometers above the ground, they are completely overshadowed by the Earth. Looking above theses satellites down on the Earth, the planet takes up one's entire field of view, and even looking edge on toward the horizon, more than half of the sky is blue marble. To understand the size of the Earth from home, you can try doing this exercise. Gaze out to the horizon. The farthest point you can see is about 5 kilometers (~3 miles). By comparison, from one edge to the other, the Earth spans about thirteen thousand kilometers. That means that you would need to see a whole eight thousand times farther to see around the Earth to the back of your head.

Even as one flies away, and the satellites fade from view, the Earth is still so imposing that one cannot see the whole thing. But once you get far enough away, almost no experience is more breathtaking. Amongst all the stars and nebula, there is not a single place that humans have found that is quite like our home, the one celestial body we can't see with a telescope. Just seeing it in a large dome is life changing. It is a testament to humanity, being the first animal privileged to see all the rolling oceans, mighty storms, and lush green continents that make our lives possible, all in one image. It is an experience that could only be aptly described using every word of every language, yet be articulated with none of them. And seeing the real thing could only be more magnificent than that. It is my sincere wish that all people one day might be able to gaze upon Earth in its entirety, with all of its history and all its majesty.

But in order to attain the fullest appreciation for our home and ourselves, we will need to leave the Earth behind, and fly away towards the edge of what we can see. In order to get a sense of scale for everything, we will use a universal constant to describe the vast cosmic scale. This is of course based on the fastest anything can travel within the universe, the speed of light. Anything without mass in the universe moves at light speed, one of the most accurately measured constants in all of science. The technical number is 299,792,458 meters every second. That's fast enough to circle the Earth over seven times in a second, which is about the same amount of time it takes light to get from the Earth to the Moon, which is our closest natural companion in the whole universe. As such, we say the moon is one light second away from the Earth. Fun fact, that means that if every planet were lined up end to end, a photon could cover that distance in less than a second.

As we begin to fly out, the Earth and Moon begin to shrink closer and closer, until they appear to be a single object. Eventually, the Sun would come into view, and would look around half a degree thick. From above the solar system, if one could see the orbits of the planets, only now could they see their circular pattern. But none of their surfaces would be visible. Even Jupiter, large enough to fit a thousand Earth's inside, looks like a shiny dot from this distance. Notably, this far out the constellations still appear identical to the way they would look from on Earth, which might make one think that they are infinitely far away. As we pass by the Kuiper Belt, we can see how lopsided the orbits of Pluto and its companions are compared to the flat disk formed by the eight major planets. This is the distance the New Horizons spacecraft took 10 years to travel, moving at 58,536 kilometers per hour (36,372.584 mph!), the fastest a spacecraft has ever been launched by human beings at that point. From here we can see the distance from the Earth to the Sun—eight light minutes—as well as the distance from the farthest planet Neptune to the Sun, about thirty times that. Once we travel about a hundred times farther away, around thirteen light hours, we come to the heliopause. This is the point where radiation from the Sun is counteracted by interstellar radiation from the rest of the galaxy, and since the Sun is traveling, this bubble appears elongated. From here, the orbits of the four inner planets get lost in the Sun, although all eight of the planets themselves have become invisible at this point. By now we have past the two Voyager spacecraft, the farthest machines to ever be launched by humans. We will need to travel a thousand times farther out before we see the next structure, this time a spherical cloud of ice. This is the Oort Cloud, the home of comets, and the extent of the Sun's gravitational influence. Finally, one light year away from home, we fully exit our solar system, and now the constellations begin to distort.

In many ways, constellations are like clouds. Everybody sees them slightly differently, and since the sky doesn't belong to anybody, they are all correct. But the patterns we see in the night sky are completely arbitrary, not just by opinion, but also from perspective. The stars that create those patters actually have nothing to do with one another in real space; they might even be separated by half a galaxy, not even born in the same cloud of gas and dust. The stars are just so ridiculously far apart that this doesn't become apparent until we travel far enough away. The nearest star system to the Sun is the trinary Alpha Centauri, just less than four and a half light years away, and almost three hundred thousand times farther away from the Sun than our planet. But we're just getting started.

Skipping past the stars in the galaxy faster than light speed, we pass by thousands of sparkling dots. The Sun and the solar system are nothing but a faint memory at this point. The planets are no longer visible, nor any of the other trillions of the mountain sized dust grains that surround them. The Sun is now assimilated into the sea of stars all around it. Then, about three thousand light years farther out, a shimmering cloud looms on the horizon. This is the same cloud that bands the night sky most brilliantly in the southern hemisphere of Earth, but only from the outside can we see our Milky Way Galaxy in all its glory. Of course, no human made machine has even come close to reaching this distance to take pictures. For now we must rely on measurements of how gas and dust are distributed throughout the Milky Way to create a model for our galaxy. Creating a good model has taken decades of pain staking observation and meticulous measurements of all wavelengths of light, but now we have a fairly confident idea of what the Milky Way looks like. Our galaxy is a flat disk, just like the solar system. It is a spiral galaxy, with a bright thick bar through the center. The spiral arms of the galaxy are prominent and blue, bustling with new stars and supernova remnants. In between them are dim areas of faint gas and old red stars. As the arms of the galaxy move in and out of the gas clouds, old gasses are brought closer together, and new stars are born. One can no longer distinguish any but the brightest of stars. The Sun, a third of the way from the center, is lost in huge gravitational ripples filled with gas and dust.

Looking at our galaxy edge on, one would be able to see several more features. For one thing, the disk is actually dissociated into a thick inner portion and a thinner outer region, indicative of the same gravitational collapse that flattened the inner solar system. From end to end, a beam of light travelling the Milky Way would take one hundred thousand years to cross. However, instead of being flat all the way through, the galaxy bulges toward the center. The bulge is the brightest portion of the Milky Way and many other galaxies, and is of course the location of the galaxy's heart, ours being a four solar mass black hole. Like the Sun, the galaxy has other bodies orbiting around it as well inside of a spherical halo. Here smaller galaxies revolve and possibly merge with the Milky Way. These objects can be small globular clusters, balls of gas with just a few thousand dim red stars. They can also be large enough to be considered galaxies in their own right, like the Large and Small Magellanic Clouds, also viewable from the Southern hemisphere back home. These objects will likely be devoured and assimilated into our galaxy, like countless ones before. Altogether, these objects contribute to a sea of stars one hundred billion strong, ranging from all sizes larger and smaller than the Sun. If we were in the 1920's, this would be the end of our story, with fewer details as well. But we can go bigger.

The Milky Way and its satellites are just a few of the objects that lie within our local group of stardust. They consist of just one of thirty groups of galaxies that all interact with one another due to gravity. This gravitational connection is how galaxies grow larger, as they drift together over hundreds to thousands of millions of years. The Milky Way too is on a collision course, with the only larger local galaxy, Andromeda. This brings us to the peak distance the biological human eye can see into space, thanks only to Andromeda's sheer size. As we continue to fly out, we can see Andromeda about two and a half million light years away from home, enough distance to fit twenty-five Milky Way Galaxies. From this distance, all the star communities begin to shrink into smudges. Only now can one get a sense of the terrifying magnitude of empty space in the universe. By the time the local group starts to look like a handful of sand, we show the audience the rest of the observed galaxies, just a few billion more. At this point we must shift our vision ever more red to see these galaxies. We can see long streams of galaxy clusters, and the Virgo Supercluster where we reside. While these filaments may tug on each other gravitationally, galaxies on the whole are no longer bound to one another. This far into space, dark energy overcomes gravity, and galaxies and clusters are spread apart by an ever-greater amount. From this distance, galaxies appear to sprinkle the sky more tightly than the stars inside of those galaxies. One can see dense areas of galaxy clusters, where immense red elliptical galaxies form as smaller galaxies collide and disperse their star-creating gas into intergalactic space. One can also see the field galaxies; mostly blue spiral star systems drifting millions of light years from dense clusters. However, one can only see where these galaxies are located, the individual structures are indeterminate from this scale. Rather, every galaxy appears just as a speck, each one a mere glimmer of light comprised of billions of stars, each star possibly being surrounded by dozens of planets, each planet perhaps having moons as well. At this point, the universe looks nearly uniform, with long dotted streams of light surrounded by empty voids of darkness. Neither the Earth, nor the Sun, nor the Milky Way, or even the Virgo Supercluster are visible any longer. Instead, all the matter from millions of light years appear to be the size of a flea, collected into a single point. At the farthest reaches of the cosmos, the farthest and oldest galaxies come into view. These are the quasars, young galaxies emitting so much energy from matter falling around their cores that they are visible clear across the universe. Then, the space surrounding each galaxy goes dark.

Let's take a moment to pause and reflect on the finite speed of light. A finite speed is the reason why it takes the fastest thing in the universe years to cross galaxies. If it takes time for light to travel, which means that by default, we only see things as they were in the past. Even reading this page, you aren't seeing it as it is now; you're seeing it as it was several nanoseconds ago. This doesn't matter much here on Earth, but it does in astronomy. Specifically, it means that objects appear to us now the way they looked years ago. This is the reason why the brightest quasars are only seen far away, when smaller galaxies were still merging into larger ones, thus making it easier for their central black holes to eject matter from their poles. From their point of view, the Milky Way might look like a quasar. Beyond these primordial galaxies, the universe looks dark because we are seeing it before the first stars even formed some few hundred million years after the big bang. Stars in the night sky look today the way they looked decades or centuries ago. The Andromeda galaxy we see today is how it looked two and a half million years ago. If we look back far enough, we should be able to see the birth of the universe. And we do.

Far beyond the farthest quasar lies the baby picture of the universe. It is the background radiation from when the universe was first opaque enough to allow light to move through it. In other words, it is our cosmic horizon, the farthest back in time we can see using light. This radiation is almost perfectly uniform, distributed evenly throughout the universe just over three hundred thousand years after the Big Bang. Even though the universe itself is nearly fourteen billion years old, due to its accelerated expansion, our horizon spans ninety three billion light years from end to end. What's more, since the universe is expanding and cooling, it means that galaxies are flying out of our cosmic horizon. In other words, the visible universe is shrinking. This is often why we refer to the present as the "Golden Age of Astronomy", since we can still observe distant galaxies. The radiation that marks this boundary is known as the cosmic microwave background. Every time you hear static on the radio, and every time you see black and white fuzz on your television, part of what you are experiencing is the light from the early universe. It has been so redshifted from losing energy to expansion that it has gone clear across the magnetic spectrum from gamma rays to microwaves. Beyond this cosmic wall, no light was able to travel through the universe; as a result there is no telescope that can be built to see farther away using electromagnetic radiation. To look farther, we must find a way to observe something else that was radiating the universe—perhaps neutrinos or gravitational waves—although it isn't clear how this can be done.

This is the end of our tour, but it is not the end of our conversation. Remember that the universe is made up of more than just space; we also need to know what time it is.

The Cosmic Clock

Of all the naturally occurring phenomena of nature, none is simultaneously as pervasive and mysterious as that of time. What even *is* time? We know that it passes. We know that it travels at different rates for different observers. Sometimes it feels like it zips by, other times it feels painfully slow. Massless objects don't even experience time, and everything else grows old because they do. But despite its enigmatic behavior, it is a part of the very foundation for how we understand the universe. One does not set up a meeting without both place and time. Speed and acceleration are based on changes *over* a certain timespan. People are paid based on the number of hours they work, and exams are given only within certain amounts of time. Time is a location just like any in space, the only difference perhaps being that we cannot go backwards in time. At least without traveling faster than light, which probably means it can't be done. The concept of time is so ingrained into our daily lives that it becomes almost impossible to define, but it is imperative to understand if we want to comprehend the nature of the universe. So what's going on here?

Things that occur in regular cycles generally define lengths of time. The very first clocks were the heavens themselves. Ancient cultures around the world didn't have watches or clocks; they only had the cycles of night and day, the passing of the seasons, the length of a year, and the phases of the moon. This is where the modern calendar comes from as well; a day is how long it takes the Earth to revolve about its axis, a year the amount of time it takes for the Earth to orbit the Sun, and a "moonth" is based on the amount of time it takes the moon to orbit the Earth. These original clocks also dictate the circadian rhythm in plants, animals, fungi, and even cyanobacteria. It is even ingrained into our DNA, allowing people to keep an internal sense of time completely isolated from the outside world. People barely had a need to keep exact amounts of time until the industrial revolution gave birth to the hourly wageworker, but it did have the positive side effect of allowing scientific measurements to become more accurate. At the dawn of the twenty-first century, humans have even found out how to use the regular atomic behavior of cesium atoms to create the most accurate clocks in the universe. The definition of a second has been redefined both by energy states of cesium and by the amount of time light is calculated to travel in one light second. Yet no matter how well we measure it, it still doesn't shed light on what time is.

Time on the cosmic scale is very different than space. From our perspective, we are at the center of the universe, but this is completely arbitrary. When we look out into space, we see a universe that is nearly uniform in terms of matter and energy distribution. There is no sign that our perspective is any truer than any other, quite the opposite actually. From what we can see, no point in the universe is any more "central" than another. Rather, every point in the universe is its own center, each with horizons ninety-three billion light years long. This should sound reasonable, after all Andromeda also sees our galaxy the way it was in the past, as is the case for distant galaxies. This was also one of the ideas I most remember from my cosmology professor. From what we can tell from present data, the universe may very well be infinite, we just cant see galaxies whose light hasn't had enough time to reach us yet. People often describe the big bang as the universe expanding from a point. But the universe is infinite, which means that all points in the universe were infinitely close, not that the universe was in a single point. Savvy?

However, time isn't quite as arbitrary. Or perhaps it is more arbitrary depending on your perspective. Every single piece of matter in the universe is the same age, the age of the universe. There are of course exceptions for particles moving near light speed or around black holes, but as a whole the universe is the same age everywhere. More importantly, time moves in the same direction everywhere in the universe, forward, always slipping into the future. However, as Einstein taught us time could still be altered in the same way space can, given enough matter and energy. It is very curious how time can be woven into the fabric of the universe and yet remains so curious. Time behaves even more strangely and becomes even more abstract under extreme conditions, like those found inside black holes or just after the big bang. As we have discussed, the only way to understand the high levels of gravity taking place at the subatomic scales of these events is through a theory that binds quantum mechanics and general relativity. With the limited knowledge and equations currently under our disposal, time becomes both infinite and nonexistent, neither real nor fake. To be fair, this is only really true for singularities, points of infinite density. The insides of black holes around their singularities can be fairly well described without a unified theory, just as particle physics can describe the universe up to several quintillionths of a second after the big bang.

One final strange oddity in the nature of time comes from looking at different temporal scales. That is to say, how can one decide what a "long time" is? A single second might seem short, but it's still a billion times longer than a nanosecond. Particles that decay in less time would easily consider a second to be an eternity, just as we might think of a billion seconds. It's all relative, just as with sizes. There's no such thing as a long or short time without a comparison, just as there is no such thing as large or small without objects of different sizes. This is one way to understand the seemingly miniscule timescale of events that occurred after the big bang. A lot may have happed in the first three minutes of the universe, but a lot also happed in the first few nanoseconds as well, and from the perspective of a quark at the beginning of time, a universe that is three minutes old is still a billion times older than it was when the first atoms formed. On the same token, we may think fourteen billion years is a long time, but in a universe that is a quintillion years old, a billion years will seem insignificantly brief.

So since time is so poorly understood, let's take a step back and look at our place in time with a more objective lens. Specifically, we can understand our place by looking at past events and how long they took, this time using full Arabic numerals. The first atomic bomb was detonated in 1945, around 70 years ago. The light bulb was invented near 1879, around 140 years ago. Galileo largely developed the scientific method until his death in 1642, 400 years and 20 human generations ago. Humans first discovered agriculture 10,000 years and 500 generations ago. The first Homo Sapiens appeared in Africa 200,000 years and 7,500 generations ago. Our last common ancestor with chimps lived around 7,000,000 years ago. The dinosaurs died and allowed for the development of mammals 65,000,000 years ago. The first multicellular life arose during the Cambrian Period around 600,000,000 years ago. The first solid signs of life occurred 3,500,000,000 years ago, and the first proto-cells may have formed earlier than 4,000,000,000 years ago. Finally, the beginning of time was 13,820,000,000 years ago, and it only took this long for the universe to begin to understand itself.

So don't let anyone ever tell you that human beings aren't smart. Despite all of the backwards and stupid people who incite conflict and hoard power, as a species we've accomplished quite a bit for only being around for such a short amount of time. It's not really their fault either, after all every one of us is born a blank canvas, naive to the billions of years we've existed as disordered cosmic dust. We also don't live very long relative to much of anything else in the universe, and we have so little time to relearn our cosmic history and pass down our knowledge to our children before we die. Yet despite those limitations we are still able to mold the universe to our whims. All of this should be an attestation to what we can accomplish in such short amounts of time with only our minds and opposable thumbs. Not bad for some hairless apes living on a speck of dust made up of specs of dust, huh?

So how do we know all of these things as a species, and how did we discover them? Well it wasn't easy, but after just a few thousand generations, humans have pretty well figured out how to categorize the ways we learn. In modern terms, we call this learning school, and the different fields of study are the ways in which we understand the world. So with this cosmic perspective to our lives, let's take a look at just what these different fields of study truly are, how we use them to understand the world, and how they are really just different flavors of the same thing, as with all else in the universe.

Part II: Pillars of Knowledge

4 | History: The Power of Knowing What's Going On

"That brain of mine is something more than merely mortal, as time will show."
— Ada Lovelace, The Enchantress of Numbers

"Human progress is neither automatic nor inevitable... Every step toward the goal of justice requires sacrifice, suffering, and struggle; the tireless exertions and passionate concern of dedicated individuals."
— Dr. Martin Luther King Jr., The Great Dreamer

Students of the International Baccalaureate program have a unique opportunity to challenge their minds through diverse and tedious schoolwork, complete with standardized exams and papers graded in Switzerland. In addition to the traditional in class midterms, quizzes, final exams, and essays, students who want to earn an IB diploma are also required to complete a two semester philosophy course, write a personal interest research paper, participate in a public service project, and for every major school subject; complete an Internal Assessment and several standardized essays and exams that are given to high school students all around the planet. In other words, it was hell. In particular, the higher-level history course was often associated with heavy anxiety. Luckily, we had teachers that were so good at their jobs, that processing and analyzing two hundred years of world history somehow became manageable. So much in fact, that history might have been one of my favorite courses by the end of high school, after chemistry and physics of course.

There's something about history classes that makes them incredibly boring, at least on the surface. When one thinks about learning history they probably think about memorizing dates and events, because dates are concrete and easy to test. This of course has the side effect of causing history classes to emphasize legislation, wars, and other events that have exact locations in time, written records, and other quantifiable information. "*Event X happened on this date, lasted for Y amount of time, and started for Z reason. Some people liked it, and some didn't.*" This sort of structure is almost inevitable for courses that only have a few months to teach hormonal young people the history of the human race. But this makes it hard to remember that history isn't made of singular dates and events. Far from the fractured view people are taught in schools, history is a continuum, an endless stream of events that occurred at the same rate they do today, filled with people living their lives in the same mundane way as everyone else, day-by-day, moment-by-moment.

History is a story; it paints a picture of strange worlds and foreign people who we will never meet. The fact that we have history serves as constant proof that the world is a changing place. More importantly, it shows us the ways people just like us reacted to the world they found themselves in, when our history was their present. Dates and events are important to know, but they are only useful if one knows the context in which those events took place, and their ramifications. This is where the student comes in, filling the gaps based on the overall cultures and mentalities pervading through the societies that made it into the history books, and this is how we were tested on our knowledge of history. During our three timed essays—the type of exams where they cram all the students into the gymnasium for—we were given primary sources such as letters, photographs, or propaganda, and separate prompts about various domestic and international policies and revolutions for each respective essay, and given a certain amount of time to write about their implications, using as many dates as we could muster for whatever prompt we were given. But we were never graded solely on how many dates we memorized. We were scored based on how critically we understood the few dates that were important to the event in question.

This exemplifies the overall structure of the IB program, teaching students how to think instead of giving them things they don't have a reason to think about. In class our teachers would sometimes joke that we were learning how to become dictators, but in a way that is sort of true. As education improves every generation, more and more people are able to study how people come into power, and how they are able to keep it through interactions with their subjects, the people who give them that power. I think this is an incredible thing, for it's better that everyone be royal, not just a minority. More importantly, when greater numbers of people know how to become dictators, there seem to be far fewer actual dictators. Demagogues can't exist in a world of equally intelligent people.

Humor aside, the study of history is perhaps the most unique and beneficial human enterprise. History defines what it means to be human. No other animal can live outside of the present to the extent we can. Think of it this way: how does a squirrel learn to hide acorns? How is it that birds can teach themselves to fly with just a little help being fed, and why can reptiles, fish, and arthropods be born completely alone and still know how to survive, while humans take two decades to fully develop? It all ties back into intelligence and genetics. Animals with the most complex brains take the most time to grow and be nurtured by their parents. This is part of the reason why nurturing young is mostly a mammalian trait, right along with the production of milk. Higher cognition depends on longer development and more resources. But humans take this a step farther, because at the end of the day, all animal behavior finds its roots in instinct, which come from genetics. Animals are born with the coding that produces the neural hardware for each animal to survive in its environment, but most don't have quite enough hardware, time, or resources to build upon knowledge from previous generations. Humans do, thanks in part to their capacity for memory. Not only can we learn knowledge from our parents and genetics, but we can also modify and enhance old knowledge before passing it on to the next generation. We do this through communication, at first through demonstration, then through vocalization, then through print, and now via wires and electrons. Modern advances in record keeping allow humans to transcend the need for new ignorant humans to be born over and over every few decades, and instead allow for every following generation to learn from the lives of every past generation. With this trait, a human could notice wood floating, and get the idea to make a boat. They could teach their child this, and when they grow up they might figure out how to carve the boat properly in order to minimize drag. Then that person's child could get the idea to add a sail, on and on through the generations until someone started building ships to go to outer space.

History is the reason humans have everything they do: homes, pets, furniture, food, electronics, water for drinking AND bathing, and essentially everything else touched by human hands. Or more accurately, the study of history and the ability to comprehend it are the reasons humans have everything they do. History itself is merely everything that has happened, and nearly all of history happened before there were any humans around to study it. In that sense, math, science, literature, and every other school course is really just a specific type of history class. Science especially is nothing more than meticulously recording history through observation. Archaeology, paleontology, and geology are all used to learn about history before the invention of writing, before the existence of humans, and even before the Earth itself. Astronomy too is like a type of history, after all looking into space is literally seeing the distant past. Speaking of which, don't forget that our cosmic horizon is shrinking. That means we are technically losing history, so we had better figure everything out sometime within the next trillion years.

Of course, with so much history, so many people to teach it to, and so little time to learn it, it's no surprise that some knowledge doesn't get passed around to everybody. This might happen for any number of reasons. As is most obvious in the schools of developed countries, some historic events just don't have anything to do with our daily lives, at least not without some connection to some landmark event. Unfortunately, what's more common is that most people don't learn any history, and so they just have to take the world at face value as if it had always been the way it is. Even people who go to good schools might have such poor teachers that they don't retain any of the history they learn. Or perhaps they have an astounding education program, but live in a place that rewrites history for some ulterior motives. History can be invented just as easily as it can be recorded, and without any conflicting evidence, a person could live their entire life in a world of fantasy and myth. This has actually been one of the cornerstones of repressive regimes, reweaving history to suit their twisted perspectives. This is especially terrifying, because a sufficiently powerful and corrupt government could undo two hundred thousand years of discoveries, and cast themselves as gods who created time immemorial. All they would have to do is raise one generation without outside influence. Fortunately, people are too stubborn and curious to live their lives with a wool over their eyes. But such behavior can only be prevented if we think critically about everything we are being told about the past. One must always be cognizant of who is telling the story, where they come from, what their background is, how reliable they are, and what their biases might be. Anybody can make history, just as anyone can make it up. Moreover, it's crucial to remember a tenant of history: that the people who record it are human beings just like us, and sometimes they make mistakes just like we do.

Let us not forget about the majority of humans who don't give their voices to the pages of recorded history. Many of us forget how much of a privilege it is to be able to read and write. As a matter of fact, out of the hundred trillion people who have lived thus far, probably just a few million had more than a fifth grade reading level, which statistically is equivalent to zero percent. A person could be the most knowledgeable human being on the planet, but if they never write anything down, nobody will ever know what they were thinking. The most they could hope for is to have other people speak about them after their death, but stories about people have a tendency to either be embellished or bedeviled depending on how much others liked them. That's how fairy tales get started. This is the problem with oral tradition, it can only converse so much detail. Of course, even primary sources straight from the horse's mouth might be skewed by a lack of perspective. Only people who don't have to spend all of their time acquiring food can learn to write, and many only did so if there was a need to write anything down. This meant that the first scholars couldn't exist in hunter-gatherer societies—even though they likely had more free time than agriculturalists—because everybody that needed to converse was within earshot, and trade was largely merit based. Written history could only occur after agriculture gave humans excess food that they could store, and people needed to record how much food everyone got, usually getting higher wages as payment. Thus writing became a skill of the upper class, and as a result, history is written by the rich and full.

This can be seen in many history textbooks as well. Recorded history starts in Africa or the nearby Middle East. Five thousand years after agriculture was developed and five thousand years before Internet, the self-recorded human tale always begins near Egypt or Mesopotamia, with Egyptians, Sumerians, Akkadians, Assyrians, and Babylonians. This makes sense, because the first humans came from Africa and the Nile is peaceful, as is the Tigris-Euphrates river system between Iran, Iraq, and Syria. Civilizations likely started everywhere as soon as people got settled into or near a base camp with fresh water, it just so happens that there were people in those areas first. Although it should be noted that China has one of earliest written histories as well, dating back to the Shang Dynasty three thousand years ago.

But just because the elites of society do the writing, it doesn't always mean they have the best perspective for what was happening in society. This is especially true if the people writing history happened to be royalty as opposed to farmers or laborers. This is ironic considering that many Kings and Queens later on were actually genetically inferior to the people they ruled over, because as it turns out, inbreeding is bad for your health. They were closer to purebred mongrels than human beings. That being said, the royal class was probably more likely to hire scribes than to do any writing themselves. Of course there were several amazing rulers who commanded people of vast lands and seas, and you can tell because their people let them stay in power the longest. Then those regimes got to spread and overwrite the histories of conquered lands. In this way the world can end up teaching one perspective over all others, so long as those in power stay in power. That's another tenant of history: when you don't please the masses, the masses don't let you keep your throne, plain and simple. Another pit fall of human made history is that it tends to center mostly on humans, which inflates our egos and gives us a false sense of superiority over the world we depend on to survive. Textbooks rarely mention how an El Niño might affect crop yields that determine whether a kingdom will survive winter or not. Historians rarely write grandiose epics about the plagues that killed millions of people, they would rather focus on the military skirmishes that claimed a few thousand pawns. We can even see examples of historical hypocrisy in the United States war for independence. The colonists fought on the grounds that "All men are created equal", which of course didn't include women, men who didn't own property, nor the African slaves that made up over a fifth of the population. At least the ideas of equality have been a positive relic of the revolution. Yet despite the hubris, all these biases reflect another tenant of history: we are all selfish, but that also makes us all the same. In this way history truly does create empathy, and this allows us to see why we learn it at all.

Let's take a common example to see why you need to know your human history. How did The Second World War happen, why did it happen, and what do we have to do to make sure such a thing never happens again? This is a decently fair question; after all we would very much like to generally avoid mass genocides and international murder if at all possible. A whopping sixty million people were estimated to have died during the war, nearly half of who were from Russia, where noted asshat Joseph Stalin was flipping sides like he worked in fast food. What an inconceivable amount of death and suffering. Sixty million young lives, just thrown out at foreign countries like a dozen apes hurling feces at one another. Sure it was only about one tenth of one percent of people to ever live—the majority of which were savagely murdered by bacteria, weather, and stupidity—but never before had so many people died in such a short period of time in such a gruesome way. Moreover, those other far more dangerous things don't go on holiday from killing people just because we're shooting at each other and blowing ourselves up. Death will come, regardless of whether we bring it upon one another. So again, we would very much like to not repeat the whole world war thing if humanly possible.

So why did World War II happen? Well obviously it was history's favorite super villain Das Fuhrer Adolf Hitler, the absolute dumbest human being to ever rule a sovereign nation, right? Any bloke on the street knows that. Except there's a problem. Namely, how did that crazy fascist lunatic get enough people to follow him to the point of mass murder through war and genocide? This is where history becomes fun. Not learning about genocide, but being like a detective at a crime scene or a news reporter witnessing glory. This is how history should be taught, but sadly it also requires you to know what order things happened in, and that requires dates, which requires thinking, which is hard to do. But there are countless ways to go about finding clues and answers that have been lost to the sands of time. We could discuss ideology. Many describe the war as a battle of good versus evil. This is valid, except people change their minds every day and nobody has a consistent definition for what a "fascist" is, beyond someone who tells others to blindly follow them without thinking for themselves. Mussolini sort of just started using the term to define his political faction and never told anyone specifically what it meant. These days "fascist" is simply just synonymous with "funkhead" when used colloquially. The Italian people also strung him up in the middle of the final year of the war. So neither he nor his ideology made him a popular dictator, if such a thing exists. So war didn't happen just because Hitler was Fascist, even though he was clearly a "fascist".

Maybe it was because the German people are inherently violent and militaristic; they do have a history of starting Great Wars and beating up the French. One can date the German predisposition for inciting conflict all the way back to the Second Reich that unified Prussia thanks to Otto Von Bismarck, back when people were serious about their facial hair. Germany even organized the Berlin conference, which saw the European powers unjustly draw borders along the African continent to decide who could colonize where, the so-called, "Scramble for Africa". Indeed, some very respectable historians contest that Germany was destined to start World War II since the end of the Franco-Prussian war in 1871. However, solely blaming German militarism or German culture for both global wars is too lazy of an argument. All European world powers had a culture surrounding the "glory of war". What's more, Hitler was born in Austria-Hungary, not Germany, which used to be a world empire until it also became Germany, since borders are always arbitrary and they don't matter at all. Hitler called for ethnic purity, not just a phony national pride in Germany.

Many historians associate the rise of the German National Socialist Party as a response to crippling debt, which was caused by the First World War and compounded by the worldwide Great Depression. Inflation was so rampant in the country during the 1920's that people needed barrels full of cash in order to buy bread. Germany was mostly blamed for the Great War by the victors, and so had to burden the costs of the war, even though this was clearly false. World War I was incited by the assassination of the Austro-Hungarian Archduke Franz Ferdinand by Serbian Nationalists on June 18, 1914, and Austria-Hungary was the first country to declare war a month later on July 28. The First World War could have been kept a small skirmish in the Balkans, if it weren't for the alliance systems that bonded certain European states to one another. Germany was allied to Austria-Hungary, and Russia to Serbia. Russia even mobilized two days before Germany on July 30, who didn't declare war until France did so on August 1. If anything, Germany was reacting to conflict in the Great War.

Every colonial European power was acting like a sadistic toddler with a shotgun, clamoring for new land and enslaving the old natives. This is probably why two wars that started on a tiny peninsula sprouting off of the Asian continent spread across the planet to people who really didn't care what all those crazy white folks were yelling about. Just ask Joseph Conrad what Leopold II of Belgium was doing in the Congo. The answer is chopping off human hands, or as he writes, "The Horror! The Horror!" I think they wanted some trees or shiny rocks or something. But hey, that's why they got to draw the maps yeah? The point is that national identity and imperialism was pervasive throughout all European cultures during the nineteenth and early twentieth centuries. It led to naval arms races and assassinations of Archdukes. All of Europe was to blame for the First World War, and by extension, the Second. Germany rightfully became resentful after the Treaty of Versailles at the end of the war decimated their reputation, and Hitler channeled that feeling of resentment to trick his way into power. Even after he began annexing territory in 1938, the other world powers did nothing more than bat an eye and slap Hitler on the wrist.

So all European colonists were a bit demented, as were all human beings before cable television and home movies. Hitler was certainly extra crazy, but just being German is too broad of an explanation for why that was. If you were to ask my parents, who work with criminals in gangs and rough neighborhoods, they would say that it was bad parenting. My parents were slightly stricter than I would have liked, but that may not have been a bad thing. As a matter of fact, how we grow up undeniably shapes how we see the world. Living with neglectful or hurtful parents is probably the strongest link to crime and unlawful behavior. However, for the special case of Adolf, I think we need a more universal understanding of what motivates murder to grow from a single gunshot, to a global pandemic. The point here is that there are countless ways to attack the problem, and they should all be acknowledged. Probably the most important tenant of history is best articulated by John Green, "the truth resists simplicity".

Despite all of history's complexities, and the countless lenses one can look through to study it, I have always found two lenses especially clarifying: resources and information. More specifically, one view of history can be charted by looking at the availability of food and knowledge within and around country borders. Ideas and opinions are formed from thousands of factors, from the culture we are born in, to the people who have power at the time, to our awareness of the natural world. But at the end of the day we are all human beings. Be it prince or popper, we all need to eat, sleep, and be safe. Otherwise we die, and then all of our ideals and opinions don't matter anymore anyway. As such, the main drivers of human history are usually based on the things that keep people from not dying horribly. Usually this is food, but it could also be the technologies that increase standards of living. People might fight over their various ideologies, politics, and customs, but they are rarely driven to kill, let alone go to war, over opinions. If one group of people doesn't like another group of people, they would probably prefer to just not talk to them rather than risk their lives in combat. This is true of all animals as well. Predators like lions, tigers, and bears, can get along with animals like dogs without conflict in captivity so long as they are well fed. Even animals in nature will only hunt when they need to eat, because hunting puts them at risk of being attacked by their prey or by another predator. In short, living things generally want to stay alive, but most living things need to eat other living things to survive, and their food sources also want to stay alive. This means that getting food is inherently risky and prone to conflict, which itself is bad for health. This extends to the human animal as well; we only fight when we see conflict as the most prudent way to avoid dying. Otherwise, why risk your life in the first place? Hence, hunger and the need for resources are prime motivators for war and conflict. This is also supported by archaeology, which suggests ancient humans only killed off neighboring clans during covert raids, minimizing casualties and enabling them to gain more resources. Of course, this correlation between war and hunger gets muddled later on in human history, when the people who declare war go from being citizens who participate directly in the conflict, to generals who oversee combat, and finally to governments who never set foot on the battlefield at all. This is probably one of the heaviest criticisms of war following 1945; governments go to war, but only by using soldiers as pawns for their strategic geopolitical gains.

The need for resources also ties together the motivations for numerous historic events. It gives credence to both the Japanese and German expansions that led to World War II. Many of Hitler's speeches even reference the need for *lebensraum,* living space for the German people. This coincides with the strong nationalist sentiment for Germany to be self-sufficient, and not depend on any other countries for support. In a similar way, Japan too needed to support a growing population and increased industrialization. Japan might have even been more strongly motivated to create an empire seeing as they invaded Manchuria in 1931 and Mainland China in 1937, over a year before Germany annexed Austria and Czechoslovakia. It just so happens that instead of collaborating with other states, these hyper nationalist counties decided to take land from other people and get rid of them if necessary. Access to resources and territory also played a role in the war as seen through relations with the Soviet Union and Nazi Germany. Perhaps the greatest miscalculation made by Hitler during the war was enacting Operation Barbarossa to invade the Soviet Union on June 22, 1941. In hindsight it is clear that this was a blatantly foolish mistake on Germany's part. Not only did it open up a second front for the Germans to fight, but that second front was also against Russia, and it lasted well into the winter. Hitler clearly wasn't a student of history, taking lessons from Napoleon instead of Genghis Khan. Speaking of Napoleon, the French revolution that ended in his rise to power was spurred by hungry citizens living under a fat aristocracy.

But going back to Germany, at the time it might have made sense to invade Russia. After all, by 1941 Stalin had already been in power for seventeen years, and his five-year plans— involving collectivization of grain, great purges of his political opponents, and secret police—had already led to millions of deaths. Specifically, collectivization had forced farmers to give up their land and crops to the government for public distribution, decreasing motivation for farmers and leading to the execution of many, thus leading to mass starvation. Some even liken this starvation with genocide, especially in the Ukrainian Holodomor. But unlike Hitler's genocide—which was based on ethnic cleansing enacted only near the end of the Third Reich after hard labor was no longer tenable and the war was looking bleak—Russian deaths under Stalin were solely due to intentional and rampant famine. It was so intense that corpses lined the streets and people resorted to eating children and pets. Stalin was basically Hitler. Both were totalitarian crackpots with dreadful facial hair and work camps. The only differences between them were that Stalin eventually got nuclear weapons and Hitler actually liked his own people, as long as they were Arian. At least Hitler didn't starve people who were Catholic and German like he was. Stalin did nothing but preserve himself. So by all accounts, Hitler may have been justified in believing he could take the Soviet Union quickly, even though this was far from true seeing as how the Soviets would be the ones to capture Berlin and end the European theater of the war.

Studying history based on resources also helps answer one of the most controversial topics in history: how the so-called West came to hold so much influence over the world. Specifically, it can help us answer the question of why people of European descent dominate the culture throughout the planet. We can analyze this question through resources, and the availability of land with those resources. One possible reason natives in the Americas didn't develop guns germs and steal was that they didn't need to. Granted, they also didn't have China to give them gunpowder in 1200 CE. However, throughout its history, the United States in particular has been infamous for having plenty of resources, so much so that modern obesity is a serious national health risk. Native Americans didn't need guns and "civilization"—the latter of which is a completely arbitrary concept—because they had plenty of what people already wanted, food. The Europeans, on the other hand, were no doubt starving. England is just a tiny island, and France, Spain, Germany, and every other European state are petite in terms of landmass. In order for them to deal with a growing population, they either had to trade or colonize, because they were hungry. And while seventeenth century Europeans certainly immigrated to the United States for free religious practice and autonomy, the vast majority in fact just wanted to live somewhere with a lot to eat, and brought their foreign beliefs and religious intolerance with them.

Land mass and resources would also explain why the United States has grown into such an economic powerhouse. Simply put, it is enormous. The state of Texas alone is larger than virtually any European state. It also has territory along a wide range of altitudes and latitudes, giving it more varied Sun exposure, and therefore giving the country many diverse ecologies. Lastly, its borders span an entire continent. Being a whole ocean away from other countries makes it much more difficult for enemies to attack the country, and the fact that its population is united under one flag makes it unique compared to other powerful nations such as Russia and China, who share a continent with wholly separate states along every border. Being allied to Mexico and Canada certainly doesn't hurt either.

This also leads to another tenant of history; collaboration is the most powerful tool at humanity's disposal. The United States of America is very unusual in that it is one nation, but is also technically made up of smaller, miniature governments. We even call them states; a term that anywhere else in the world would mean a completely separate country. This structure contrasts from groups of small sates that don't always work together, such as in Europe, but also from countries with larger land mass and more people than the U.S. that don't have as distinct internal borders, such as China or Russia. The U.S. is also infamous for spending a lot of time with internal bickering, usually about giving more people liberty. But when the nation faces a threat, more and more diverse people come together to fight for a common cause. Unfortunately, this still leads to the same strange sense of nationalism that occurs throughout the world when people think fake lines they've drawn on maps should dictate the way they treat each other. But that's the main difference between the United States and the European Union. You might live in a country that is a part of the European Union, but you are a citizen of your country first. On the other hand, I was born in Colorado (which is also larger than several whole nations), but I can go to school and live anywhere in the country with nearly zero trouble. First and foremost, I (like all of us) am a proud citizen of the United States of America.

~ *Cue Stars and Stripes Forever* ~

It's that very subtle difference in perspective that makes our country so strong. It unites over a third of a continent with people living in entirely different environmental parts of the planet. Granted, it was founded by elitist Christian European misogynists that owned other human beings as property and killed off the natives with disease, BUT the idea of liberty has been so pervasive throughout the country that people have never given up on trying to prove the founders correct by immigrating here and fighting for the freedom Europeans had denied them for centuries. Over time, being American has grown to mean including people of all races and creeds, united under the banner of liberty, justice, and equality. And even though we treat poor countries like a helicopter parent treats their children, and even though we treat those who don't speak English like savages, people still move here, face prosecution and hate, and even learn a new language, all because we happen to be the winning team at this given moment in history. But that's what makes us so powerful, we always choose being the best over being intolerant and we allow more and more people to become innovators. We also have plenty of landmass, which means we can hold more innovators.

Our large size may make it difficult for tolerance to spread, but it also makes it easy to fester, since like-minded people tend to live together. This happens everywhere, but unlike on other continents, like-minded intolerant people are forced to be citizens under the same flag as the people they dislike, and they have a strong enough sense of nationalism to keep it that way until people forget their differences and just learn to get along. This juxtaposes us with some large countries that are far more diverse than we are, but also from small countries where everyone looks exactly the same, which is boring. All we need to do is stop being hateful to people that want to do the jobs we don't want to do, realize that other countries have better equality than we do, be a safe place for people to enhance our economy, and allow people from all backgrounds to represent our citizens, which themselves come from different backgrounds. As a result the people replaced by foreign labor can become innovators and be free to pursue their passions, enhancing everyone's lives, making everyone have less hate, and live happier lives. If we take advantage of all the people living here, give them opportunities, and give them a stake in the development of the country, then the United States will continue to be the greatest, freest, happiest nation on the face of the planet that it claims to be!

USA!! USA!!!

Now, there may be concern about people who work in the country and send money to their families outside of our borders. This is only a short-term problem, if one at all. Again, those people are providing services for native citizens, and the things they develop and produce are still used by those citizens. People don't' work in a different country if they can make more money at home. The money they make comes out of the pockets of wealthy citizens, and into the pockets of developing global citizens. It's essentially free charity. It's no different than getting the same good or service from a native citizen, except with the added bonus of lowering the global wealth disparity that plagues the world's citizens. People will always travel where the money is, and flow away from where money isn't. People will be lifted out of poverty, be able to innovate new technologies that better the lives of all citizens throughout the world. This will spread information and knowledge amongst the world's population. More people will be able to make more things to be sold to more customers around the world, and people will see wealthy countries as allies and fellow human beings, spreading democracy and peace across the planet. People in oppressed countries will associate wealth with equality and freedom, and want the same freedom for their citizens, making them fight to ensure good working conditions with fair wage.

This global freedom will create more innovators, speeding up progress at an even faster rate. People will live longer, healthier, and without fear of being blown up. Everyone will treat everyone else as a human being, borders will dissolve, and educated citizens will all have a say in the laws they follow globally by voting on secure servers on the Internet from home, and have a say in the cultures they celebrate locally, by actually talking to each other, like human beings. Anyone will be able to go anywhere and live whatever lifestyle they wanted to with however much money they felt like earning anywhere on the planet. People would be poor by choice, not because of where they were born. Rich people would earn—but never hoard—their money, because they created something that other people need or want to buy, without inheriting fortunes from mommy and daddy. Nobody would be hungry. Everyone would work to benefit each other, ergo themselves, and make sure everybody had food to eat, so that nobody would have any excuse to not contribute to society in some way. Everyone would be friendly to everyone else, and people will stop killing each other. Innovators and scientists from across the globe will cure cancer, create cheap solar power, merge quantum mechanics and general relativity through research conducted by building worldwide particle accelerators, and everyone will be able to eat printed bacon without killing an animal or polluting the planet. THEN, after over two hundred thousand long years, we will all treat each other and nature with respect and dignity, and finally be able to do kickass space things, and we'll celebrate Earth Independence Day by watching supernova explosions with fancy space beer in our fancy space cruise ships. Unless you're a huge jerk who enjoys causing pain and suffering, then you'd be left behind on Earth like the backwards primate you are, or worse yet, solitary confinement on an asteroid. And no space beer for you! So just behave and act like the 13.8 billion year-old pile of particles you are. People are far more likely to give you money if you're useful anyway.

U.S.EARTH!!! U.S.EARTH!!

~Music slowly fades ~

Again, there's no single correct way to study history. Every individual could tell the same story a different way. But the nice thing about viewing history in terms of resource allocation is that it universalizes the histories of people around the world and during all times in history. And when one looks at the entirety of human history, one thing becomes abundantly clear: nobody has a single clue about what is going to happen in the future, very little clue about what happened in the past, and a general disdain for thinking about the present. We all just grow up in whatever form the world happens to be in, and all we can do is look at the way things are, and hope that everything else we are told is indeed true. People constantly try to make predictions about how the world is going to be, and their predictions are never fully correct. Who among the ancients would have guessed that diseases would be cured and modern agriculture would be able to sustain billions of people? If they had known that there would be ways to manipulate nature into making electronics, I wonder how many of them would have wasted their time conquering some land that would just be taken over again long after they died.

Who would have guessed that people are far more efficient when they work together? Sometimes history has a way of seeming glamorous and thrilling, but that is usually a consequence of being selective in what details one looks at. More often than not, learning history makes one glad to be living in the present, and want to race to the future. No matter what, going to the past always means more stupidity and conflict, all because fewer people are connected across the globe. But no matter where you come from, you still share the same history as everything else in the cosmos, and you are especially shaped by the actions of every human being who has ever lived. Every generation is the same play, only with different actors and new props. We are all members of the same team, the same species with identical strengths and weaknesses. We must be able to understand that we could easily have been born in each other's shoes, and be thankful to learn from the mistakes of our predecessors. Understanding history takes collaboration, not conflict.

By definition, the study of history will become more difficult as time moves forward. But as more information becomes available, minor details will become lost in the fray, leaving only the most universal concepts to be studied. There will be more opportunities for people to specialize on certain details of history, so that even though there will be more history to learn, there will also be more scholars to understand and summarize those histories for the general populous. This is of course assuming that the trend of increased knowledge and connectivity seen throughout all of human history continues as it has for the last two hundred thousand years. We all know there is no future without a past to tell us how to live in the present, and that is the core reason why we must study history, and do so through as many different lenses as possible. It doesn't matter how old you are, be it ten or a hundred; if you only base your decisions on events that happened to you in your lifetime, or the last hundred years of your family, or even the recent short millennia of human history, you're only scratching the surface. Without knowing the full depth and fortune of our several billion yearlong history, one can never hope to be anything more than a child.

5 | Mathematics: The Power of Encoding the Universe

"It is impossible to be a mathematician without being a poet in soul."
— *Sofia Kovalevskaya (1850-1891)*

"Pure mathematics is, in its way, the poetry of logical ideas. "
— *Albert Einstein (1879 – 1955)*

Few topics in academia inspire such a dichotomy of emotion in the same way mathematics does. For many young students, math is the key topic that decides what path they follow as an adult. Even amongst collegiate students of science, it is often not the actual scientific concepts that instill fear during midterms, but rather the math used to describe them. The neural rigors of mathematics can even lead to a sense of hubris among scientists, with the biologist being ridiculed by the chemist, who is teased by the physicist, who is scoffed at by the mathematician, all based on how well they work with numbers and equations. What is possibly most tragic is that math is also one of the primary reasons non-scientists are afraid to learn about and understand the discoveries made by researchers. But like any topic, I tend to believe that an understanding about what math is and how it works serves to quell fear, and perhaps even inspire a means to use it.

Math is the language of cold hard logic, the lexicon of nature. The tongue of the cosmos is the most difficult to translate, but also the one that people are most universally fluent in. If different fields of history are defined by the period of time and the perspective they choose to look through the past, then different fields of mathematics are defined by the ways they manipulate numbers, and like all schools of thought they are not mutually exclusive. Rather, many of them rely on one another, and more simplistic ideas are used as the foundations for more complex ones. You have algebra for rearranging equations and finding new laws of physics, geometry for understanding shapes and angles, calculus for understanding changes, and statistics for understanding probabilities. You can also crossbreed old maths into new ones, like linear algebra for vector spaces instead of scalars without direction, or dynamical systems and chaos theory for situations where slight changes lead to hugely different outcomes. Mathematics is critical to understanding the true elegance of the universe and the deepest beauty of nature. It is the difference between merely looking at a sunset, and understanding the interactions of scattered photons through the atmosphere creating a vast palate of the electromagnetic spectrum.

I know that for many students, the most dreaded aspect of math exams comes in the form of word problems. More often than not, they always seem to ask the same question: *"Suzy has three pears, and Johnny has four apples. Calculate the velocity of the Sun at noon, given that pancakes are on the roof."* By the way, this is generally the same format taken by most higher-level science exam questions as well. They can all be solved in much the same way, by first writing down all the given parameters, everything you know, and the parameter the question is asking for. I feel like this qualitative reasoning is a wise strategy for any real world problem as well. First understand what you know, and then it becomes much easier to deduce what you need to know with greater clarity. And if you were wondering, the answer to the example question is purple. Why? Because aliens don't wear hats.

While you might not believe it, all humans can do math. Anyone can take a sum by adding together quanta, like adding blocks or money. But, you can also add together infinitesimal amount of anything within a given range by taking an integral. Similarly, if one wants to do the opposite, and look at a ratio of how two things are changing, one can use the derivative. It may not seem altogether obvious how these two are opposites; you might just have to prove it to yourself. One could also think of a derivative as riding along a slide, and an integral as filling up the space underneath the slide between it and the ground. Anyone can do math, it just takes practice, and that may be because the universe we are made of can be described mathematically. Math is simply everywhere in nature. It is how we are able to calculate the masses of stars, galaxies, and black holes, how we are able to balance our budgets and budget our economies. We use it to tell time, cook, create art, and build architecture. It is even used to discover planets, not just outside the solar system but also within. Urbain Le Verrier calculated the location of Neptune based on the irregularities in the orbit of Uranus in 1846. The Fibonacci sequence is seen in flowers, pine cones, seeds, spiral shells and much more. Alan Turing even used maths to understand how stripes, dots, tentacles, limbs and digits develop in animals. But no other mathematical feature is as ubiquitous and famous as the number pi.

Pi is nothing more than a ratio, between the perimeter and the diameter of a circle. But it is universal. Take any circle, measure the length of its outside, and divide that by how wide it is, and you will always get π. Moreover, because circles are the most perfectly balanced figures, and the two-dimensional analogs to spheres, π is an integral number of the universe. Spheres are more often than not the de facto shape of celestial bodies, and the shape of the observable universe. Even if objects aren't spheres, they're often circular disks. Not only that, but circles also describe phenomena that occur in cycles. Just like a ball moving on a hoop, anything that occurs in cycles requires the use of π to understand, from quantum waves to pendulums and springs. Since both light and sound are waves, music and entertainment depend on an understanding of π. Interestingly, the length of rivers divided by their distance is on average close to π as well. Not only that, but π also plays a role in statistics, namely in the Bell curve, also known as the normal distribution. The Bell curve is used to describe how far a specific data point differs from an average, and these separations are usually described using standard deviations, the same way scientists describe uncertainties in data. One standard deviation more or less than the average will include 64% of the total data if it is normally distributed, which happens to be nearly equivalent to 2 divided by π. Finally, perhaps the most interesting aspect of π is that it is irrational, a number that cannot be expressed as a fraction because it goes on forever. Most curiously, the decimals of pi form combinations of numbers that never repeat, and they go on infinitely. This means that if one were to convert the numbers in π to words, every phrase, name, password, and your entire life story could all be described with nothing more than a circle. Its no wonder π has its own holiday.

Despite all its uses, math still has limitations. A weakness of math is that it is actually too precise at times to be used by physicists and engineers. In the real world, quantum fluctuations ensure that nothing is ideal. As a matter of fact the majority of the time the true versions of equations are never used to do real things such as construct buildings and bridges, or send satellites to distant worlds. But don't lose all you're hope on scientists. Approximations aren't dangerous; you just need to how much precision is good enough to get the job done. It's good enough for government work, and it's good enough to send hairless apes into space. This limitation is also quite artificial. We observe that at the most fundamental level, all matter behaves probabilistically. God is a gambler, you might recall. In systems as vast and complex as molecules, brains, and weather, made up of so many particles, it becomes difficult to model them mathematically. Luckily, humans have clever inventions that can do math for them. You might even have one in your pocket this very minute. With the advent of quantum computing, a whole new world of computational power will be unleashed on science and mathematics, making jobs like meteorology and neuroscience far more mathematic and precise. Approximations are needed all the time. Cows are spheres, as we would say.

Many people often think that math and science are synonymous. But while scientists do rely on mathematics, both fields recognize that there are stark differences between the two. The differences between math and science might have been best described in Richard Feynman's lecture on Mathematicians versus Physicists. Feynman is often revered among the scientific community for his uncanny ability to simply explain complex concepts, and because on top of his ability to speak he was also quite humorous. According to him, mathematicians aren't grounded in reality; they don't know what the real world implications of what they're saying. Due to this they are also overly general. For instance, a mathematician would try to devise an equation applicable to any number of n-dimensions, to which the physicist, who only needs the specific case, would reply, "well that's very nice, but I only need three dimensions". That is, until the physicist tries to solve equations for a ten-dimensional string theory, which again demonstrates the need for mathematicians. On that same token, the lack of a need for mathematicians to conform to reality gives them a certain liberty to explore the frontiers of logic, developing equations which might not find practical use by scientists for centuries, if ever. This contrasts with a need for physical understanding, a trait that is often the difference between scientists and mathematicians. When Einstein was developing the mathematics of general relativity, a professional mathematician David Hilbert saw his work, and tried to race Einstein for a solution. I believe that the reason he lost was because Einstein had his imagination. Beyond mechanical logic, one must also have a vision for what they want to attain. Einstein had spent years imagining warping space and time whilst traveling on a beam of light. Hilbert never had a chance with logic alone. But he did come close, and did so in about the same time. That should be a testament as well to how cold hard logic can be used without prior knowledge, but with a firm foundation alone.

When one imagines a scientist, they might often think of a chalkboard filled with equations. Indeed, long strings of letters and symbols are usually seen as synonymous with genius. But while complex functions seem intimidating, it's important to remember that they are nothing more than abbreviations for words. Every mathematical symbol can be seen as an idea, and in that way long strings of equations are nothing more than compact essays written in logic. When thought of this way, it becomes much easier to understand how scientists develop their equations.

Let's try out a hypothetical exercise to show how you can experimentally verify Newton's law of gravity and make an equation. Assume gravity is the force of Earth pulling on an apple. You measure the force of gravity by measuring the weight of an apple. Assume a precise scale that can measure how much the apple is pressing into it. Presumably if you cut the Earth in half, the apple would weigh half as much since half the Earth would be pulling on it. You can also just cut the apple in half to confirm this. When you double the height of the apple from the ground (with a longer scale), the weight is one quarter. When you triple the height from the ground the weight is one ninth. You deduce that as the separation between the Earth and the apple increases, the weight of the apple goes down by the distance squared. Suppose you guess the weight of the Earth based on how much rocks weigh and how big you think the planet is, with its size based on measurements from Eratosthenes in Ancient Greece, who could make a decent guess based on the height of the Sun at different altitudes (not that Newton necessarily used this information). Now you have an equation that says the force of gravity from the Earth onto an apple is proportional to every bit of mass on Earth times all the mass of the apple all divided by their separation squared.

$$F \propto \frac{M * m}{r^2}$$

But there's a problem. When you go to test your fancy new equation by measuring the weight of all your philosopher stones, you notice your guess based on those masses is different than what you measure. Alas. But wait, if you take the force you measure (on the left side of the equation) and divide it by the one you calculate (right side of the equation) you notice that every single time, no matter what the measurements, you get the exact same number. That means all you need to do is slap that constant into your equation (call it G), and now your equation works every time! What's more, since everything in the universe made of the same matter on

Earth has mass, your equation is valid for apples, stars, and even galaxies (after making certain estimations on their shapes of course). Now you have Newton's Universal Law of gravity.

$$F = G * \frac{M * m}{r^2}$$

Well done, and thanks for GPS, cell phones, and pretty pictures of space!

Hooke's law with springs is similar. As you stretch a spring, the force you measure is proportional to how much you stretch it. Stretch twice as far and the force pulling on the spring in either direction also doubles, that's it! When you make measurements on different springs, you can always divide the force by the distance you stretch, and you get a number that represents the properties of that spring. Voila! You just derived an equation that states that the force of a spring is equal to a constant multiplied by the distance you stretch or compress it!

$$F = kx$$

If you know that work (a change in energy) is the sum of all forces over a certain distance, you can make a new equation describing the energy in a spring (after making more measurements to be sure your constants haven't changed). It would look like:

$$E = \frac{1}{2} kx^2$$

Good work! As a bonus, you can use this equation not just for macroscopic springs, but also as an approximation for the behavior of molecules and other objects that behave using simple harmonic motion.

Due to how much influence mathematics plays in science, it leads to one of the greatest philosophical questions of the natural world. It is the question of whether math was discovered or invented. On the one hand, there is strong support that mathematical concepts are innate to the universe, seeing as how we are able to use them to describe nature in the first place. The universe has always worked the same way, long before humans ever existed, which would suggest that we are merely unraveling the mysteries of mathematics over time. Of course, much of our mathematics doesn't obviously describe anything in reality, which suggests that humans develop mathematics from our minds, and it only happens to describe the world. However, an understanding of numbers has occurred gradually throughout our evolution. Number knowledge is observed in numerous animals, even fish and insects. Furthermore, humans use the parietal lobe in the middle of the brain to process numbers, which implies early development. From this perspective, mathematics would seem to naturally derive from our minds. This would seem to make sense, seeing as how the same laws of nature described using mathematics are at play inside of our brains. So while we can certainly create some maths, much of it is a logical translation of nature's voice as well.

But how did math arise? Some of the first signs of counting are seen in the Congo, through marks on bones. Many animals also seem to have a basic understanding of quantity, however only humans could develop abstract number theories, especially after gaining access to more nutrition via cooking and agriculture. Moreover, counting things with greater accuracy was likely only necessary to keep track of resources and food supply. It may well have been that the first great mathematicians were tax collectors, business entrepreneurs, and politicians, earning extra food and divvying it up as fairly as possible throughout the village. Clearly this would become more difficult as the villages grew into cities, empires, and nations. It is important to note that complex mathematics isn't necessary for survival. Some indigenous cultures don't even have number systems. One native Australian group only has words for "one", "not one", or "many". Instead of calculus, they use their brains to memorize songs throughout countless generations that give them internal maps of their environment. Areas with great potential for agriculture gathered people, and caused them to be fruitful and multiply. Around the world, areas with more food had more people, more mouths to feed more taxes to collect. Large populations necessitated mathematics, which in turn drove prosperity. Keep in mind there were no written symbols for ideas (i.e. words or books) quite yet, although those would be soon to follow. Mathematics may have been the origin of literature, but also monarchy, serfdoms, and wealth inequality. Everything is a mixed bag, but we're working on that.

Numbers became associated with status, as large numbers were thought up for great rulers and all of their subjects. Measurements became standardized to create large buildings and infrastructure to keep everybody safe, and geometry was developed to correlate the heavens with the Earth. This did such things as optimizing Sun exposure, and create monuments like the pyramids of Giza, which mirror the belt of Orion, demonstrating how humans have a tendency to see things, "as above so below". The Greeks found prosperity just north of the Egyptians from the Mediterranean Sea and further developed number theory. Axioms about angles, shapes, and divine ordered pairs were dreamt up and abstracted by great philosophers such as Plato, Pythagoras, and not

Aristotle. Music and art were developed throughout the world, made of pleasing mathematical combinations of sound waves and brushstrokes that added like water waves. Numbers went from being tangible things like stones and wheat, to being divisible pieces that could be broken into various fractions. There were numbers within numbers within numbers. Ideas developed about infinite numbers, and indivisible numbers. Just by drawing in the dirt, people were able to logically deduce ways to fill shapes, chop them up into pieces, and maximize the efficiency of farmland. Eratosthenes used geometry to find the size of the Earth, using the height of the Sun on the same consecutive times of the day. Those calculations translated to more modern architecture, farmlands, and population distributions. The map could now be turned from a sphere, into a flat sheet of paper. By this time, if the world was ever flat before, it was no longer. A flat Earth would get you lost at sea or in hungry jungles.

Then the Romans came, straight out of the Italian heel. They were less articulate in the language of nature, but very successful at acquiring land to export food from. But the powerful Roman Legion needed numbers as well. How else could a century of sentries march to the battlefield? This is why only simple things can be described using the rather inconvenient, albeit fun, Roman numeral system. Thankfully, useful math theories were still being developed in Arabia, India, and China, which also had vast numbers of people to govern. There, god-like levels of numbers began to arise, as people created ridiculous planet sized structures using individually shorter times, scales, and pieces. They created many from less, and that gave birth to the modern Arabic numeral system, which would be used for several millennia thereafter.

Then zero came, and finally there was nothing. With zero, the whole universe became describable using maths. With the ability to put a zero within and behind numbers, humans gained the ability to become infinitely small and infinitely large, the most crucial ability when studying the hearts of black holes and the birth of the universe. No Roman army on Earth could do that. With a firmer grasp of numbers and the universe, science, technology, and economy boomed within the Eastern borders whilst the west was forced to collect land to feed their citizen soldiers. Arabic numerals had the capacity to determine the most accurate understanding of the shape of the Earth, how it spun on its axis, and how the Earth revolved around the Sun, a millennia before Copernicus in Europe had to give to his fellow Christians a stern talking to at the dawn of the sixteenth century. Numbers were replaced by variable symbols, allowing for mathematics to become a sort of ethereal glue that could be used to bind logic into complex equations using algebra to rearrange them. Fibonacci was born in Italy, and many people like him saw the utility of our now everyday numeric system. Of course, the imperial and brutish Roman system had already been in place for a thousand years or so, and it took several generations for people to trust a complete overturn of their primitive mathematics. Italian cities even banned our number system altogether, because of hateful mistrust and inability to adapt. The Catholic Church had considered interest in Arabic numerals one of their many sins, which was a cute idea, but it also wasn't possible to maintain using the quantized roman numerals, nor did their primitive ideology promote economic growth in an increasingly global market.

Thus the Middle East experienced a golden age while Europe had been stricken with a medieval dark era, as the Roman Empire crumbled and the church oppressed the people and the land became scarce along the tiny peninsula. From the eighth through the 13th century, the Abbasid Caliphate, centered in Baghdad, developed upon algebra, trigonometry, and geometry, which would become the basis for Isaac Newton's calculus in England. With their fluidity, Arabic numerals even began to overturn the use of the ancient Chinese calculator, the abacus. When this new mathematics was finally adopted in the west, markets began to flourish, and as gunpowder was imported from China, Europe began to regrow. Unfortunately their militarism and obstinate religions didn't change very much. The Europeans became great navigators, allowing them to emerge from poverty thanks to trade and conquest. In 1679, the philosopher Gottfried Leibniz developed binary, perhaps inspired by Daoist concepts of Yin and Yang, of oneness and nothingness. He did away with all numbers, and replaced all of them just with ones and zeroes organized into placeholders for other numbers. Two became 0010, 4 became 0100, eight became 1000, nine became 1001, and forty-two became 101010, each first digit determined by doubling the previous number, starting with one.

When the Big European Island that needed ships the most found coal in their dirt, the industrial era began. Europe harnessed the power of coal and took over the world, and the added prosperity was driven and enhanced though calculation. Laws for thermodynamics, chemistry for agriculture, biology for crops and ecology all developed. Taxes modernized to be able to extort resources from colonies and keep track of captured and purchased slave numbers. Ada Lovelace died before turning 37 in 1852 London. But before she did she also created the first computer algorithm using advanced number theory and algorithms based on Boolean logic. A century later, the mid twentieth century saw the first computers based on the work of her correspondent, Charles Babbage, and he gained the title "Father of Computers". By the time the twenty-first century hit, humans were living in a global digital community made up of ones and zeroes. Mathematical theories led to the likes of General Relativity and Quantum mechanics, and for the first time ever, humans were making predictions and revealing truths about nature, before humans even noticed they were happening. Math led science, and the universe grew from the size of a continent, to a globe, to the Milky Way, to the Virgo Supercluster, through the Cosmic Microwave background, and into the heart of the early universe immediately after it was born. With the next steps in human understanding, computers are being devised based on the observed quantum pandemonium, in order to enhance our technology from one or zero, into one and zero. With more powerful computers, we may calculate new ways to live and prosper, to travel to the stars, and get a better Internet connection.

So in short, logical, concise, and inventive mathematical minds are to blame for the entire world's prosperity, and some of its strife. Everything is a mixed bag. But I'm definitely of the mindset that simulated war games on computers are more fun than their real life counterparts. In video games, you can ride a bicycle into a tank and still win, if not just respawn. But more usefully you can practice tactics and refine your brain. Math in general is quite good for your health.

6 | Science: The Power of Observing the World Around You

"If I have seen further than others, it is by standing upon the shoulders of giants.
— Isaac Newton, who showed how the heavens go, despite believing heaven was a real place

"Science knows no country, because knowledge belongs to humanity, and is the torch which illuminates the world. Science is the highest personification of the nation because that nation will remain the first which carries the furthest the works of thought and intelligence."
— Louis Pasteur, who showed that life can't spontaneously spawn from meat and wheat

Imagine for a moment what is was like to be the first human. When you look out at the world, what do you see? The world seems to be divided into two parts, above and below. In the below is a vast landscape that seems to go on and on for an infinite length. On it you would see a diverse cast of living things from plants to animals to fungus. Above you would see a constantly shifting blanket, half blue with a life-giving disk, half pitch dark and speckled with glimmering lights. You would see countless varieties of living creatures, all perfectly suited to their environments. Beyond that, all of existence would seem perfect, elegant, and uniform. Life too is evenly balanced, as all things that live eventually die and return to the Earth, until new life rises to take its place.

But despite the seeming perfection, there are still many mysteries that reveal the true chaos of the universe. Sometimes the ground splits open as volcanoes erupt molten ashes that release the fury of hell itself upon the Earth. The skies will occasionally grow dark as winds rage and oceans drain from the sky. Amidst the dark clouds, hot flashes of fire crash down from the heavens, striking the mighty fury of god down to Earth. Even when the ground is calm and the sky is clear, the grim reaper's breath might still creep into the unsuspecting, and rot the flesh from the inside out. If none of these grim fates find you, then you might fear for the worst fate, being terminated by one of your brothers or sisters. Volcanoes, earthquakes, floods, tornadoes, lightning, hurricanes, plague, and war, how can one reconcile the uniform majesty of the universe with the wanton destruction that takes place inside of it? For 99.9% of human history, 99% of human beings had zero idea. The best they could venture to guess was that all of their fortune and misfortune was the response of some divine power that was judging them based on their actions, no more or less moral than they could have been with so little knowledge.

What if I told you that the answers to all of those questions surrounding the mysteries of nature could be understood, that we could read the mind of the gods? *"Heresy!"* you might interject. What if I said that diseases were caused by microscopic life forms that outnumber every plant and animal a million to one? *"Nonsense!"* And what if I were to say that all life on the planet developed from those most simplistic forms of life? *"Blasphemy!"* Or what if instead I were to tell you that volcanoes are nothing more than cracks in a thin shell around the molten rock that is our planet, heated through the energy of radioactivity and the energy of a quadrillion rocks colliding? Or if I said that very rock is just one of a probable trillion planets around one of a trillion stars in one of a trillion galaxies in a potentially infinite universe that has been constantly expanding for thirteen billion, eight hundred million years? *"Sacrilege!"* But it's all true, and it paints a story far more glorious and awesome than any folktale.

The thing about humans is that we are experts at making generalizations. With limited data, our assumptions are almost certainly false, if not very situational. Science too, is essentially made around assumptions, but extremely good assumptions. Science is based on making precise and repetitive measurements, testing something over and over again. As we make more observations, more measurements, we become more confident in our generalizations. The end goal of science is to be able to describe all the vast complexity of the universe in a precise, elegant way.

We notice ways in which objects always behave, and call them laws. Building upon countless factual, repeatable observations, we derive theories that describe the way the world works. The universe expands, life evolves, matter is energy, and energy warps space and time, causing gravity. What's more, scientists are always looking for new questions to answer. They are aware that there may be more to the picture that they are missing, and they actively seek to disprove themselves and one another. Through this method our understanding becomes ever more firm, and we are able to use what we know to develop new technologies and enhance the way we live. That's the power of science.

Science has become such a powerful force for development that it has nearly become a religion all its own. People have so much faith in it that you might hear science enthusiasts using the word as a noun, justifying claims with statements such as, "because science!" as if science was some authority figure that should be trusted at face value. People even use the word colloquially to validate their non-scientific claims, with such statements as, "There's a science to it", which often describes when something is methodically well understood. Science truly has become a cultural phenomenon in the twentieth century. But while I truly love this common passion for knowledge, I can't help but feel that the true meaning of science is lost to the general public, and as a result may lead to a denial of scientific claims due to the word's casual use.

It's important to remember what science truly is at its heart: an abbreviation for a process, known as the scientific method. Science isn't an authority, nor is it something that should be accepted as a mere source for blatant fact that everyone should just accept. Science is a process, and simply accepting any claim because "science" says so spits in the face of the method itself. If scientific claims were obvious, it wouldn't have taken us 199,800 years to realize that the world isn't 10,000 years old, or that you should wash your hands with soap. The gist of the scientific method can be seen in any elementary school classroom or with any normal child. It starts with nothing more than a question, of how something works and why something is the way it is. You might have a guess in mind based on your previous experiences, and in order to test your prediction, you make observations. After multiple people through multiple different experiments have independently made enough observations, and they all point to one idea, we begin to develop a theory of the world. Nobody will mind if I try to prove a scientific theory wrong, because I usually can't. Scientific claims are grounded in what people actually see is happening around them, not based on what people think about the things they see.

Both means of understanding are crucial, and only one can give you better agriculture and television. But they can both help you make decisions on how to use those things. However, the distinction between a professional scientist and an inquisitive mind lies in how rigorous these observations are. Anyone can draw a million conclusions from a single observation, but a million observations leaves only one conclusion. That's what science is all about, and all you need to do it is to look a little more carefully than normal. You need to do it for yourself, not just listen to what others think.

This is why the modern scientific method didn't develop until Galileo Galilei began his work in the sixteenth century of the Common Era, just at the end of the renaissance and ushering in of the era of enlightenment. He himself followed in the footsteps of Leonardo da Vinci, another harbinger of curiosity. Often considered the grandfather of modern astronomy, at least because he was the first person to stumble upon the use of lenses to see far away things, Galileo was just beginning to revolutionize human perspectives of the universe around the same time Spain was colonizing South America and plundering their silver. You'll recall that I said that was only twenty generations ago, and I can deduce that claim scientifically. We can assume many people in 1600 CE had children in their teens. Since *Romeo and Juliet* takes place in a similar time, where Romeo was 16 and Juliet was either also 16 or just 13 (depending on the source), we can assume they are a fair approximation of what circumstances people in Italian, Spanish, and English cultures were living under. 2000 minus 1600 is 400, and 400 divided by twenty is also twenty. Science! Isn't it truly astounding how much we were able to learn about the world in such a short time? One can only imagine how much we could have accomplished had we started conducting scientific research sooner. Shakespeare might have been writing plays for a live studio audience broadcasted across Europe.

The story of Galileo also highlights another critical aspect of scientific thought, which is that it requires freedom of speech. Galileo was put under house arrest as a heretic, because most educated people at the time ascribed to the geocentric model of the universe given by the Catholic Church. He was labeled as an apostate for claiming that the Sun was at the center of the solar system, on the grounds that this view was in direct conflict with the bible. They cited several verses, including Psalms 104:5, which claims that, "the Lord set the Earth on its foundations; it can never be moved", just one of many tragic flaws in that bronze aged text written by dessert dwelling sheepherders. Unfortunately, these conflicts between ideology and reality are all too common, and while it is good to challenge scientific claims, I find it absolutely criminal to think of all the people who were condemned for preaching the truth, despite the widely held theology. Although I can't say these outcomes are in any way surprising. Communicating the complexity that science has laid bare has and will always be the greatest challenge. Luckily our neural hardware supports the capacity to accomplish just that.

I'm not surprised it took so long either. Science is hard to do. Anyone can do it, but proper science isn't something that can be done by a peasant, or really anyone else who needs to work during every hour of the day just to put food on the table. Many of the first scientists were nothing more than wealthy bankers and politicians who had extra time on their hands and money to build scientific equipment to run experiments. It's no wonder why the likes of Hooke, Boyle, and Newton all came out of England. They all lived in the wealthiest land and were born with silver spoons in their mouths, likely with some association with politics and the church, back when those two things were always synonymous. From these earliest pioneers came the most fundamental form of science, experimentation. Before that, the closest people to experimental scientists came out of Asia, or even the Islamic golden age of the eleventh century, before they banned mathematics as works of the devil. Before them, only philosophers really had the time to worry about the deep metaphysical questions beyond the juvenile mathematics that they had. But unfortunately rigorous observation and finely controlled experimentation are the only ways to make definitive predictions about reality. This was true for several centuries, until theoretical physics began to lead the way during the nineteenth century. Once the experimental physicists were able to derive laws of nature based on observation, theoretical physicists such as Einstein were able to use mathematics to predict depths of reality that wouldn't be experimentally verified for another century after he died.

If experimental physicists were more often than not rich and bored, theoretical physicists are often the ones who daydream the most. Experimenters could buy or make whatever equipment they needed, but all theorists needed were peace and some paper. But only after we had some good facts from experiments could philosophical daydreaming turn into mathematical daydreaming. People can do experiments and theory in great numbers, but for creativity to blossom one often needs a healthy amount of time for quiet introspection. In addition, it has been individuals, or small groups, which have made monumental advances to human understanding. Too many minds can actually stifle progress. The only real difference between scientists and commoners is that the vast majority of experimenters were born rich. They didn't have to go to war, they got to stay in school, or at home, and just waste away their time thinking and dreaming about how the universe goes. Historically, wealthy people do the best science, since wealth leads to a higher standard of living. It's a shame more rich folks living in the twenty first century don't conduct scientific experiments. Most of them are just fake skinned airheads. But regardless of where scientists come from, one cannot exaggerate the importance that both experimentation and theory have on the scientific method. Theories are useless without some sort of experimental verification, just as experiments should always lead to more accurate theories. When unified, these two methods give scientists the power to do something psychic mediums can only generalize about, predict the past and control the future. Nature's laws are true for all times and places. With Newton's laws, one can send rovers to Mars and speed satellites out of the solar system with gravity assists. With knowledge of kinematics, a detective can piece together a car crash, and with some chemistry know-how deduce a homicide case. Of course, those kinds of things might be mitigated if everyone had a personal phone camera. As a matter of fact, portable video cameras have probably contributed just as much to world peace as the polio vaccine or water purification, and those inventions only exist thanks to an understanding of cells, chemistry, and quantum mechanics. It's thanks to the critical scrutiny of science that humans have come to understand the world well enough to come out of the dark ages and revolutionize the way we have lived for a thousands of years.

Some of the most complex concepts discovered with science are so abstract that they are difficult to explain. One of the most common ways to explain difficult concepts is through way of analogy. But scientists must remember to be cautious with analogies. For example, my description comparing electrons to balloons isn't a common interpretation. We are taught that electrons are "smeared" into clouds around the nucleus, and so I personally envision a thick membrane of charged energy. We also now know that the classical "solar system" atom with balls orbiting a nucleus isn't accurate, even though it's still far more accurate than the old plum pudding models they used in the late nineteenth and early twentieth centuries. However, if you imagine the electron orbiting in a perfect circle around the nucleus, if it moved fast enough it would look like a hoop, and if the hoop was spinning fast enough, then it would appear to be a spherical shell. This may also be a viable way to envision an electron as a point-like particle as well, just a dot going in circles going in circles. Again, be cautious, recall the wave-like behaviors particles exhibit also. This phenomena is probably best explained when electrons are thought of as excitations in matter fields. Their location is uncertain because electrons can be thought of as popping in and out of existence from an electron field, like ripples in a pond.

In general, the majority of people seem to support studies made with the scientific method, so long as it has tangible benefits to their lives. Now, I would personally argue that understanding how the world works is always an enriching and beneficial endeavor, but I'm the kind of person who thinks in the long term with reference to the distant past. I understand that this isn't true for many people, who would prefer to look with great detail at the relatively near past and future, rather than generally throughout all of time. For those people, advances in medicine and technology take precedence over research that attempts to answer more metaphysical questions about where we come from, how we should live, and why we're here. The reason, I believe, is that these questions arise inherently from our ability for abstract reasoning, which is thanks to our highly advanced neurology. Since these questions are inherent, they are asked by all people throughout all of time, which means everybody has an opinion on them. These curiosities are written into everything we do, from our art to our literature, and of course through our research. This is why the most controversial discoveries made by scientists are the ones that starkly contrast our previous dogmas and philosophies. These discoveries become so disputed, that merely mentioning them becomes taboo in certain circles, leading to people who will vehemently deny them despite any and all evidence in support of those ideas.

In my experience, the three most controversial discoveries made by science are the ones that answer the biggest questions. There is the big bang, which explains the history and future of the universe as a whole, where we all reside. Second, there is evolution, which explains the history and future of life, the thing that unites us with microbes and distinguishes us from rocks. Third, there is human accelerated climate change, which is caused by our vast intellect as a species, and which threatens to put an end to all of our advancements. Speaking to you not merely as a scientist, but as a human being who understands how we've come to these conclusions and how they best describe the world we live in, I can tell you that all three of these claims are indeed true. They are not to be feared, but to be understood and celebrated as a demonstration of our intellect, the trait that has allowed our species to survive long enough to colonize the globe. These things are not a matter of belief, but a matter of observed reality. So how do we know?

Let's begin with climate science, which has only become a matter of interest just over a century after the industrial revolution, near the end of the twentieth century. The issue has been heavily politicized beyond nearly any other scientific claim, as well it should, considering how it affects the whole human species and could lead to the end of industrialized civilization. Folks who don't understand how humans are accelerating climate change started by denying that the planet was getting warmer in the first place. Back then it was called global warming, which admittedly sounds quite alarming. After the temperature visibly rose year by year, certain parts of the planet began to see unusual droughts or floods, storms began to increase, carbon dioxide levels were measured to rise, and the oceans began to expand and acidify, those people stopped denying that the planet was getting warmer, and instead denied that humans were causing global warming, and the colloquial term became climate change.

People often claim that the weather is and has always been unpredictable, and that the climate has always changed and fluctuated over time. If they don't deny that the planet is getting warmer, and that human activity has caused carbon dioxide levels to rise since the industrial revolution, then they deny that there is any correlation between the two. This argument is mostly just given by politicians, whose only way to stay in power is usually by defending big oil companies and catering to an uneducated public. If their power and influence rest on denying how humans accelerate climate change, it's no surprise that they defend the skeptical stance vociferously. While the weather is quite unpredictable and variant, it's critical to emphasize that local weather and global climate are not one and the same. In fact it is quite narrow minded to say that just because the weather is still cold in your local neighborhood that the Earth—whose surface is some tens of millions of times larger than the horizon—is not getting warmer. The connection between the industrial revolution and the climate isn't difficult to comprehend either. Humans light carbon on fire, lighting carbon on fire produces carbon dioxide, and carbon dioxide makes the planet get warmer, simple. To be clear, it is certainly true that the global climate has fluctuated countless times over the millennia. That's not the problem. People are sounding alarms about the issue not because the climate is changing, but because human accelerated climate change is happening so quickly. When changes occur quickly, our infrastructure is at risk, our food will die out, and the biodiversity that keeps these changes intact shrinks as animals that cannot adapt quickly enough become extinct.

From a scientific perspective, it is crucial to understand what climate researchers measure in order to understand climate change and devise preventative measures to control it. Measuring the temperature is rather straightforward, one merely has to use accurate thermometers around the planet corroborated with satellite infrared cameras. Humans have also recorded temperatures for nearly as long as we've been able to write things down, giving a third, albeit less dependable means of measurement. The temperature of the atmosphere also corresponds to properties of the oceans, specifically the sea level and its acidity. Like nearly all substances, water expands when it heats. Coupled with the melting of glacial sea ice, this leads to rising sea levels that threaten coastal cities. Now, it should be noted that as more freshwater ice melts into the oceans, the salinity of the ocean decreases, and as the ocean becomes more diluted, it melts at a lower temperature, seasonally increasing the levels of sea ice in certain regions. However, this should not distract from the overall effect of melting sea ice, which as a whole is still decreasing and threatening to expand the oceans and reduce wildlife habitats. Moreover, the danger to those who live near the ocean too is not quite as severe as ocean acidification, which is one of the less discussed threats of global warming. Oceans acidify because carbon dioxide snatches oxygen from water molecules where the sea and air meet. This turns CO_2 into H_2CO_3, known as carbonic acid. Since the oxygen hold onto the electrons that were once shared with hydrogen more strongly than the hydrogen nuclei themselves, free hydrogen ions will detach and be left to roam around the ocean, turning carbonic acid into carbonate and acidifying the ocean. This effect decreases the amount of oceanic calcium that can bond with carbonate, reducing the amount of calcium carbonate that can be used by animals that utilize that chemical to build their shells and homes. This effect is most pronounced in coral reefs, which have begun to lose their color, fading into a ghostly pale white as they die off. Since these animals mostly lay at the bottom of the marine food chain, their demise results in the deaths of the more complex organisms that we fish out of the ocean to eat, as well as the animals on land that feed off of marine species.

Measuring carbon dioxide levels is slightly trickier, especially before the industrial revolution, or even before human beings ever existed. But like anything done scientifically, these measurements have been made by several independent sources. Measuring carbon levels in the atmosphere today is relatively simple, one has to do nothing more than go outside and collect a sample of air. Carbon dioxide levels can be measured several ways, my favorite being through spectroscopy. By shining light on a sample, one can determine what's inside of it by observing what wavelengths of light were absorbed and reemitted. To measure carbon levels in the distant past, climatologists must look for evidence recorded by mother nature. Two famous methods come in the form of ice cores and tree rings. Since plants require carbon dioxide to live, one can estimate the levels of carbon in the atmosphere by observing how trees have grown over their lives. Every season, trees grow by adding new layers to their bodies, which appear as the rings you see in a tree that has been cut down. Scientists essentially measure how much food was available for the trees to eat by observing how thick their rings were each year. In addition to looking at recently living trees, one can also look at fossilized trees, which can be carbon dated just as any other fossil (more on that shortly). Obviously, tree growth depends on several factors, including availability of moisture, and so other methods are needed to demonstrate carbon levels throughout the ages. This is where ice cores come into play. Like trees, ice levels build up in polar sea ice throughout the years, and when that water freezes, it traps air from the atmosphere inside of it. By understanding how ice builds year by year, one can correlate the contents of the Earth's atmosphere throughout the ages. What we've discovered is that since we started using fossil fuels for energy, the levels of carbon dioxide have risen to over 400 parts per million, twice as much as the concentration before humans existed and over a hundred parts per million more than any of the highest levels in the last four hundred years. So yeah, global warming is a thing, and humans are accelerating it. Who knew? Actually, a lot of people, did you? If you still have lingering questions, we will continue the conversation on industry in Chapter 12.

How about evolution? People seem to get personally offended by evolution far more than any other natural phenomena. Calling people animals is most often a severe insult. We try so hard to distinguish ourselves from, "the beasts", and when we engage in our primal instincts to feed and mate, it is often punished by those who mark those actions as being the antithesis of civilization. We try so hard to assert dominion over all other forms of life. We build large homes, and go out of our way to kill off anything that dares to try and live in our space. To prove it, just look around your home. Presumably you don't see a lot of insects, fungus, or other pests nearby. If you do, many people would consider you to be poor, when in reality you might just not be selfish with your space. Although you probably don't want to live in a moldy house with cockroaches and rats, that's bad for your health. But if you don't keep your space clean, something will try and live there with you. Life is so rare in the universe, if there wasn't something for them to feed off of, they wouldn't be there. The only way we keep our homes, "clean" is by making them nearly inhospitable to any organism besides ourselves and a few select plants and pets. There's no denying that our intellect sets us far apart from nearly all other organisms, but that doesn't change the fact that we are all living organisms. I never really understood why people were so sensitive about that. There's nothing wrong with being an animal, it's just what we are. We all breathe oxygen, need to eat, and we even share basic survival instincts with our closest relatives. There is simply no other way to explain all the similarities that tie us to every living organism, and there is no other way to explain what makes us unique. Our intellect doesn't really make us superior to other animals either. It is a great advantage that allows us to adapt our environment to our needs, but that isn't what evolution is about. Evolution is about what traits happen to be good enough for organisms to keep doing the whole living thing.

But just because an idea is obvious in context, that doesn't make it obvious at a casual glance. Charles Darwin's *On the Origin of Species* is perhaps one of the most seminal books in all of history, precisely because nobody before in human history had thought to connect all the different living beings on the planet in such a way. Thomas Huxley, who was mockingly referred to as Darwin's Bulldog, described its implications most accurately when he said that, "Old ladies, of both sexes, consider it a decidedly dangerous book." It's no surprise, Darwin's findings led to the theory that shattered the perceived boundaries between humans and the animal kingdom that we had worked so hard to build. Darwin didn't have a full understanding of the big picture, but ever since he published his book, new data has done nothing but enforce his understanding that all life on Earth ultimately evolved from a single common ancestor.

Evolution occurs through the combination of two main factors: genetic variation and natural selection. Put simply, evolution is the process by which the creatures that survive long enough get to pass on their genes, and those creatures that don't survive don't pass on their genes. Each creature carries genes for various features, and their offspring have their own unique genes that vary ever so slightly from their parents. These genes are expressed externally as behaviors and physiology, through what are called adaptations. The DNA of organisms can change every generation through several means, such as insertions and deletions of new genes, duplications, transportations, and point mutations, where an individual nitrogenous base is replaced with another (e.g. adenosine is replaced with guanine). Whole chromosomes can also split and merge over time, as was the case with the human-chimpanzee ancestor, which is why human chromosome 2 is the merged version of two chromosomes found in chimps. If adaptations help an organism to not die, then those traits get passed on to *their* offspring, on and on until those traits are either not beneficial or ingrained into every descendent of the animal. In other words, if you're good enough, you get to live on through your offspring; otherwise another one bites the dust.

There's a common misconception about one of Darwin's most famous idioms to make it into common vernacular: the survival of the fittest. This idea sounds very powerful, and that's probably why most people use it incorrectly. People often take it to mean that the biggest and strongest members of a species are the ones who make it to the top. It's further misinterpreted when used to describe social Darwinism, which claims that the strong do what they will and the weak suffer what they must. The problem with this line of thinking is that evolutionary fitness doesn't have the figurative meaning of being strong, but rather the literal meaning of fitting into an environment. What helps an organism fit into its environment is entirely dependent on where it lives and what niches it fills in that environment. Bacteria are fit if they are immune to antibiotics just as prey animals might be fit if they have camouflage or wide vision. Again, if an animal can stave off death long enough to pass on its genes, then it's fit, simple as that. For that reason, the rich racist colonists were some of the least fit human beings, because they were too stupid to realize that the only reason humans can outcompete every other animal is by working together intelligently. In that same sense, the fittest humans are the ones who can solve problems and communicate well enough to gain allies and work with others.

Evolution ties together and explains nearly all phenomena which we observe of life, both in our own daily routines and when we research nature. To begin, we should first understand what a species is before we discuss how they arise. By definition, a species is unique when it cannot mate with a member of another species and produce viable offspring. This is why a horse and a donkey can mate to produce a mule, or a lion and a tiger to produce a liger, but why both mules and ligers are sterile. This is also why all humans are the same and why defining race and ethnicity is stupid and naïve. We are all human beings, because we can all have kids with one another. But clearly humans did evolve slightly various external traits based on what part of the world our ancestors grew up in. Darker skinned people were exposed to more sunlight, and people with narrow pointy noses developed in cold dry climates. But our subtle human differences are still nothing compared to the various dog breeds we have created, and all of them can mate as well. New species arise when two groups are isolated, choose a different food source, or in some way are kept from breeding long enough for their genes to stop mixing and for their offspring to adapt to a new environment, food source, etc. This is why some of the best areas to study speciation and see unique biodiversity are islands, as they are essentially microcosms for evolution. It is important to emphasize that evolution is a generation by generation ordeal, which is why animals like mammals take longer to adapt than say insects, who breed every several days or weeks, and why there are far more types of simple organisms than complex ones. For instance, there are only seven great apes, but over eight billion types of beetles. This highlights why rapid climate change is dangerous, not because changes are happening, but because changes are happening so quickly that many species are dying out before they can reproduce and adapt their populations to those changes. Speciation might be deceptive as well. Convergent evolution frequently happens when different species evolve similar features, like how whales have fins like fish or how bats have wings like birds. Sometimes what works, works.

Evolution is everywhere you look. It's why there is such a variety of dog breeds, with traits selectively bred for by humans over tens of thousands of years, testifying to how dog breeders understand the world better than creationists. It's why people need a new flu shot every year and why overusing antibiotics leads to resistant bacteria. Extremely subtle differences between individuals, from homosexuality to lactose intolerance, are all products of evolution that may have helped our ancestors reproduce. Hypothetically, in the case of drinking cows milk instead of human milk, which is already a strange concept, those that could digest lactase could have had access to additional nutrition that those who are lactose intolerant wouldn't. More controversially, homosexuality could be the expression of genes that when present in a heterosexual individual make the opposite sex more masculine or effeminate. This would be a tenable explanation for why homosexuality is inherited despite those individuals not reproducing. Masculine qualities that help males mate with females would lead to homosexuality when expressed in females, just as effeminate qualities that would help females find male mates would lead to homosexuality when expressed in males. It seems likely that these traits are neurological in nature and based on varying levels of chemicals associated with sex, like testosterone and estrogen, since people are born on such wide spectrums of gender and sexuality, and because those two traits are often independent of one another. Homosexuality is also observed in nearly all mammals, and may be why higher levels of homoeroticism distinguish bonobos from chimpanzees. Bonobos seem to act much more peacefully because they make love, not war. Speaking of socializing, another example of evolution at play can be seen in humanity's other favorite pastime, drinking. Most animals can't process ethanol, but doing so might have allowed our ancestors to eat partially rotten fruits and vegetables. Some even contest that agriculture began out of a desire to ferment grains. Evolution is even seen inside of us. The laryngeal nerve, which controls the nerves of the larynx, wraps around the heart's aorta when it could just as easily go a few centimeters from the brain to the voice box. It is found in animals as early as amphibians, and this strange feature is even seen in giraffes, which comically have the nerve running all the way up and down their ridiculous necks. There is absolutely no need for this. But evolution doesn't do what's best, it just repeats what works, and certainly isn't intelligently designed.

Obviously the long-term ramifications of evolution are best laid bare by fossils. In addition to the bones we all adore, organic matter can be preserved in multiple other ways as well. Insects can get petrified in amber from the sap of ancient trees, and casts of animals can be solidified in stone. But I think few would argue that the most exciting types of fossils form when water carries sediment through the corpses of dead animals, and bone turns into stone. Indeed, the visage of ancient animal bones are often the very epitome of what people think of when they hear the word 'science'. In order for paleontologists to connect fossils to modern living animals through time they need to know two main things, the fossil's age and what parts of the organism in it belonged to. The age of fossils, like rocks, can be determined thanks to the weak nuclear force via radioactive decay. Every element has radioactive isotopes, and by understanding the ratios of radioactive isotopes to stable isotopes when the radioactive ones are still free to enter a given system, one can determine the age of a sample by looking at how the percentage of radioactive isotopes in that sample now compares to the percentage before. Rocks are dated from when they solidified from magma, were deposited as sediment, or otherwise cooled off. By dating the rocks around where a fossil was found, researchers can get an age for that fossil. This is often done by looking at the ratios of potassium-40 to argon-40. In the same way, newer fossils can be dated from when the organism died, through the better known means of carbon dating. Since all living things have carbon in them, this method is popular for dating all sorts of objects, even including old literature and mummies. The coolest thing about this is that the radioactive carbon-14 originates thanks to radiation from the Sun, which is then incorporated into the atmosphere's CO_2, taken in by plants, and then integrated into the rest of the living kingdom.

The process of deciding how prehistoric fossils relate to modern organisms is a task of taxonomy, anatomy, and physiology. Paleontologists also need to understand what modern animals look like, especially by looking at their bones. They look for similar types of bones, and if they're lucky how whole skeletons fit together, in order to deduce the sorts of environments those animals lived in. They might look at the teeth to understand what the creature ate, the hips to understand how it walked, or even the patterns on the bones to see whether a creature had scales, fur, or feathers. Through ever more advanced technologies and countless carefully excavated samples, we have been able to put the tree of life back together. We even learn about the order in which each class of animal developed. For instance, we never find any mammalian bones which are as old as the oldest fish bones, which are never as old as the oldest fossilized stromatolites left behind by microbes. Thus we learn how humans didn't just evolve from a monkey-like ancestor, but mammals from a reptile-like ancestor, reptiles from amphibians, amphibians from fish, all the way back to some single cells that ate a bacterial mitochondria.

One of the best understood and most interesting family trees can be seen in whales. Whales surprisingly evolved from terrestrial animals, becoming the only mammals to fully return to our aquatic roots. One can tell superficially that whales are mammals by observing how they have fur, give birth to live young, breathe air, and have horizontal back fins unlike vertically finned fish. The most amazing thing is that we can observe the transition from feet to fins and from noses to blowholes through fossils. We've found intermediate fossils through the last fifty-five million years or so, even near the base of the Himalayas. From the terrestrial Pakicetus, we see descended species, which had features indicative of aquatic species, such as Rhodocetus, and later Dorudon. Whales today still have residual foot bones left behind from their ancestors. This is a key point. Just as humans still have the appendix attached to their large intestines, whales still retain the genes for legs. The legs haven't completely disappeared, because evolution isn't goal based. If a trait is useful, such as the lack of visible legs for whales, then they are passed on. But if a trait is useless, like the fact that whales still technically have legs, then it too will get passed on. Traits are only selected out when having them results in members of a species being less likely to survive long enough to mate. Useless traits are just as favorable as useful ones. That's all evolution is, life doing its best to not die in an ever changing world, thanks to a combination of quantum genetic shuffling paired with naturally and intellectually chosen expressions of that shuffling.

Finally, we come to the big bang. It's a rather misleading term. The word *bang* usually implies an explosion, but explosions require something to explode inside of. Let's clear this up now. When we look out into the cosmos, we observe that all the billions of galaxies are uniformly distributed. Not only that, but the cosmic microwave background at the edge of our visible horizon too is almost perfectly uniform. Finally, the expansion we observe happening in the cosmos is happening equally throughout all of existence, much like paperclips on a rubber band which is being stretched out. In other words, every galaxy is spreading apart from every other galaxy by exactly the same amount. From what we can tell, the universe appears to be infinite, if not far, far, far beyond our horizon. When people describe the big bang, they usually make it sound as if at the beginning of time everything in the universe was all crammed into one place, the so-called cosmic egg. This is even how they taught the big bang in our second semester astrophysics class. I find this description remarkably misleading, and when describing the beginning of the universe, that is a great disservice. I much prefer the interpretation given in my actual cosmology course. As time goes on, the whole universe expands outward, and everything gets farther away from everything else. Ergo, when the universe began, everything was closer together. But infinity is still infinity, it doesn't matter whether there was more or less "empty" space in the universe. Everything in existence wasn't packed into a single point. Rather, every point in the universe was a singularity, which is and always was infinite. Otherwise, you have to deal with the visualization questions, "where was the singularity, what was outside of it, and where did the cosmic egg come from?" and these questions that obviously make no sense because by definition there was no such thing as "before" or "outside" of the universe, because there was no time or space before the big bang. Spacetime is what's expanding in the universe today, and that is what everything else resides in. This is a subtle difference, but I feel like it is important to mention, because most people think that the big bang happed when the universe was literally smaller, instead of more crowded. Of course, mathematically we can't yet describe the very instant of the big bang, let alone imagine anything before it with testable certainty. However, an infinite universe where everything is infinitely close and infinitely hot would therefore be infinitely energetic. That sounds like god, but only so much as we can't describe it simply yet. So maybe god did exist at time zero, and then promptly stopped existing when time began, depending on how you define god.

So how on Earth do we experimentally verify an expanding universe? Of course, the first major discovery was to see galaxies actually flying away from us. Like all things in astronomy, the way we can tell is through light. Just like how the sound from a siren lowers in tone when the siren is moving away from you, light is redshifted when the object emitting that light moves away from you, and we can see that in the spectra of distant galaxies. We can measure their distances as well, although this isn't easy. In order to know how far away objects are, we need to corroborate several different methods and gradually work our way outward. For nearby objects, we use parallax—which is the effect you notice when you hold your thumb away from your face and close either of your eyes—except instead of your eyes, astronomers use the Earth at opposite ends of its orbit, and instead of your thumb, astronomers see stars shifting, and calculate how far away they are by how much they shift. For farther objects, like the Andromeda Galaxy, we use bright stars that vary in brightness based on how large they are overall, known as Cepheid Variables. If we know the period of oscillation, we can figure out how bright the star is, and if we know how bright it should be, we can figure out how far it is based on how bright it looks to us. For the farthest objects, we need to use some other standard candle. The most popular and reliable candles are type 1a supernova, which if you recall all the way from Chapter 1 are white dwarves that explode and always release the same amount of light since they explode at a specific threshold. There are several other more nuanced ways to measure distances in the universe, but these methods are probably the most important.

Once we realized that the universe was expanding over time, and that galaxies farther away from us were moving away faster, we could calculate the age of the universe, when everything would have been together, thus the big bang model was born. This theory had to make several testable predictions about what we should see in the universe, and all have been verified. One prediction was that we should see a point when all of the light in the universe was first able to travel freely throughout it. This is what the cosmic microwave background is. The WMAP spacecraft first got pictures of the leftover radiation from the big bang by being placed behind the Earth in a lagrange point, where it would always have its back to the Sun, and thus be able to map out the cold void of space. Another prediction made by the theory is the abundance of deuterium (a proton and neutron) in the universe. The reason for this is subtle. Deuterium only forms today inside of stars, where it is promptly turned into helium, but at the beginning of the universe it would also be created until the universe expanded and cooled enough to "freeze in" the amount of deuterium in the universe. As a matter of fact, free deuterium can only be created shortly after the big bang, and the model predicts a specific amount of deuterium, which fortunately happens to be the amount measured by astronomers. As measurements get better, cosmologists can better understand the parameters of the early universe, like the percentages of matter, light, and the cosmological constant that describes expansion.

In order to understand events before the CMB formed, we turn to particle physics. The reason that the universe was hotter farther back in time is the same reason gas heats up under pressure. As matter gets hotter, it goes from solid to liquid, then to gas, and then to plasma, eventually to gluon-quark plasma. Physicists can actually replicate these conditions inside of massive particle accelerators by smashing protons and electrons together near the speed of light. So as we go back in time, matter would have been closer together, making the universe hotter. Therefore, one of the goals of particle physicists is to build more energetic accelerators, because high energies are needed to replicate conditions in the early universe. One of the main reasons we haven't observed supersymmetry or gravitons yet may be because of their theoretically high masses. Supersymmetry is the idea that for every fermion with half spin, there is a supersymmetric partner fermion with no spin, and for every boson with full spin, there is a supersymmetric partner boson with half spin. These superpartners would be extremely massive should they exist, and so high-energy accelerators would be needed to coax them into forming. The reason physicists are looking for supersymmetry is because it is a natural extension of already observed symmetries in nature, and because it may provide insight into such things as superstring theory. So as a rule of thumb, the more energetic the accelerator, the farther back in time we can learn about.

We can also take inventory of the universe through several means. We can weigh the universe through relations we observe between the brightness of stars and their masses, measure the amount dark matter by seeing how galaxy clusters warp space through gravitational lensing, and measure the amount of dark energy by measuring the expansion of space. If the whole idea of empty space having energy sounds strange, you can experimentally verify that as well. One cool experiment is known as the Casimir effect, where two plates, set a micron apart actually come together due to virtual particles. The reason this happens is because of the uncertainty principle, which allows energy to appear out of "nothing" for short periods of time, like borrowing energy before anyone notices. Since matter also behaves like waves, that means that matter with large wavelengths and short wavelengths are constantly zipping in and out of existence on quantum scales. Between the plates, only small wavelengths appear, but outside, both short and long wavelengths can appear, meaning that outside there is more overall energy, so the plates get pushed together. Energy carried in free space is also seen on the macroscopic scale, through traveling gravitational waves. These ripples in spacetime were some of the last of Einstein's predictions to be verified experimentally, first observed in 2016 by the two LIGO detectors in Louisiana and Washington (#Nobel).

Today, the last piece of the universe's history that we have yet to understand is the same as what's inside of a black hole, the singularity. One of the most exciting prospects for that is string theory, which postulates that the most fundamental unit of energy comes in the form of vibrating strings. The different ways these strings vibrate would result in the different fundamental particles we are all made of. These units would come at the smallest possible size, known as the Planck scale, which is calculated by combining fundamental constants of nature to get a value in units of distance. The Planck length is unimaginably minute. It is as small compared to a letter on this page as a letter is compared to the visible universe. But isn't it exciting to think what we still have to accomplish, considering only 200,000 years ago we were fighting for our lives on the savannah?

Quite contrary to the commonly held beliefs that all existence is perfect and cyclic, our discoveries have revealed that change is the nature of the world. The Earth is never in the same spot. Moment by moment, the entire universe is constantly shifting, always new and fresh, all things like drifting stardust. But that's the great thing about science, that it allows us to understand this constantly shifting universe at all. It's simple, it's elegant. In fact it's so astounding that it sounds miraculous. But it's not. Scientific theories are built upon the knowledge of people who dedicated their entire lives to just staring at a single animal, a single element, and a single patch of sky at a time. They collected and recorded countless observations, and derived trends using mathematic logic. You can do it too. You just need a little bit of history, a little bit of math, and a little bit of perseverance.

Science doesn't have hierarchy among researchers, but of ideas. Take for example how physicists seem to get elevated by popular culture as being especially important intellectuals, regardless of which one of the dozens of subfields they may do research in. But there are many less glamorous fields that are just as important to research. One field that gets picked on the most as being dull is often geology. However, without geology we would never be able to understand fossils or the age of the solar system, which is why the only professional scientist to go to the Moon was a geologist. Even obscure sciences are important, like the science of feces. For many biologists, one of the best ways to understand an animal is to understand its diet. If the creature is elusive, or even extinct, excrement can provide critical information about an animal's health, its environment, and its place in the larger food chain. What's seldom expressed is how every field of research somehow plays a crucial role in helping humans understand the world as a whole, and this is why all scientific endeavors are ultimately good, no matter the field. At the end of the day more knowledge is always a good thing. What's more, it eliminates the need for authorities. Facts are facts, regardless of who says them. Lawrence Krauss best describes this by saying that, "there are no scientific authorities, only experts. Almost every major scientific discovery was at one time dismissed by a majority of so called experts. The beauty of science is that old ideas are thrown out in light of new evidence." A smart person listens to the majority, but a wise person only cares about truth.

Science does however have bureaucracy, but people are both open and assertive when new ideas are under consideration. This bureaucracy isn't based necessarily on merit, although the most experienced scientists are often trusted more if they have made important discoveries. More than merit, ideas are based on probabilities, on what is likely to be true given available evidence, what things are 99% almost certainly true, and what things are possible, but only .001% likely to be true. Some near certain truths include the following: the Earth has air and circles the Sun that circles a four solar mass black hole. Bacteria live in and on your body, and both you and bacteria are made of squishy water sacks. Burning requires oxygen, and when you burn carbon, they combine to make carbon dioxide. When you do that in air, the air heats up and makes your planet look like Venus. Things that are almost certainly untrue include leprechauns, fairies, mediums, astrology, and ghosts. These things are not impossible; they just have nearly zero evidence to support their existence. Science allows us to deconstruct ourselves into our most fundamental constituents, and every day we pry deeper and deeper. But in between every discovery and every data point, the interpretation is entirely up to you and me.

Even scientific geniuses can be wrong, and if they weren't there wouldn't be any new science to do. Take for example Newton. Sir Isaac was they lawmaker of scientists, his pivotal role being almost entirely thanks to discovering the same laws of gravity that allowed future students to send people to the moon, allow us to find destinations quickly with GPS, and scientists to land a car sized rover on Mars with a flying crane. He didn't do this out of thin air however. He used existing data, and used his ideas to refine old laws. One of his most pivotal revisions was to Kepler's third law, which my very first astronomy teacher taught as the, "most important equation in all of astrophysics". It takes the form:

$$P^2 = \frac{4\pi^2}{GM} a^3$$

This equation relates the time it takes for one body to orbit another (P) to the distance from the center of that body's orbit to the farthest edge of that orbit (a), also known as the semi-major axis. Both of these parameters can be measured with telescopes, and by using this equation one can find the mass of the object that body is orbiting around (M). This equation can be used anywhere in the universe, and is the primary way astronomers measure the masses of objects in space. For the solar system, for which Kepler derived the original equation with just P and a, P would be one year, and a would be the distance from the Sun to the Earth. But Newton still wasn't completely correct. Even though his laws can be used for low speed space flights of only a few thousand miles an hour, it still doesn't describe what gravity is, and it assumes that gravitational force travels instantaneously. It wouldn't be until Einstein that we came to understand the nature of what gravity is. But even now we still struggle to understand gravity at the atomic scale. This is why science is so exciting; there are always more questions to answer.

Science is also one of the most innocuous of the disciplines that students are taught in school. In history, students are required to interpret events and memorize dates. Language arts require students to think critically about things other people have thought and written in order to help them articulate their own thoughts. Math is cold and rigid, and while it is based on a sound foundation of logic, it too requires disciplined memorization of rules. Science, however, inspires creativity in the most tangible way possible. A teacher may be able to read a book or tell a story from history, but only in science classes can students actually prove to themselves first hand how the world works. In science class, you become the scientist, you can conduct the same experiments that people did centuries ago, and see what they saw. Being able to solve mathematical equations by yourself is incredibly rewarding, but being able to use equations to describe real things that you can touch and see is the most powerful thing a human being is capable of. With science, one uses repeated observations, follows trends, and with that knowledge, actually predicts the future.

It's important to remember that this power to predict the future is directly proportional to how much data a scientist has, how many measurements they have made, and how meticulous and accurate those measurements are. This forces scientists to constantly be aware of the world around them. Any average person can go outside, look at a leopard, and come to the conclusion that its spots can't change. But a scientist will go out and follow a leopard, take notes about its life and write down what they see. A scientist would observe a leopard at birth, and see that its fur looks completely different than when is has grown up. They would take pictures or make sketches of the leopard, and share it with others. A scientist would keep careful catalogs of the leopard's parents and cubs, and notice the different locations of each generation's spots. Someone could take these observations a hundred years after the scientist has died, and compare their findings with current leopard populations, and find that leopards in the past may have had twice as many or half the number of spots as they do today, and come to the conclusion leopards do indeed change their spots all the time. Like any discipline, science requires lifelong dedication, and while an understanding of all the disciplines is crucial, only scientific reasoning has allowed humans to create antibiotics and vaccines that decimate diseases, and telescopes that let us peer into the very birth of the cosmos. You can do it too. You just need a little time, dedication, and faith in your predecessors.

To demonstrate the power of science, consider this example: The moon is tidally locked with the Earth, which means that you can only ever see one side of it at any time. One could go outside from Earth everyday of their life and only ever see half of our nearest neighbor in space. Now, I could tell you that there are unicorns living on the backside, and you would have no way of proving me wrong. You might add that there is no air on the moon, but I could refute by saying that unicorns are magical, and so don't have to bother with air. One could feasibly create an entire religion based on the unicorns in the sky (which incidentally could also be a great name for musical group). People could go centuries or millennia believing in such a thing, until people actually built rockets and sent people and satellites to look at the moon, and see that it looks remarkably the same as the other side; dry, with extreme temperatures, and completely devoid of enchanted equidae. But if you were to go to the moon and actually look around, you would find the far side to have a thicker crust; no dark patches that you can see from Earth (known as *maria*, the Latin word for seas), and that can tell you about the history of the Moon, the Earth, and the solar system in general. Not quite as good as unicorns, but definitely a close second.

In this way science seems to demystify things, but it really doesn't. It has always been recognized that the Sun is what allows crops to grow, and as such it is naturally revered in all societies as a bringer of life. In many agrarian cultures, the Sun was considered to be the ruler amongst a divine pantheon of gods, if not one of the most prominent deities. The most revered gods are more often dictated by what kept certain civilizations alive and what killed them the most. In the same way, the moon often plays a prominent role in mythology, being the only large bright object visible in the night sky (even though it's up during the day half of the time). Thanks to telescopes, we can actually look at these objects, and understand the complexity that makes them no less majestic, but far from perfectly divine. Now we understand the Sun not as the god Ra, or as a ball of fire carried by Apollo, but as a medium sized star no different than the specks of light in the night sky, a ball of super-heated charged plasma powered by nuclear fusion that comprises nearly all the mass of our solar system. The Moon is nothing more than a dried up rock, but its cratered surface teaches us about the history of our entire solar system. Fantasies are fun, but living a life without a full scientific understanding of the world is like going to a play with a blindfold or an opera with earplugs. But there is no reason for faith and science to be irreconcilable. Scientific discoveries can't disprove the existence of a higher power. If you do happen to believe in a personal deity, think about it this way; religion is what happens when people assume they know what the gods are thinking, science is what happens when people take a step back and observe what the gods have done. That being said, nothing breaks the laws of nature. Our world is consistent and knowable, and thinking otherwise is nothing more than cowardice.

So how do we reconcile our faith with new discoveries? You might ask me, how do you know that people turn into plants and back into animals when they die, but their consciousness doesn't survive? That's a completely reasonable and absolutely valid question. I could say that when plants and animals die, they get broken down and eaten, most often by fungi, and the fungi transfer nutrients and water to plants while the fungi get the sugar produced by the plants via photosynthesis. That's great, and tidy, but someone could still say, "well that's a nice story and all, but do you have any proof? Have you actually seen this happen?" and answer of course would be no. I'm an astrophysicist, I study stars, and many of us mostly use data from telescopes in orbit while we are at a computer. What's more, without access to the internet, photos, or documents, I would have no way of proving that people actually built robot eyes in the sky or spend entire days watching plants grow and die. In some ways scientists have more faith than the average person. It's just not blind faith. We have to assume that the data we're getting and the people who report them are true, and their interpretations are neutral and testable by anyone else. Furthermore, while it's true that scientific claims have to be able to be replicated by others, it's also true that average people simply don't have the time to test every scientific claim. But if other people can independently test those ideas, and verify interpretations, then that topic is put higher on the shelf of, "we're pretty sure this is true". Nobody has to rederive Newton's laws, even though they could. Nobody has to prove chimpanzee intelligence after Jane Goodall and several others have already lived among them. Instead they may study the capacity of chimp intelligence to see how they interact socially, while also respecting their intelligence enough to not put them in cages. You need to be able to trust their data, and build upon it. It's no different than learning carpentry from your grandfather, volunteering to build new houses, and then becoming the project manager for a cathedral or fancy school. Ideally, a scientist has wisdom and experience, the same as any expert we learn something else from.

I don't know many people who have actually had a one on one conversation with a god in their native language, they mostly just ask them for things like food and health. People trust public figures to interpret books written by other people, and those are the ones who apparently spoke to their god. Scientists also put faith in their fallible human peers, but only if given ample evidence presented in a way that somebody who grew up on a desolate island (or at least someone who understands Fourier transforms) could replicate. All they would need is sufficient background information based on observation and data. This may seem shocking, but coming to terms with our imperfection and constantly hunting for patterns is what leads to cars, electricity, and the internet, where a global community can discuss, review, and analyze observations that researchers so painstakingly took, and where experts can record videos and address the public face to face. At the end of the day, science and religion both preach trust and faith, both in nature and of each other. The bonus with science is that it also comes with a side of scrutiny and a heaping helping of critical thinking, as opposed to divine authoritarianism.

So in summary, space has been expanding for almost fourteen billion years, life on Earth has and continues to change every generation over the last four billion years, and in the last few hundred years one of those organisms has rapidly heated the planet by lighting dead trees and dinosaurs on fire so that they can eat frozen pizzas, or whatever. The universe is vastly more complex and beautiful than anyone could completely fathom. But by building upon past knowledge and working together, we can put the pieces together, find trends, and use them to make life a thrilling adventure for everyone. If you happen to observe something that doesn't fit into the present scientific theories, have the data to back it up, and can still explain all previous observations that support that model, great! You might even get a Nobel Prize.

But if you just don't "believe" in things such as evolution, climate change, or the big bang, because you like being difficult, then that makes you no different than the Catholic Church that imprisoned Galileo, nor any different than Aristotle, who was wrong about everything because he didn't bother to take actual data. That line of thinking is lazy, dull, and potentially quite dangerous. It keeps societies from progressing, and if you choose to reject the scientific method, rather than improve upon it, then you should lose all your privileges to modern medicine, grocery stores, radio, that little device in your pocket that gives you access to all human knowledge, anything that plugs into the wall, and television (that means you have to get off of your couch to watch sports, and also you don't get a couch anymore). It's perfectly acceptable (and encouraged) to disagree with any single scientist's claims, in fact that's the first thing professional scientists do. But when the vast majority of them all agree on one thing, you can be fairly confident they know what they're talking about, and disagreeing about anything just because it makes you uncomfortable is a waste of everybody's time, especially yours. In the immortal words of Hank Green, "There's nothing here to argue against. It's a process, not an ideology", and that is why we love science. Unless it's a result from a brand new study that hasn't been peer reviewed, in which case you should criticize it, heavily. So can we please just move on now? I would like a space resort on the Moon sometime this century.

Lastly, if you are one of society's productive, busy scientists, and this is the only chapter you had time to read, here's a nice little inside joke as your reward:

How do you organize a party in space?
(Answer omitted to conserve paper)

And for everyone else, who are also likely busy but kind enough to read this far:

A custodian, a scientist, and politician, all friends since grade school, go to a bar as they do every Friday. When they all receive their first round of drinks one night, their usual, trustworthy bartender is out sick, and is being replaced by a new, young trainee, and he accidentally mixes up their order with different (but equal strength) beverages. The politician asks to see the manager and demands that his friends get their right drinks. The scientist goes along with the politician in order to avoid making a scene. But unfortunately for the politician, the custodian had already finished the gin and tonics they were given, because somebody happened to vomit all over their pristine office space an hour ago.

7 | Language: The Power of Immortality

"The limits of my language means the limits of my world."
— *Ludwig Wittgenstein*

"Because without our language, we have lost ourselves. Who are we without our words?"
— *Melina Marchetta*

Language is the most exalted of all evolutionary adaptations, allowing a mind locked inside of the body to escape through a voice of precise sounds translated through the air or through complex symbols. In the case of humans, it allows us to communicate with far more nuance than through light, odor, or one pitched whistles and clicks. All of the most inspirational expressions of human will are literary. From The Divine Comedy to the Gettysburg address, the aspects of our history we most respect and revere are the ones that are best articulated through writing and speech. This come as no surprise, many animals can communicate through sound, but alphabets are one of the few things that Homo Sapiens have that no other creature does. Whether it be oral tradition, pictographs, epic poems, plays, speeches, letters, or business emails, words are the format for which we understand our existence and share it with others. Words can build worlds, and everything in the universe has at least one, even if we don't know them yet. I understand we've been discussing some pretty heavy stuff in these last few chapters, so in this one let's try and have some fun playing with our vernacular, savvy? Here's an example, written in the language of math:

$\sqrt{-1}\ 2^3\ \Sigma\ \pi$, and it was delicious! Do you like pie?

What we learn from observing hunter-gatherer societies and the origins of humanity is that language itself is rather new. Many of the oldest markings of humanity come in the form of silent art and painting. As they say, a picture is worth a thousand words, and the cave paintings that date all the way back to the advanced Egyptian hieroglyphs attest to that. But the majority of history in between Egypt and the Paleolithic has been mostly lost as oral tradition, as have many cultural remnants of hunter-gatherer societies since the sixteenth century. For Native Americans, Australians, and Africans, there was never a need to write anything down, which unfortunately has meant that their historical perspectives have been overshadowed by the longevity of Eurasian writing.

Just like with mathematics, it appears that agriculture provided the need for writing. When people had a surplus of food, they began to congregate and settle the land. People were allowed to specialize their jobs, and people began to earn wages in the form of food. In order to keep track of things, people had to write and record names, property, and such in order to regulate income. In this way advanced language has always been an aristocratic privilege granted to only a few individuals, which is why today many of the only written records of ancient cultures, from Japan to Italy, come from some sort of noble class. Even into the twenty-first century, literacy rates have become one of the most important factors for marking a successful society. Even in the United States, I have seen young people my age incarcerated with third grade reading levels. This in no way is meant to diminish those who cannot read or write, but the fact that there are so many people who lack that ability to me is tragic. Linguistic skills provide people with a voice, a way for them to express their intellect in a way that resonates with others, and I lament all the stories that are lost because so many of us don't have that voice. But as proof of our progress, never before have so many children grown up learning how to read the minds of others and write down their thoughts for others to read in kind.

Writing is cathartic. In many ways, the ability to put something from the storm of neurology in your head down into a concrete form is symbolic of solidifying one's own ideas about the world. Writing can help you understand things that you can't express simply by speaking, and therefore help you understand yourself in a deeper way than any illiterate person could. Sometimes not thinking is more difficult than thinking hard, and so it's quite beneficial to get all of your thoughts out of your head as a way to clear the mind. Studies show that writing and speaking about anxieties can calm you down and even improve your test scores, perform better in sports, interviews, presentations, or do anything stressful. Most importantly writing can help people cope with great emotional trauma. The next time you fall in love, feel the need to do something violent, or lose someone or something close to you, consider writing an essay in response. It doesn't even have to be coherent; you can just grab a piece of paper and a writing utensil, then jot down the first things to come to your mind. Once you've finished, you could publish your work, throw it away, or simply burn it. Just the act of writing alone can provide a powerful outlet, beyond even what a drawing or sculpture can allow. Unlike more abstract forms of art, writing requires you to assign concrete labels to your thoughts, and while there are certainly aspects of the human condition that are beyond words, the very act of trying to describe the indescribable seems to me uniquely virtuous. The reason for this lies can be seen in how we use language as a form of social cohesion. Possibly the most difficult thing about being a human being is dealing with other human beings. As adept as we are at perceiving outward traits, body language, and recognizing social patterns, there's only one way we are able to read each other's minds, by speaking. It doesn't matter if it's vocal, or through written symbols or complex hand signs, speaking is the only way to know the thoughts of others. As such, speaking is the only remedy for conflict short of mutual annihilation. The pen is mightier than the sword as they say, although nowadays the keyboard is mightier than the pen.

Languages evolve in much the same way species do. When people become isolated, experience different climates, with different flora and fauna, their languages adapt to represent the world they find themselves in. Over time small variances in dialect become new languages with wildly different diction and syntax. But at the end of the day, words don't really mean anything. At least out of context. You only understand what I'm saying right now (I hope) because we have agreed to the meanings of some arbitrary symbols as representations for thoughts and experiences. But if you were to go somewhere else in time or to some distant land, these symbols would have zero meaning to you, nothing more than specific curves and lines drawn out by ink. Language is as fluid and arbitrary as human consciousness, and as such any number of different words can be used to describe the same thing. It doesn't matter so much what you're saying, but whether anyone will understand you. This is true of people as well. Names and labels only identify things as much as we want them to. A rose by any other name would smell as sweet, have just as many thorns, and be composed of the same molecular structure.

One of the greatest examples of the arbitrarity of language comes from the thought experiment of Ludwig Wittgenstein's beetle in a box. In it he asks us to consider a room full of people, each having a box. The content of each box is sealed from the outside world, and only the owners know what is inside of them. Each person in the room refers to whatever is in their box as a 'beetle', even though there is no way for one person to know the contents of anyone else's box. It could very well be that each person has something different inside of their box, and they just happen to use the same word to refer to what's inside. For this reason, it is impossible for anyone to use the word 'beetle' in any meaningful way in conversation, because nobody is using that word to describe the same thing and there is no way for anybody to know what anyone else means. Wittgenstein used this experiment to illustrate how we can never adequately describe our subjective experiences. We might both say that the weather is cold, but a person from Russia probably has a different idea about what constitutes 'cold' than someone who lives in the tropics. The same person may very well define cold differently throughout the year as well; spring feels blistering compared to winter, just as fall feels frigid after summer, even though spring and autumn tend to have the same temperatures. This concept extends to nearly all sense perceptions. I don't know if my red looks exactly the same as your red, or whether the pain I feel is equivalent to the pain you feel. According to Wittgenstein, words can't describe subjective experiences, only concrete signs that are visible to others. We understand pain by looking for the symptoms of pain. But I would extend this to all words, even those that seem more concrete. I could say 'desk', but the type of desk you picture depends on the types of desks you've seen. I could say, 'glass desk', but you're still free to interpret the shape, size, and what's on it. This is the difficulty of language. We must agree on the meanings of words in order to use them, and if we don't we might as well be lying to each other. We need to be able to understand what others are saying to use any word meaningfully; otherwise it's just a beetle in a box. Whereof one cannot speak, thereof one must be silent.

Even though many languages have fallen out of common use over time, some of them have been resurrected thanks to science. Scientific names have adopted a historic title in order to standardize them and to honor human history. The exception to this is to have a law of nature named after a person, which I find is rather arrogant, but perhaps rightly so. After all, nobody breaks nature's laws; we only learn how to work around them, much like man-made political "laws", albeit more universally. But many names do have history behind them, and there is even a whole bureaucracy for the naming of different scientific discoveries. As it turns out, the word quark is from James Joyce's *Finnegan's Wake*, and the word Lepton is Greek. The 88 official constellations are also Greek, although every culture throughout history has unique star shapes of their own. Interestingly, star names are often Arabic, like our number system. Fossils are given Latin names so that they can consistently describe their evolutionary status, and to compare them to modern fauna. Chemicals also abide by standardized nomenclature rules. Chemical names do however tend to sound snobby and pretentious at times because they have too many prefixes and suffixes, which is bad because we need to eat certain chemicals to survive. However, because scientific names are often complex and usually inconvenient, language evolves over time to use colloquial terms for scientific words. This just means that people can make up words to describe nature, and revive old ideas to honor them. I honestly think it's more fun when names are made up. People's names are usually only used for complex ideas that describe how the natural world works. Personally, I think it would be more fun to have a creative title for a law—General Relativity sounds more robust than Newton's Law—but my name is too long for students to have to memorize, so perhaps that isn't my calling.

The greatest injustice of science was too an error in vocabulary, by calling its highest form of understanding "theory", the same word that under any other circumstance would be equivalent to "hypothesis". Whoever had the dreadful idea to use two words that are so nearly identical to describe ideas of such contrasting gravitas should be banned from every scientific textbook. The National Academy of Sciences defines scientific theory as something that "refers to a comprehensive explanation of some aspect of nature that is supported by a vast body of evidence." It is something we understand so well, that any further observation or evidence will almost certainly build upon, but never refute it. A hypothesis is equivalent to a guess. After making observations you start to get an understanding. After corroborating many repeated, tested, criticized, and independent observations of seemingly unrelated phenomena that all support the same idea, then that hypothesis is no longer uncertain, and it is elevated to the exalted status of theory. Scientific theories are the best understandings of humanity to describe what the hell the world is doing. They are never perfect, but asymptotically strive toward perfection. That is why Evolution is a theory, and The Garden of Eden is a hypothesis, as are talking animals and humans from dust or ribs. A fair guess, but long since left behind in the dustbin of other good ideas that couldn't explain reality.

Since language is the medium in which we understand the world, the words and phrases you use matter. The other difficulty that has arisen around science is that of jargon. Just as with business, law, or any other nuanced discipline, the specialized language used by scientists can often sound like gobbledygook to the average person. This does have the great benefit of making us sound much smarter than we are. For instance, between these two titles, can you tell the difference between the real published research paper and a fake?

1. **"Calculations of pKa's and Redox Potentials of Nucleobases with Explicit Waters and Polarizable Continuum Solvation"**
2. **"Observational Effects of Quantum fluctuations on The Interstellar Medium and the Production of Zeeman Split Polarization"**

As I have stated, communication is one of the most difficult aspects of science. Nature is complex, and understanding just one of her aspects is difficult enough to do by yourself. A person can become a brilliant engineer or mathematician, revolutionizing their field of study, but all that effort is useless if other people can't understand their work well enough to use or improve upon it. This becomes increasingly true as scientists begin to specialize more and more, to the point that only a dozen people can understand each person's research.

At many universities, science departments have special seminars and colloquia, where guest professors come to speak about their research. Professors attend, as do graduate students, and undergraduates are encouraged to attend if possible. As an undergraduate, there was not a single talk in which I understood more than a quarter of the presentation, simply because the vocabulary was so far removed from everyday speech. From my interactions with the graduate students, this discontinuity is rather common, even PhD candidates are often lost by much of the nuance of current research. This isn't a matter of basic knowledge of astrophysics, but rather a natural consequence of how much we have accomplished in our endeavors to understand the natural world. I could give you a basic overview on the features of the Sun, how it shines and the various cycles and processes it goes through—from the p-p chain to coronal mass ejections— but I could only do this qualitatively. Only a helioseismologist could do something like analyze measurements from the SOHO spacecraft to give you a detailed description of how solar temperature, pressure, charge, density, and rotation fluctuate over time, and actually give you equations that approximate the behavior of the Sun. Not to mention the fact that their expertise depends on whether they focus primarily on the interior or the atmosphere of the Sun. That type of specialized understanding requires dedication, but as a result it makes that knowledge increasingly difficult for researchers to explain their work in a way they can readily communicate to their collogues, let alone to the public. This is only compounded by how introverted many intellectuals have to be in order to do their research. The negative impact of nuance on communication extends to all the sciences, and even to all jobs, but communication is especially important for scientists, because no matter how we make a living, we are all part of the natural world. Solar research impacts everyone living under the Sun, just as medical research affects everyone who is alive. For this reason it is crucial to bridge the gap between scientific research and public life. This can be done through vocabulary; scientists should strive to not lose their layman diction, and be able to understand their concepts well enough to explain them in a simple and engaging way. In addition, everyday people should strive to have great scientific literacy. If nothing else, it makes you a very interesting person to talk to at dinner parties.

Different words have various implicit cultural and historic connotations, and effective communication dependents on all sides understanding how the other side will interpret the words they use. As a teacher, I have to think about this concept heavily. One of the most important class management skills I learned was to use positive correction rather than criticism. When a student makes a mistake, it is your job as a mentor to correct those mistakes, but the way you do it has tremendous consequences for how effective those corrections are. If you are cold and demeaning, or if you shame someone for their mistakes, that hinders growth and keeps the student from wanting to pursue self-improvement. The difference between criticism and positive correction is the difference between, "You're an idiot" and "Here's what you can do to improve". What helps even more is if you include something to the tune of, "you did this really well, but here's something you didn't do so well", beginning with a complement before a critique. I find this rule to be demonstrably important in everyday interactions as well, especially with people you disagree with. Being respectful and forcing yourself to begin with compliments always makes people more receptive to criticism than just outright slander. There's no difference in the desired outcome, only in the path chosen to get there, and it is your language that carves that path.

The impact of language is also seen in the power of profanity. Profane words don't just express immediate frustrations, but they also demonstrate how sometimes the most simplistic ideas are the most powerful. Take for instance the common issue of political correctness. The very fact that we censor ourselves to any extent shows how simple words carry great weight. Whether those words are meant to disparage specific people or a general group, it is important that individuals are cognizant of how their words affect others and they temper them accordingly. Likewise, it is important to remember that conversations are two-sided, and that half the power of a word is given by the speaker, and the other half is given by the listener. When we hear others speak to us, we also have the power to choose to be offended or to take those words as water off a duck's back. This is especially important when listening to advertisements and government propaganda.

It is important to understand how propaganda has been used throughout history, especially during wars, and how influential they can be over public perception. Sometimes these influences can be as harmless as comical, but it is incredibly fascinating to notice how powerful something so subtle can be. For instance, have you ever noticed how movie villains always seem to have foreign accents? Influences can also be overt, like the banners created to instill a cult of personality around totalitarian dictators like Mao and Stalin. Mao was especially skillful in this regard, even publishing a little red book with his quotations that was required reading for all Chinese citizens. The United States even shares a similar history in bolstering support against their enemies. Internal slander is usually only apparent during elections, but foreign propaganda has also been very effective at dehumanizing our enemies during war. Even Disney got involved in creating war propaganda cartoons during the Second World War. This isn't to say that bringing citizens together is a bad thing, but it should be noted that some groups have to be labeled as enemies for propaganda to work properly, something that is very dangerous to do. It isn't just limited to wartime either. Anyone who has listened to any kind of advertisement knows that word choice can have a powerful impact on whether your product sells or not. Words can easily be used to deceive and misdirect. A powerful wordsmith can brainwash various specific people into changing their lives with the power of tact. This was best described by Winston Churchill as the, "ability to tell someone to go to hell in such a way that they look forward to the trip." The exact same event can be made out to be either good or bad, with nothing more than the diction and tone used by the speaker. This is why one should be weary of where they are getting their news. A five person shooting could be described as a massacre by the media, which might paint a far different picture than reality. An inconsistency in the way we use our words can dramatically warp a person's point of view. As a human being, you should take everything you hear with a grain of salt.

One of the greatest testaments to the power of language is found in works of literature. Every biography read is another life lived. You can learn about one person's story in detail, or the story of villages or nations through the eyes of people that were alive at the time. Books are portals through time as seen through human eyes, just the same as telescopes allow one to see the universe unfolding. They reflect the cultures and humanity of those who wrote them. Poetry for example reflects reality by surmising emotion into rhythm. But poetry doesn't have to be short like a haiku; Edgar Allen Poe's prose poem *Eureka* discussed cosmological concepts before Einstein and Hubble were even born. Fantasies do this too, but fiction allows for more imagination and freedom, and requires more thinking on the part of the reader for deep understanding. Take for example Bram Stoker's *Dracula,* or Mary Shelley's *Frankenstein*. On the face, these works are nothing more than monster stories. But look deeper, and one finds that they reflect the very ways we define our own humanity. *Dracula* is a work about defying appearance stereotypes, especially gender based ones. Dracula appears classy, but is really a murderer, defying villain stereotypes of being brutish and heinously ugly. Instead of analyzing outward appearances, *Frankenstein* tackles questions about what it means to act like a human. By bringing the dead back to life Dr. Frankenstein effectively becomes god, but is ultimately fearful of his creation, casting it out as an abomination. Despite this, the monster becomes well read, and attempts to find solace in love, asking Frankenstein for a wife. In the end, the doctor destroys the second monster, and in response the first one kills Frankenstein's fiancé. By showing us the evils of both Frankenstein and his Monster, Shelley forces us to question whether our actions are predetermined, and whether the fault in our actions is to be blamed by the creation, or the creator. She even draws parallels to Paradise Lost—the story of Lucifer's fall from heaven—as one of the only books the monster reads. Both of these famous tales bring to light intricacies of humanity that wouldn't be possible in reality, which is why they stand the test of time beyond simple autobiographies. You can't judge a book by its cover. After all, did you expect vampires to appear in this one?

The most important thing about language is that it unites us. People have known this for millennia, even one of the oldest myths revolves around language: The Tower of Babel. In this grand tale, all of humanity once spoke the same language, and they attempted to build a city high enough to reach heaven. As punishment for their arrogance, Yahweh gave all the people different languages and scattered them across the Earth. Despite the fact that several countries have been able to develop space programs without making god mad, this story does hold a central truth; that the more people are able to work together, the more they can accomplish. But on that note, I'm not sure how simply making people speak different languages affects their ability to build towers. I'm pretty sure they can do that with blueprints and mechanics. Regardless, I often ponder how many wars and genocides would have been prevented had each party spoken the same language and dialect. A major reason why many colonial natives were considered "savage" was likely because the invaders couldn't understand them. It was only long after the Inca and Aztec were wiped out did we learn about those great empire's rich culture and advanced astronomy, agriculture, and architecture. If they were able to communicate, they might have been able to trade with the Spanish. Although likely not, since Europeans had guns. What's more, the fact that many natives didn't have written language also hindered their reputation. Those who can't read or write are often seen as barbaric. Of course the problem with that is that *barbarian* is a term that originally referred to anyone who didn't speak Greek, and very often these people would be enslaved to the Greeks after conflicts. As a result the word has now become synonymous with a lack of civilization, even though that entire concept is tenuous at best. After all, the majority of human beings today are barbarians.

The origin of our words chart the very history of humanity itself. Language is fluid, and readily adaptable for what people need to understand and communicate. The fun thing about language is that it's always evolving. But as a consequence, it means that the words we use come to take on new meanings over time, and just as when lifeforms evolve to the point where they almost become unrecognizable, so too to distant languages become difficult to translate. This is perhaps best exemplified in the dichotomy between the romance languages and the languages of the orient. For an English speaker, learning Spanish is one thing, but learning Mandarin is a completely different ball game. Although as an English speaker, I feel as though our language provides a certain freedom that isn't seen anywhere else. This is because English has been around the world, and has been influenced by a global community, which is why it is such a good case study for analyzing how languages evolve over time.

There's a lot of history behind the fact that I speak English today, although it could easily be summarized in ten minutes. Modern English begins at the dawn of the fifth century, when the Romans left Britain and the Angles and the Saxons came to replace them, bringing their useful everyday vocabulary, with words like *house* and *loaf*. Then the Catholics came and added to the Latin left behind by the Romans, with words like *martyr* and *font*. Then the Vikings came, after which the oldest known work of English literature, *Beowulf* was written. The pagan days of the week became commonplace as well, with words like *"Thor"sday*. Then William the Conqueror came and added some French into the mix, until the 100 years war, which left English as the dominant language of Britain. Of course, one cannot discuss English without mentioning the patron saint of the language, William Shakespeare. During his career as a playwright he wrote 17,677 words, one tenth of those he just made up. If you've ever uttered, *critical, monumental, majestic, frugal, countless, hurry, pedant*, or *champion*, you can thank Shakespeare. He demonstrated the flexibility of English as a language of beauty and emotion, as well as one capable of great abstract thought. His works were so revered that astronomers too decided to honor him, by naming all the moons of Uranus after characters in his plays, including one from *The Tempest* named Ariel. Then the King James Bible was written, adding abstract metaphor and ethics into English. Then as the enlightenment developed, science went from Latin to English, and words were invented for anatomy and natural laws, solidifying English as the language of intellectualism, religion, and poetry. Then the crown began colonizing the planet, and further assimilated new words from the Caribbean, Africa, Australia, India, and of course, North America. English began to develop around the world, taking on new dialects everywhere it went. Dictionaries had to be invented, as lexicographers began telling people what words meant and how to spell them. The Oxford English dictionary was born, which never really stopped English from growing, but did provide a standard of speech. At this point, we had books about books, with words about words. New words were added in the Americas by all of its new immigrants, like the Germans with *kindergarten*, the Dutch with *cookies*, and the Italians with *pizza*. It's lucky for English that America became a wor(l)d power, influencing technology and entertainment thanks to its doctrine of free speech, which helped English grow more and more. Then the Internet came, words abbreviated and, FYI, English became the language of the digital age (LOL). Now let's see how languages, English especially, evolve over time.

One way words develop is through errors. Words that have been misused are not always corrected and thus often evolve in either their meaning or their spelling. This often occurs when words are misheard or misunderstood. One form of error is known as *catachresis*, an about face in meaning or connotation. Bill Bryson discussed this phenomenon in his book *The Mother Tongue: English & How It Got That Way*. Once upon a time, "counterfeit" meant legitimate, "brave" meant cowardice, "garble" meant to organize, "crafty" was a form of praise, a "harlot" was a mischievous man, and to "manufacture" something was to do it only by hand. For a more comical example of how expressions can shift through time, here's an excerpt from Charles Dickens' *Bleak House* (1853), "Sir Leicester leans back in his chair, and breathlessly ejaculates, 'Good Heaven!' ".

Another way that languages can attain new words is through adoption from foreign languages, when words from other languages are preserved and used as part of the old language. English especially is famous for pilfering words from around the globe. As I have mentioned, we can go from Greek à la Latin à la French à la English, and words from each language change slightly in spelling and meaning as they pass through lands. They can exist in two or more related forms, such as canal and channel or amiable and amicable. We have also Anglicized some words to make them sound more English. For example the native word raugroughcun was modified into raccoon.

Words can also die out, and then become commonplace again, or just take on multiple meanings. Bologna isn't just a type of meat, it is also an insult. The word "neck" was once a measure of distance, which is why the phrase "neck of the woods" exists. What's more interesting are words with no equivalent in English, and should maybe be added. The Russians have *Toska* to describe an aching soul, Indonesians have *Jayus* for jokes so bad that they are still funny, and then there's *Ilunga* from the Tshiluba dialect spoken in the Congo, for a person who is ready to forgive and forget any first abuse, tolerate it the second time, but never forgive nor tolerate the third offense. I find it interesting that although these words don't have a singular English equivalent, one can still articulate them using multiple words instead.

But again, if you need a word for something, and you can't find it in a foreign language or your own, you can always just make one up. One of the best catalysts for the creation of new words comes in the form of new technologies and medicines. Physiologists for example have some of the most fun terminology. Instead of heart attack, they get to say myocardial infarction. Words can be created for any specific purpose, even if it's just to mess with people. My favorite example of this is when the U.S. Army took a survey in 1974 of soldier's food preferences, listing one option as "funistrada". This wasn't actually a food, but it ranked a higher preference than lima beans and eggplant, simply because of the way it sounded, and probably because lima beans are gross.

People speak English all around the world, and many languages have influenced it in turn. So whose language is it anyway? Not understanding what people are saying is very obnoxious. If people are going to work together as a global society, eventually it's going to just become more convenient for everybody to speak the same language. The big question, of course, is what language should everyone use? Should it be Mandarin, the most spoken native language used by over a billion people? Should it be the second most used native language, Spanish? Or should it be English, because despite having the third most native speakers, is the language with both the most secondary speakers and as a result the most used language in the world? Or perhaps Arabic should be the global language? It's the second most common second language, is centuries older than any romance language derived from Latin, is the basis of the western numeral system, and the language used for the names of stars. It's important to respect the culture and heritage of every nation and recognize the beauty of their languages, but it is arguably more important to remember that the language you speak isn't the same as the language spoken in your homeland or by your people a few millennia ago. Also, if people need to learn a new language to make a living, they will readily do so. Having a global language doesn't mean that people can't still speak the tongues from their homeland. In fact, being multi-lingual is widely regarded to have a strong correlation with higher intelligence. In an ideal world, people wouldn't bicker about which language is best; they would all just shut up and learn how to speak several, so that later they could bicker with anyone in the world face to face. All people could speak one global language, but also be multilingual in native and personal interest languages.

Possibly the most amazing thing about language is that it allows people to express broad ideas with fine-tuned vocabulary. For instance, in order to analyze and test students about literary works, teachers often task those students with finding themes and motifs within book. This is because very often books are nothing more than huge swaths of words defending an idea that can be summarized in just a sentence or two. For example, I could easily just post something online for the world saying, "Everyone just shut up and be nice to each other", but instead I took the time to write a detailed (perhaps even mildly entertaining and informative) piece of prose to better support that perspective. Now wasn't that nice of me? And we still have so many amazing and interesting things in the world to talk about!

Language is also the only way one can articulate abstract concepts like love. One can display affection, and that too is quite powerful. But everyone knows that the mind is the true path to a person's heart. Take notes, if nothing else here's why you should be stoked to go to English class and learn about literature everyday. One can do any one thing at a time with their body, but through writing, one can paint a landscape of several ideas, any of which can be personal, but all of which should be sincere. You can simply say more by writing down your thoughts, and it is more convenient for the other person, and more impactful because you can spend more time planning when you write. With concepts as complex as love, wrath, sorrow, or joy, only a wordsmith could ever hope to bear the gravitas of those tiny words.

Language is one of the highest exalted abilities a creature can have for this exact reason. Only the most intelligent beings can do it, and among them none are more complex and enthralling than that of human beings. We have the potential to communicate a whole lifetime of events, upward of an entire universe, using nothing more than some lines and vibrations of air. So what's love got to do with it? What's love but a sweet old-fashioned notion? Who needs a heart when a heart is so easily broken? Answers to such complex questions often vary from person to person, but just think about how nice it is to not have to talk to someone, and just know what they're thinking, while simultaneously being willing to tell them anything. That is what makes up the substance of true love. This is part of the reason you can love objects and pets as much as people. Objects like beach balls, and pets like rocks. You can tell others whatever you want, and either they won't be thinking anything, or they will be decently good at telling you how they feel. Simple pets don't think much, but they are very good listeners. And that's one of the biggest differences between someone you hate and someone you love; how and if you ever communicate. This is especially true if they're in a whole other country and you only ever see frozen images of them. In that case it's much easier to dehumanize people. But the second you are in visible contact with someone, everything changes. They become tangible, and visibly human. The only way to for people to be crass to another person within earshot is for one of them to already feel harmed by the other person or for one of them to just not care about the other. If you talk to them, look at their eyes. You have to either feel some sort of kinship with them, or seriously believe they are a threat to your existence. If you have a full conversation, then you might still disagree with their entire worldview, but this is the critical turning point. There are three primary outcomes after this. Following just a handful of conversations, if two people fundamentally disagree about the world, those two people will either dedicate their lives to proving the other person wrong, give up and expunge the other person from their life and never listening to what they say again, or take measures to end the other persons line of thinking by killing them. If neither of them kills the other, and they instead choose diplomacy, then they are left with two outcomes. If they are very poor at either listening or articulating their thoughts, then they may go their entire lives disagreeing. This is fine; plenty of rivalries have this sort of dynamic, which may be the most powerful form of friendship in some cases. But if this option isn't maintained, it can easily devolve back into hate. Finally, the last option, if neither people get lazy and try to murder the other, and they are both excellent at articulating themselves (or are at least persistent enough to try) then they will be able to describe the world through their eyes. Two people could grow up on completely different planets, and if they are able to put words to their experiences, and give each other a mile in the other's shoes, they can only grow to love each other more and more. Language is why animals love and hate at all, and why humans do it more passionately than any other creature. This may be why the best marriages are between best friends. You have to be willing to tell the other person about your life and listen to what they are thinking. That's the only way to be with another person for extended amounts of time; you have to constantly articulate your thoughts to each other. You have to always keep talking, because you can only hate someone you rarely have a conversation with. The best way to have a conversation is to have mastery over many words, and being able to write down those ideas gives them power and longevity. Of course, if people just assume everyone is as complex and worthy of respect as they are, then they can skip this whole process and just be nice to each other. All that other jazz is only necessary if you want to fall and then also stay in love, which you can do with whomever you want.

More than simply knowing things through history, math, or science, language is important because it creates bonds between living things. Language can be used to unite the worlds of not just people, but also animals, the most intelligent of whom are like mute children. Human communication with animals has become extremely well documented in recent decades. We all know that our pets are very responsive to us. Some dogs can distinguish between nouns and verbs, and can be trained to recognize hundreds of words and commands, most notably Border Collies. Of course, several dog breeds serve with law enforcement and rescue teams as well. But that's not all. Any pet owner can testify to the responsiveness of their companions, including rats and pigs who solve puzzles, and even cats, who despite being aloof have nearly twice as many neurons as dogs. This ability isn't just held by our mammalian relatives either, pigeons demonstrate the aptitude to understand basic math and can respond to human commands well enough to perform tasks for navigation as well as entertainment. Even more remarkable are the parrots, some of which are actually capable of human speech. Of course the intellectual capacity for communication expands beyond the animals we choose to keep company with. Prairie Dogs have been observed to communicate in a way that describes different predators and people using objectively specific calls, demonstrating their ability to recognize physical features and reference them to others. Dolphins are famous too for their ability to communicate with one another using clicks and whistles, make future plans, invent new swimming maneuvers, and recognize themselves in mirrors, demonstrating self-awareness. Elephants display similar abilities, but instead communicate with low tone infrasound over long distances. They even share the same physical and emotional languages as humans, as they have been shown to cry during traumatic events. But no creature communicates with the complexity of our closest relatives, the great apes. Chimps, orangutans, and even gorillas have all been taught sign language, and have shown the ability to hold basic conversations with human caretakers. Koko the gorilla is just one of the most famous case studies, exhibiting a vocabulary of over a thousand words, ten times that of one of the most well spoken birds, Alex the parrot. She could also combine the words she knew to create new more complex ideas.

It is important to remember that some animals think deeply as well. Apes can talk to us through sign language, and they know when we've abducted them and when we tear down their homes. They also know they can barely do anything to stop us with our poisons, weapons, and traps. They live their lives in hopeless submission. Thankfully, many people make friends with them, and at least try to treat them as autonomous adult animals that survived millions of years just fine without us. So maybe if we could unite the human species with language, we could unite all life by finding some means of communication. Perhaps one day we could even teach apes astronomy, although they usually sleep through the night on instinct. As we can see, there has been a progression of language in the animal kingdom over time, and instead of drawing a line between humans and animals, observation has clearly demonstrated that language lies on a spectrum, on which we are just recently the most advanced. And just as with language, we can see that same progression in the ways that we develop our culture.

8 | Culture: The Power of Being an Earthling

"I think there's just one kind of folks. Folks."
— Scout, To Kill a Mockingbird by Harper Lee

"You know what I want to think of myself? As a human being. Because, I mean I don't want to be like 'As Confucius say,' but under the sky, under the heavens there is but one family. It just so happens that people are different."
— Bruce Lee

This might sound controversial to you if before reading this book you were one of those people who thinks that humans are completely separate from the rest of the universe, but as it turns humans do not have a monopoly on anything, not even culture. Nothing we do is unique amongst animals; we just do everything with far greater complexity. If you thought that humans were especially vile for our history of war, just take a look at chimpanzees and ants. Ants even take captive slaves after a conquest. If you thought that murdering babies is absolutely amoral (which it is 99.9% of the time), take a gander at lion prides. Infanticide is a common practice in the animal kingdom as just another means to guarantee that your genes get passed on and your competitor's genes don't. Hell, shark babies kill their siblings while still in the womb. Even rape isn't off the table. Dolphins are notoriously horny even around humans, and otters will engage in coital actions with baby seals to the point of murdering them.

But don't get the idea that our animal nature is just a whirlwind of violence and murder. Elephants, apes, whales, and even birds all exhibit sorrow and mourning in the face of death. Elephants shed tears, and even try to burry fallen loved ones by covering their corpses with twigs and grass. Like elephants, crows, magpies, whales, and apes all huddle in groups around their kin, mourning their passing and refusing to leave them behind. Mother chimps have even been seen carrying the corpses of their infants for days after their passing, and apes will actively protest when their human caretakers try to take away a fallen member of their group. But animals don't just extend compassion to their next of kin. Remember Alex the grey parrot? His last words were, "You be good, see you tomorrow. I love you", the exact same thing he would always say before his best friend Irene left for the night. In addition, dolphins have been seen protecting humans from shark attacks. Any horse owner knows that their steeds are basically family. Take for instance the case of Clever Hans, the horse that was able to fake being a math genius by reading the subtle body language of his owner. Even crows, who have recently gained the sour reputation as birds of death, show humans compassion. There's a girl in Seattle Washington who gets gifts from crows in return for feeding them, including trinkets, shiny objects, and buttons! And of course many humans already know how compassionate our ape cousins can be, just ask Michael Jackson and Bubbles (just don't get any ideas that apes actually make good pets). Although, non-pet animals can have pets of their own, like Koko the gorilla. She even blamed one of her cats for a sink that was torn out of a wall. Just like humans, animals take credit for good behavior, and blame others for bad behavior.

That's not all. Sure animals exhibit both violence and altruism, but those emotions are often associated with our more primitive nature, to the point that some people actually believe that absolute morality is instilled into our very being. But we still have the edge in cleverness right? Animals don't have a sense of individuality or freedom do they? Not so fast. As it turns out, many animals know that we keep them in cages, and some of them don't like it. Take for example Fu Manchu the Orangutan, who picked the lock of his enclosure to escape every night using metal wiring he kept hidden in his cheek pouch. Or how about the gorilla residents of the Los Angeles Zoo, Evelyn and Jim, who have worked both collaboratively and independently to coordinate their escapes. But possibly the most famous escape artists in the animal kingdom are the ones who don't even have spines. Octopuses easily escape their enclosures, can solve puzzles well enough to unlock their aquarium jail cells, and even slime around buildings as if going for a casual walk. So while we might like to gawk at the animals we keep in zoos, it's important to remember that even though we might be taking care of them in prison, that doesn't change the fact that they share the same curiosity that has driven human exploration since our time on the savannah. Sy Montgomery, author of *The Soul of an*

Octopus, put it best, "Astronauts don't go into outer space because they don't like Earth, they just want to see what else is out there." Of course, even though some animals exhibit human-like intellect, by and large they are still more ridiculous than we are. I've see cats, penguins, fish, raccoons, chickens, dogs, turtles, and even spiders all chase laser pointers. I've only seen people do that after a few hard drinks.

All of these observations of the animal kingdom drive home one universal truth: humans are animals, plain and simple. There is nothing special about our behavior, and nothing we do is unique. However, that still doesn't change the fact that the human species as a whole IS unique. Our highly developed frontal cortices allow us to pursue the animal endeavor into depths unfathomable to any other creature, and this is nowhere more evident than in our culture. Unlike history, maths, science, or language, culture doesn't really have a satisfactory definition; insomuch that I can't definitively think of a defining feature that puts culture into a single box. History is just what happened in the past. Mathematics is nothing more than finding logical connections between quantifiable phenomena. The scientific method is simply the process of taking careful observations, finding trends, and using those trends to predict the future, with the most precise measurements yielding mathematical relations. Language is certainly quite nuanced, but whether it's through vibrating air, written lines, the way you move your body, or the way your neurons fire, language is still just how organisms communicate information. Culture derives from and influences all of these things, and the complexity of culture is directly correlated to the complexity of the animal. The cells in our body work together to perform the task of surviving, as do the organelles inside of eukaryotes. However, neither the simple organisms made of single cells nor the intellectual birds and mammals have the complexity necessary to develop art, organize religions, or to just party like Homo Sapiens. So, since "culture" is so difficult to define with just a few words, instead I thought I would define it by looking at the many ways it plays a role in our daily lives.

What is Culture?

Culture is why the only president in United States history to be elected both three and four times also held office during the greatest blights to face the global community; the great depression and the Second World War.

Culture is the reason ancient Rome associated wearing pants with barbarism and a lack of civilization, but only until they expanded north.

Culture is why people can become world famous celebrities for doing absolutely nothing useful besides being strange, interesting, or just distracting.

Culture is why entire textbooks are dedicated to the lives and exploits of "great" men such as Alexander, Plato, and Dante, while the women of the time are marginalized to the footnotes of history.

Culture is why even after the advent of new millennium technology women are only famous if they look a certain way, rather than for their accomplishments.

Culture is why men are often encouraged to copulate with multiple people, yet somehow an equal number of women are shamed for doing the same.

Culture is why people are afraid to go to outer space after only 32 space related fatalities, even though some 20,000 people died along the Oregon trail.

Culture is why Athenians and Spartans never considered each other "Greek" until after they went to war with both with the Persians and then with each other, much like the how the American colonies didn't fully consider themselves "The United States" until they declared independence from Britain and had a civil war.

The spread of culture is why the Silk Road and the Indian Ocean Trade Company were far more impactful than most wars in shaping human history.

Culture is why people think Europe is a continent, even though if one looks at a map they can see that it's much more like a glorified peninsula coming off of Asia, which happens to lie on the exact same tectonic plate as Europe.

Culture is why people only remember wars in history rather than the events that led to them.

Culture is why generations of people rule while others are oppressed, and why some people wear plastic bottles as shoes while others have closets full of leather.

Culture is why people who only eat hamburgers and pizza use vitamin supplements as opposed to eating a better diet.

Culture is why we wear the clothes we do. We don't even wear anything we need to except when it's too damn hot or too damn cold, then clothes are basically mandatory. How else could hairless apes that evolved to survive naked on the African Savannah cope with the dessert of the Middle East or tundra of Colorado? We get negative temperatures out there, but nobody freaks out about it because we have animals and plastics with fake jewels and real jewels and hot coffee and huge buildings where electricity and gas make it feel like spring. What I'm saying is, clothes are smart. They reflect what a person looks like in their mind. Folks can dress one way this day or another way a different day, and you can tell what they are thinking based on their clothes. Even so, heroes and monkeys can both wear suits.

Culture is why people who practice the Sikh faith are profiled as terrorists, despite originating in India and having nothing to do with the Islamic faith, and why people with ethnic backgrounds will often, "act white" in order to fit in. It's also why people make snap decision judgments about each other, and those decisions in turn shape culture.

Culture is why a generation of southern voters in the United States went from being part of the Democrat party to being part of the Republican party after the civil rights act was past in 1964, and why the party that originally emancipated the slaves lost the black vote.

Culture is why countries have state animals, most of which make no sense. The emblem of the U.K. has a lion and a unicorn, one of which is fake the other of which is native to Africa and the Middle East. Belgium's state animal is also the Lion, which as you might recall from *Heart of Darkness* is wrong on so many levels.

Culture is the reason that people can't downgrade lifestyle, even at the cost of the environment and other people.

Culture is why Aristotle taught Alexander the Great and influenced deep thinkers centuries after his death, even though he was wrong about EVERYTHING. Except maybe the idea of catharsis, which he probably couldn't explain.

Culture is why economics is so mind numbingly complex and dauntingly nuanced.

Culture is also why Alexander the Great is so revered throughout history, even though his empire crumbled immediately after his death.

Culture is why people are executed, maimed, or ostracized for committing adultery, even if their relationships are abusive or unsatisfying. It is also why people face the same persecution if they live under Sharia law and then decide they like Buddha better than Muhammad.

Culture is why Santa Claus has a part time job selling cola products, and why Coke used to contain cocaine in the early twentieth century.

Culture is why people get married, and also why shiny pieces of carbon are so expensive, despite being the fourth most common element in the universe and also possibly raining from the atmosphere of Uranus.

Culture is why some people are scared of eating bugs, even though they are highly nutritious, already end up in bread, and are enjoyed as delicacies around the world.

Culture is why monkey see, monkey do. Chimpanzees (which aren't monkeys) are observed to be able to learn how to use mechanical devices after researchers trained an individual in a population the trick to using them, and the researchers saw the knowledge disseminate throughout the population, even past segregated enclosures.

Culture is the reason people have lawns full of useless grass instead of lawns full of native rocks and fauna. If you have a big grass lawn, shame on you. You're wasting water, and grass is a dreadful carbon sink. It takes a thousand square meters (a quarter acre) just to absorb a few days' worth of a person's carbon footprint, not including the gas needed to cut them. Slavery was also a byproduct of culture. Atlantic slavery specifically could have arisen from consumerist culture, since the main crops they grew (pure sugar, tobacco, and coffee) weren't actually useful for keeping people alive. This is true for sugar because for millennia, people got their glucose from wheat fruits and animal fats.

Culture is why the United States still uses pennies (and the freaking metric system, but I already covered that). They cost money to mint and are way useless, so the government is literally subsidizing a small copper coin because of its cultural status. If people feel bad about losing Lincoln, just print a penny on a nickel, on account of Lincoln freeing slaves and Jefferson owning and having children with them. Although nickels are also kind of useless as currency.

Culture is also why we put Andrew Jackson on our money, who basically got famous for destroying banks and appropriating land from the natives whilst killing them off.

The internet has a culture as well, one that results in people eating raw cinnamon and dumping buckets of ice on their heads to fund amyotrophic lateral sclerosis research. It's important to not take people too seriously, especially on the Internet, where any buffoon or child can anonymously make hateful comments without fear of repercussion. However I do believe there is a silver lining to foul online behavior like trolling, cyber bullying, and YouTube comments, namely that over time people started responding by actually becoming more respectful, and when I watch celebrities reading mean tweets or hear colloquial jokes about not reading the comment sections of videos, it gives me the impression that people on the internet are more cognizant about treating each other as human beings, as opposed to just usernames with an opinion. We should spend more time appreciating that the only reason we can communicate globally is because of underwater cables that could circle the globe twenty-two times.

What the Internet most shows is that human culture has no borders. As Mr. Rogers would put it, we are all neighbors, and this is no more evident than in the greatest, freest, fattest nation, the good ol' U.S. of A. Yet somehow people always seem to forget that. It's very strange how people fear cultural blending, especially in the United States, the supposed great melting pot. Nobody who lives in the United States is actually descended from the people who originally lived there, and even the Native Americans technically still just emigrated from Africa. Then Europeans came, which led to a genocide of the natives due to disease and war, they imported about five percent of the Africans in the Atlantic slave trade, and then Chinese, Irish, Jewish, and enumerable other immigrants eventually flocked to the country in hopes of getting wealth, fleeing prosecution, or just not dying. This even continues well into the early twenty-first century for people of Islamic faith. For goodness sake, Albert Einstein immigrated to America during the Weimar Republic crisis, just before the rise of the Third Reich. Every one of these groups were ostracized for not being "American", despite all of the actual Natives being killed off and the

United Sates only taking up a small fraction of all the land in America. This mentality is so pervasive that America is synonymous with United States. People often don't realize that Brazilians, Cubans, Puerto Ricans, and yes, Canadians, are all Americans as well. And how nice is it that they think the states are so safe and prosperous that they want to immigrate and contribute to our economy and make it more prosperous? Except Canada, I think there are too many guns and too little universal health care for them to want to move south. Actually, many people are afraid to immigrate to the United States because of how common gun violence and obesity are. Ironically, this is even true in Africa. According to the new South African born *Daily Show* host Trevor Noah, African mothers tell their children, "Be grateful for what you have, because there are fat children starving in Mississippi."

Culture evolves every generation, as demonstrated by Elvis Presley's hips, the Earth Environmental movement, and student protests around the world in Tiananmen, Mexico, all over post-soviet Eastern Europe, and the Middle East. It seems that for every generation, as kids grew up in a larger and larger world—that went from a patch of land, to a continent, to a globe, to a galaxy and far beyond—people have been much less likely to kill each other. After all, young people often have enough energy to care more about improving society. What humans do, better than any other life form, is take what's bad about the world they grow up in, that the previous generation was too afraid or unable to change, and find solutions to those problems by using the wisdom of the past with eyes unclouded. Of course, this doesn't always happen. Sometimes people get into ruts of living the exact same lives their parents and grandparents did, without realizing that it could be better. Normally, culture is why we call ourselves "human beings", and it is how we decide what is not a "human being". But here's a question, if we are made of the universe, then are all things equally cultural? Probably not, stupid people are far less interesting than planetary nebula, and they usually only live adhering to one culture. So it probably depends.

Finally, culture is why people throughout all time get together to hold feasts, festivals, play music and games together, and why all people gaze in wonder at the night sky with reverence to their ancestors, and despite all adversity are able to celebrate life regularly. It is why we are all the same, why we are all ridiculous, and why we never realize it.

Art

Culture is the soil from which art blooms. It comes in many forms and has many influences, and is almost as difficult to define as culture itself. Art can be found in the form of architecture; from The Eifel Tower to The Taj Mahal, The Roman Coliseum to The Pyramids of Giza, from The Forbidden City to Vatican City. Although separated by time and space, all these beautiful structures reflect states of human affairs and a longing for beauty and eternity. They demonstrate our need to make a permanent etch in history. But not all art is built to last; sometimes it is made to be fleeting, and sometimes all that matters is that it existed at all.

Though seldom recognized as such, cooking too is a form of art. Master chefs spend just as much time crafting their dishes as they do actually cooking them. In that sense, the cost of a dish usually correlates with how much artistic endeavor goes into making it, in addition to the rarity of the ingredients. On the same token, the cheapest food is also the most likely to be made through animal cruelty, which makes eating a difficult balance of cost and morality. Sushi chefs and professional cake bakers go through nearly as much training and apprenticeship as a medical doctor before they are allowed to become even a sous-chef. Training is more important for individuals who need to prepare potentially lethal dishes like puffer fish. It should come as no surprise that food is one of the most common archetypes for distinguishing cultures, or why so many social bonding experiences happen over the dinner table. Cooking really is one of the few things that only humans do, and it likely played a key role in allowing our brains to develop to such a degree. Plus it's nice to spice up something that we all have to do every day anyway.

Temporary art can also be expressed through the ways we present ourselves to the world. Whether it's the clothes you wear, your hairstyle, cosmetics or prosthetics, the way we dress provides us with everyday artistic license. Entire industries revolve around clothing fashion, which is a particularly fascinating thing for an animal to do. When you remember that humans are really just hairless apes, it becomes quite remarkable to observe how we use plants and other animals to cover up our "indecency". This shouldn't be seen as a negative however. In truth the sheer multitude of clothing styles and beauty products is a testament to human creativity. The fact that we have so little hair also means that we become much more creative with the hair that we do have, trimming, curling, straightening, and cutting it in a way that suits us just right. And just like peacocks, the clothes, makeup, and hairstyle we choose reflect our social status and mating profile. What's more, with the advent of plastic surgery, people can modify themselves to even a higher degree. With advances in genetic research, there may even come a day when people can choose to have designer babies born with perfect symmetry, musculature, and bone alignment. Many people are put off by artificial body parts, but many people also find confidence and freedom in being able to surgically modify their bodies. While I do have to say that excess botox and fake calves are more gaudy than aesthetic, I don't see any problem with grown adults changing their bodies if they wish. After all, I personally participate in one of the most time-honored traditions in the history of body modification: tattoos.

Ink and jewelry are some of the most powerful statements that can be made of oneself. This is especially true of tattoos, since they are essentially permanent, given the long and uncomfortable option of laser removal. Jewelry isn't usually permanent, but there are certainly plenty of cultures around the world that do permanently modify their bodies by putting rings through their ears or around their necks. While the longevity may vary, both of these forms of art share the unique property of directly involving pain. Again, this is most true for tattoos. While having a needle put through your flesh certainly isn't painless, it's not quite the same as having one or more needles repeatedly plunged into your skin for several minutes at a time. That's really what makes tattoos such profound statements of one's humanity. A tattoo is something one endures, intentional pain that allows people to make a statement about who they are and what they stand for. Many people probably associate tattoos with criminals and thugs, and there certainly isn't a shortage of gangsters around the world who show their allegiancea through ink. But the history of body art spans tens of thousands of years, with calligraphy tools found in the remains of prehistoric hunter-gather societies and the remnants of tattoos found on ancient mummies. More than a symbol of defiance, tattoos have been worn proudly by great warriors and explorers, as signs of fertility and power, as well as for branding people as slaves and criminals. I too take great pride in being a part of this ancient tradition of human expression, a permanent reminder of my history and conviction. This wasn't always the case either; when I was young I was quite opposed the idea of getting ink. My two most affirmed statements were telling my mother that I would never get tattoos, and I would never have children. Thus far, I don't have any offspring, but I do have two prominent pieces of art forever etched into my skin. It's strange, because I feel like even in eighty years, I will be happy that I took several hours and some mild pain to etch a permanent reminder of my life into my arms. It makes me afraid that one day I'll have children, and have to raise and teach them what my parents and the countless wonderful people I've met have taught me. I suppose every tattoo is like an important memory of all those people, but I still grimace at how people can so easily express certainty in the present with no knowledge of what the future might hold. As it turns out, predicting the future isn't easy. If only we had more empirical data...

While some forms of art resist the passage of time, and some acquiesce to the flow of time, others attempt to capture snippets of time. This type of art can find many incarnations, including painting, photography, and cinematography. Visual art can even come in the form of propaganda, reflecting the state of a nation or political faction. Advances in technology have only created new avenues for people to record history and present it in an artistic way. Where before images could only be recorded through the likes of Van Gogh's *The Starry Night* and Picasso's *Guernica*, now captured images can do the same, like the Migrant Mother and the Times Square celebration at the end of World War II, not to mention the thousands of documentaries that have been created. Photography is even the chief way astronomers inspire us with images from the Hubble telescope, showing how the wonders of the cosmos are arguably the most majestic images one can see. After all, it wasn't until we saw our planet as a pale blue dot from the Moon and Saturn that we truly began to grasp our astonishingly minute place in the universe. Of course, visual art doesn't have to literally represent reality. Just as many if not more pieces of art have been created to express the wonderful internal complexities of an artist's experiences. This is clearly evidenced in the wide variety of genres found in literature, drawings, pictures, and film.

Finally, for every piece of art celebrating life, there is another piece reflecting on death. It's impossible to say how many artists contemplate death while crafting their work. One might only see death represented superficially in works of horror, but that is far from the case. Perspectives on death are as diverse as the number of lives lived. As such, art doesn't need to have gore, skeletons, or anything macabre in order to reflect mortality. Who knows how much beauty has been given to the world by artists who had just been through an existential crisis? But regardless of how we react to it, nobody would deny that death is a topic that almost always provokes deep thought. Between the coffins and complex rituals and burials, humans tend to be quite strange when faced with their end. Death itself has even become somewhat of an industry in the developed world, and for many people dying actually costs money. This is rather inconvenient, since most people are too uncomfortable with their mortality to even address it and include it in their life plans. As a matter of fact, quite a few people are thoroughly convinced that they will never die.

Religion and Grappling with Reality

Let me begin by saying that religion is a beautiful thing. Quite literally, it brings peace in the face of certain death. As a general rule, religions throughout history have served as safeguards to destruction. By appeasing to some divine power, people hoped to receive better crop yields, protection from natural disasters, and victories in battle. Religions unite people and give them a sense of purpose. Without a doubt, it's certainly understandable that one would need a higher calling in a chaotic world where death can come at any moment. This would have been especially true for those living in areas where calamities were common and before the days of satellites and modern medicine. Praying is also probably good for the health, as is meditation, which is essentially the same thing. An occasional peaceful break from reality is important for one's sanity.

There have been no less than hundreds of unique and diverse systems of faith throughout the ages, each one reflecting the story of the people who lived by them. Here we will define religion simply as a person's belief in a higher power or influence beyond themself. In the twenty first century one can find churches everywhere they look, for all kinds of religions, which is great. Mosques, Cathedrals, and Temples everywhere have been a source of peace and prosperity for countless human beings looking for hope, truth, and vindication. That's why people need religion, and faith, because at the heart of every religion is the same divine light: hope. Life is horrifying, mostly because it ends, rather abruptly at times. Religion helps us find peace with mortality, and inspire us to treat one another as family, even if we come from different homes. It unifies us under a holy banner.

It just so happens that nobody can decide which flag to wave or why we even have flags at all, and they change their allegiances like cutthroat pirates. Most of the time, people are merely guilt tripped into their religion by family and society. I hope there will always be safe places where people can go to find themselves, no matter the shape of their faith. That said, it would be nice if we could save some land by consolidating a few of them, five religions and dozens of spin offs seems a tad crowded. As a matter of fact, just five major religions are still rather suffocating at times, just ask any Jew, Christian, Muslin, Buddhist, or Hindu.

Religious history is so fun! It has so many twists and turns; you never know what's going to happen next! It's like quantum mechanics on steroids, which is basically what the homo sapien brain is anyway. For example, out of the five major faiths, Muhammad (ergo Islam) came last, even though it professes to derive from Abraham, who was the first Jew. Abe dates back before Christ and Buddha, but not before Mesopotamia and Egypt, and completely separate from American and African native shamanism, Aztec ritual sacrifice, and anything to do with the rest of Asia. The first mainstream for of Christianity—Catholicism—was spread in part by the Roman citizen Saul Paul. Rome was ruled by one of the sons of god Emperor Augustus, who was oppressing people as usual in his empire, in such areas as the Dead Sea, right by where Jesus was born. As one of the earliest versions of Christianity, Catholicism is slightly more senile than the versions of Christianity that came after reformation, as many older religions tend to be. It's much more "cake or death!" than the more modern Christianity spinoffs, spouting off on how crackers are literally the flesh of Christ and how you'll burn in hell for doing anything at all.

The first gods were in all things for hunter-gatherer societies. The animals, the wind, the rivers, the stars, planets, sky, and Earth all had a soul, and had a god associated with them. After agriculture began, and cities began to develop, kings and queens began to be seen as the decedents of the gods. This was true for polytheists like the Egyptians and Chinese, as well as monotheists like the Europeans. Even Mesopotamians had Gilgamesh, who was considered semi-divine, but was probably just a king ruling citizens that were either gullible or just wanted someone to rally behind, as per usual in human history. Being some of the first people to settle into cities, Mesopotamian and Indus River civilizations had polytheistic religions based on erratic gods, which in turn were based on the erratic environment in which they lived. Eventually this would develop into monotheistic Judaism when Abraham made the covenant with Yahweh after he asked Abraham to circumcise himself, because in the dessert genital mutilation was one of the only ways for people to be trendy. Across the pond and farther south, several religions developed over the millennia around the Inca, Aztecs, etc. The Aztecs had a particularly destructive version of religion. Not only did they need to make sacrifices to appease their gods, they also had four times as many apocalypses as the Hebrews. Similarly, they both had a flood, but they also had three other worlds that their gods created, and then promptly destroyed. One of these worlds was ended by raining fire, another by raging hurricane winds, and another by jaguars. Across the pacific, the Japanese also had their own creation story. Their mythology begins with Izanami and Izanagi creating the Japanese islands. Lastly, in tandem with

Europe's history, the Greeks had a mythology beginning with a war between titans and the gods they named the planets after.

Religion commonly arises in the face of great hardship. Why did the African slaves become some of the most devout Christians? I think it was partly to assimilate to white culture, because they had their own religions in the motherland before Europeans came ashore, like Akom, Vodun, and Islam. After several generations of having families separated and being denied any education in America, Christianity was the only option available to them. Another major factor was likely because of the suffering they endured. When you're treated like second-rate manure and sold like cattle, the hope of salvation must sound infinitely appetizing. The African slaves may have even resonated with the story of Exodus and the escape of Hebrew slaves from their Egyptian masters. This would have been the same reason why Jesus was so influential. In his time, the Jews near the East Mediterranean were facing oppression from a tyrannical Roman Empire, and Jesus preached against the rich and the powerful, which must have instilled hope in his fellow Jews. This suffering may also be why so many people "find Jesus" in prison, although it may just be a two-faced lie that criminals make to appear to have morals. Or maybe the criminal justice system legitimately puts the fear of god into them.

Religions have also given us some of the greatest historical mysteries of modern archaeology. Stonehenge was probably a weird religious thing, it's basically just a church with big bricks instead of small ones, which makes sense because Stonehenge happened way before Romans or Christians invaded Britain—or even most of Europe—and most of the other places where people were doing their own weird religious things. The Egyptians also left behind many great relics of their faith, at least the ones that weren't stolen by grave robbers. They left behind the pyramids, the sphinx, and of course the mummified remains of their divine rulers and their pets. Interestingly, they had a cardio-centric view of humanity, and so they buried people without a brain and with their hearts, which means they literally threw away the actual spirit of their pharaohs before they got to the afterlife.

Although very few polytheistic religions remain, their legacy lives on. Gaia, Ra, Jupiter, Zeus, Thor, and Quetzalcoatl are still used in popular culture or are at least heard of in some way. Generations of people used to give sacrifices and live their entire lives believing in these deities, and they did so for centuries or millennia at a time, completely isolated in time and space from one another and living in totally different worlds. And now in western culture we just put them on cartoons or in live action movies with explosions. It makes you wonder why we revere so many figures now, and what our decedents will think of us in (hopefully) a thousand years when they're sailing through the Orion Nebula. Now, just because people decided to wrap all of those things together into one god doesn't make them right, it just reflects how human understanding about the world has grown. We've gained the wisdom to understand that the living organisms, weather, and all the matter in the cosmos are connected, and the evolution of the world's religions reflects this. People just up and ditch their religions if a more powerful state comes along with a different one, just to get the prestige and to make foreigners like them. Other times, rulers decide to adopt a religion as a means of gaining social cohesion and solidifying their rule, just as England adopted Catholicism, or how Africa adopted Islam, and then Christianity. Just to be clear, England eventually ditched Catholicism later on and came up with their own church, and they were able to do that thanks to the reformation.

Christianity is sort of a one size fits all in terms of religion, which makes sense since it's pretty lazy spiritually. All the other religions require pilgrimages, mandatory passage readings and body alterations (although a lot of Christian sects do that as well). If you're a Muslim, you are contractually obligated by Allah to go to Mecca before you bite it, and if you don't live in Saudi Arabia, that's just a huge hassle. You have to deal with passports, dirty thieves, racial prejudice and discrimination. Plus the poor Sikhs from India are always getting confused for followers of Islam, which is even more inconvenient. This whole religion thing is starting to get confusing. Unless you've chosen Christ, then you can basically be forgiven for any of the stupid things you do free of charge. That may be one of the reasons why Christianity is so popular in the West; it's the religion of the busy bee capitalist. You just gotta accept Jeezy Creezy as your metaphorical father (or shepherd) and you're saved! Except there's also Yahweh, who's his father. So your grandfather? Or is Jesus Yahweh in human form? And there's a ghost too I think. I'm actually not sure how that whole family tree works. According to Dante's *Paradiso* the catholic god is a group of really shiny rings. Not sure how we're made in that image, but hey, who's Dante Alighieri to decide what god should look like? He only invented the modern scary version of hell with Judas at the bottom with Brutus and Cassius (Roman Italians REALLY hated the Greeks, a la Trojan War).

Anyway we aren't even made of rings. I think. Unless string theory strings are rings! Bada Bing, I just solved religion and the physical theory of everything. If only we had more physicists to experimentally verify god. But alas, they have so little funding. Did you know that there are billions of particle physicists living in poverty all over the third world? They had their money stolen by Catholic and Protestant Europeans, and a wee minority of angry Muslims oppressed some others. Now they spend their days drinking impure water, defecating in ditches, and living under inept governments. Those poor particle physicists. Although, blame is shared by all of us. All the Buddhists do is meditate, all the Christians, Jews, and Muslims do is pray, and all of a handful of each of them periodically send those poor particle physicists pocket change like common homeless veterans. Oh right, they also like to argue over who's version of the one true god is the most awesome and terrifying. They're all so scared of their gods that they'll actually fight if it means feeling godly, when all they have to do is give those poor particle physicists a chance to prove one of them right once and for all. Poor, impoverished particle physicists, if only they had textbooks and building materials. How can people serve god properly if they're all scared for their life that some jerk with a gun or a bomb might send them to heaven early?

As a matter of fact, people often claim that nothing can be done about natural disasters like volcanoes. Don't worry about Yellowstone, that it's out of human control. So instead of doing anything useful they just pray. How pathetic is that? Who would have guessed that humans could harness the power of lightning to create indoor Sunlight, laptops, and Tesla coils? I'm sure if we had enough data on the subject someone could figure it out. People are already learning how to control weather to increase rain in dry climates. All we need are enough fed and educated people to pay and mitigate disaster. There are plenty of poor people in developing nations who I'm sure would prefer scientific research to save humanity over walking several miles every morning to get a glass of clean water, and there are plenty of rich Christian Europeans living on the same continent as Yellowstone who would probably appreciate not dying from volcanic ash. Don't Christians like donating to charity to avoid paying taxes anyway? What's the hold up? Give those kids some geology and climate science already!

People only seem to invoke god after a catastrophic event or monumental disaster. But if you're smart and know what a volcano, hurricane, tornado, or earthquake are, then you're much more likely to make it out alive than if you completely luck out and thank god for it. It in no way diminishes the sheer scale and horror of natural disasters, but at least it isn't a supernova. Whole planets can't survive those close up, and even their atmospheres could get blown away from only a few light-years out. Entire cities can be wiped out by the likes of Toba, Krakatoa, Vesuvius, and Yellowstone, but at least the whole of humanity can make it through that. Life has been living through heavy volcanism for the last four billion years! Life is resilient, and people are crafty. I guess if you still want to blame god you could argue that every now and then god needs to weed out the globe to allow some chosen people and animals to continue living in his (her?) image. But that's what evolution is, and if you care enough to go that far to blame everything on god, you probably don't believe in that. What, did god spend fourteen billion years to make creations that would either sit on a rock in space and shoot themselves or go out into space and meet him (her?)? I suppose if god is infinite, then a few billion years still isn't that long, infinity is by definition always longer. So I suppose we're just a zoo for some god(s?) then, living for some seventy years? Or thirty if you lived with Jesus, or two if you got sick, because god made bacteria too. And then he (she?) kills those babies and stupid lazy people off so that they can be in heaven with him (her?). There are a lot of dead infants in heaven (as long as they're baptized). I wonder if they get to grow up there? How does that work anyway? Do old people stay old forever in the afterlife? Or is everyone twenty and good looking? I swear, this whole religion thing is more complicated that quantum chromo dynamics and vector calculus. I tip my hat to monks, priests, bishops, rabbi, and imam everywhere. But which one of you is right and why didn't your god tell everyone about him (her)?

By the way, who told everyone that the one true god was a dude, and why are monotheistic women complacent with that? Where's his wife? Probably doing all the executive work while god sits on his throne, smiting people, sending plagues, and picking favorites. Why does god smite people? Well if you ask a human person on the street, apparently everyone is either an infidel or a sinful blasphemer and sodomite. Darn, if only god just showed up and told everyone to stop killing each other in his name, instead of letting children grow up as terrorists and gangbangers, who only exist because of oppressive selfish people who claim to be devout. Whatever, dark and mysterious ways, or some other malarkey, I won't tell you what to think. All I know from experience is that some people are always mean to each other, it doesn't even matter what religion they choose (or more likely born into). Luckily everyone is nice, if not to everybody then at least to somebody. It's just so easy to forget that everyone is as equally human as we are.

Witnessing the evils of the world has a way of making a person say, "Just shut up and be nice to each other", which is in part why nearly all religions promote peace, so long as you pick the correct one. Jesus promoted peaceful rebellion against the oppression of the Jews by the Romans. He was also the coolest prophet, because one of his most famous magic tricks was making food and wine come outta nowhere. He totally picked that trick up from the roman gods too, after all everyone likes bread and wine. Buddha didn't promote alcoholism, but he did promote self-meditation and spiritual salvation by releasing the self from Earthly wants. Yahweh was famous for killing a lot of people in the Old Testament, kind of like the various Mesopotamian, Aztec, and especially Greek Gods liked to do. As a matter of fact, the Greek gods and Yahweh both got women pregnant, and then had their kids do magic tricks for a while, before viral videos were invented. But after having a kid Yahweh mellowed out. Although I suppose it could have just been one god punishing people around the world for not believing in him (her?). I just feel like it would have been easier to just tell people he (she?) was there, and explain how weather worked.

Despite the peace toting religions have, they're not above succumbing to our human desire for growth. It's always the same childish nonsense with religious people. "God gave this land to me"; right up until someone with better weapons comes along with the same story. Apparently god can't decide who to lease the Earth to, or what to dress like, or how we should dress, or how many gods to be. At least god kills everyone fairly, with plague, famine, and weather. Oh wait, never mind, humans figured out what pathogens are, how to build more stable infrastructure, and how to stop hoarding food. Sort of, there are still a few children who get into gunfights with each other instead of fixing the problems god gave them. Go ahead and read your little story books on ethics, scientists will just be over in the corner reading textbooks written by humans more curious than prophets about how god's creation actually works, and how to manipulate it to their whims. The Bible might be the best book to learn how to be kind to your fellow humans, but it can't teach you how to create an electric power plant or create antibiotics. Thank god every motel has a bible inside, because lord knows people only go there for hookers, drugs, suicide, or because they're poor. Evidently those people need Jesus the most. Although I feel like a glass of Christ's blood would be much better at taking the edge off. That's what I image ritzy hotels do, if you have enough cash to go on vacation.

So anyway, Abraham is in Judaism, Christianity, and Islam? That's an impressive amount of fame! Like Jewish Hebrews, Muslim Arabs trace their ancestry back to Abraham. Like Jews, Muslims believe that Abraham had two sons, Isaac and Ishmael. While Jews see Isaac as the Hebrews' progenitor, the Muslims trace the Arabs back to Ishmael. So although agreeing with Jews in terms of ancestry, Muslims shift the emphasis from Isaac to Ishmael. According to Muslim tradition, Ishmael helped Abraham build the Kaaba, and Ishmael's descendants became the Kaaba's guardians. In addition, while the Bible describes Abraham offering his younger son, Isaac, as a sacrifice to Yahweh (before He stopped him), the Qur'an describes the same story, but with Ishmael as the nearly-sacrificed son. Confused? I am.

The traditions of Islamic faith unify religion and politics, a concept that has long since been abandoned by industrialized nations. This is mostly the case because unlike Jesus, who was a peaceful hippy, the prophet Muhammad pulled double duty as general. So unlike Jesus, he had an empire to run, and a religion to start. Jesus didn't even seem to want to start a new religion, that didn't happen until a few decades after the Romans killed him. However, his religion is still strangely revered in the United States, mostly by people who don't know the history of their own country but still want to be patriotic. It wasn't until 1954 that the phrase "under God" was added to our pledge of allegiance, because the U.S. was afraid of the Russians, who were communist, which denounces religion, even though Russia wasn't completely atheist and the United States was surprisingly able to separate church and state decently well for 178 years. In just sixty short years people would start assuming that it wasn't always that way and The United States was always a Christian state, despite the first sentence of the first amendment of our one constitution literally saying, "Congress shall make no law respecting an establishment of religion". And if citizens choose to not say the pledge of allegiance, they face being singled out amongst their peers. Granted, there's nothing wrong with believing that Yahweh blessed your country, but it seems like a vehement damning that people, especially children, should be ostracized as unpatriotic just for exercising their first amendment right to practice another faith. Or maybe a millennia old faith is more important than a century old federation? There are plenty of nations that don't let their citizens choose their religion after all. It may also be true that in general people only pretend to care if other people share the same belief system they do, and so it isn't a problem to put "God" in a national pledge, or to even say one in the first place. It would be un-American of me if I didn't let you decide for yourself.

Many religions tend to be very awkward about sex, especially the Catholic Church. Although, deterring people from procreating does sort of make sense when you remember that many religions originate before anesthesia and proper birth control. Religion would have been a great way to regulate population back when people's livelihoods depended on yearly crop yields, not to mention that raising human larvae is arguably the most difficult child rearing in the animal kingdom. Body shaming also isn't limited to religion. Like I said, a lot of people don't like to be reminded that we are in fact smart hairless apes. So there really isn't any problem with religious people wanting to save themselves for marriage, that would actually be quite impressive in the twenty-first century, or any era for that matter. However, it is a little more than problematic when the Catholic Church promotes the spread of HIV by telling millions of Africans that they shouldn't use condoms, or when their priests vent their sexual tension on young boys, or when people shoot up abortion clinics. That kind of behavior is just naïve and childish, and it causes far more death and suffering when people have to live through AIDS, live with sexual abuse, or grow up with parents who don't want to raise them. At that point, religious sexual ignorance becomes quite abhorrent.

People who are religious also seem to be the only people to care about gay marriage, for some reason. Perhaps they get jealous from losing two potential mates at the same time. They usually cite this distaste from Sodom and Gomorra, or some such thing, out of a book that was written during a time when people didn't even have pizza or know what the Sun was made out of. They also didn't have genetics or observations of social homosexuality throughout nature, which would have actually been useful information for not discriminating against people for being in love. I personally haven't heard any viable justification for this discrimination, especially since continuing the human race isn't really a problem anymore, and marriage has historically been used for alliance building since before recorded history. People are weird. Luckily for the Catholic Church, The Holy Father Pope Francis is one of the chillest popes to date. He even has emojis, and fights climate change! The Vatican also has a prestigious astronomy program as well. They are looking into the eyes of god better than most Christians these days. Doesn't quite make up for child molestation, an alliance with the Nazis, or the Inquisition, but hey, at least they're trying. The Vatican even accepts evolution, something most Christians still seem to struggle with.

There's also a common myth that all wars were started by religion. This is clearly hyperbole. Sure the Nazi belt buckles said *gott mit uns* (god with us), but people are far more likely to fight over lunch than religion. Just think about the House of Wisdom in Baghdad developed during the Abbasid Caliphate. It was the epicenter for much of the scientific world from the 9th to the 13th centuries, where people of Islamic, Jewish, and Christian virtues all came together as scholars to study the ways of the world. That is until the Mongols arrived. The Muslims of Spain also developed water engineering and heightened agriculture yields through the study of botany, sewing the seeds for Spanish imperialism. People can survive under any religion, but if you don't eat, you die. Religion is just one excuse people can use to justify killing each other. If two people believe in a different God, then they have the option to either convert the other or just leave them unenlightened, without risking their lives. Or they could just deem each other as blasphemers who have abandoned the true way. Which one is the blasphemer? Obviously the one who doesn't have the latest weapons.

The influence of religion can also be seen in all the holy texts, which are ripe with ethical lessons that influence many works of art, like Michelangelo's *Creation of Adam* in the Sistine Chapel, and the book *Of Mice and Men*, which was influenced by the story of Cain and Abel. As an aside, Michelangelo's depiction of Yahweh sort of resembles a human brain. This makes perfect sense; since nobody has ever seen god, we have to use our brains to decide what he(she?) looks like. It just so happens that most Italians pictured Yahweh as Santa Claus. Religion also inspired the remarkable poetry of Dante's *Divine Comedy*, where he perfectly outlines each level of hell, purgatory, and paradise, all of which the Bible seems to have neglected to give any details about. By the way, are all scientists in the eighth circle of Malebolge? That's the ten-section layer of hell where Dante decided to put all of the sinners who committed fraud, including sorcerers. Their unique punishment is to have their heads twisted in reverse for trying to look into the future (symbolism!), something that scientists are quite good at, much better than actual mediums in fact. I say that they would be there because obviously we don't control where we are born or what path our lives take, so we are all just born predestined for eternal suffering or eternal boredom. That would be quite a shame, without scientists we wouldn't have GPS, or the Internet, where one can buy several thousand ladybugs for no reason for less than an hour of work. That doesn't really seem evil, just really human. Oh wait, the Catholics think everybody is born in sin, because of a naked dirt couple in a garden that never existed.

The idea of sin is completely arbitrary. Most people generally regard murder as evil, but only so long as they are a part of "our group" rather than "their group". If people couldn't justify murdering each other, then war would never happen. This probably doesn't come as a surprise to anyone who regularly practices a faith, all three of the dessert religions in some form or another regard humans as inherently sinful, which is why they are in need of salvation. But when the holy laws people use to justify their behavior are so fluid that anyone can get away with anything so long as they attribute it to some divinity, then they aren't laws, they're guidelines. This isn't a problem until two separate groups of people see each other as ideologically incompatible and heretic, and then slaughter one another, both believing themselves to be doing the work of god and both of them genuinely believing that their causes are ordained from on high. And when the winner is decided, they subjugate people for thinking differently, right until the oppressed group rises up, perpetuating a cycle of hate and misery, kind of like the Sunni and Shia. All of this, because everyone is so sure that they know the true path to salvation.

Religions also use grey flowery language that's inconsistent, yet all similar in how their various teachings are just as genuine as a fortune cookie. Case in point:

"Can the Ethiopian change his skin, or the leopard his spots? Then may ye also do good, that are accustomed to do evil." — Jeremiah 13:23 KJV

You could read this one of a thousand ways. Some used this and the Curse of Ham from the Book of Genesis to justify chattel slavery. Others use it to show how people never change. The only problem is that leopards do actually change their spots, both from birth and each generation. Also dark skinned people can get sun burnt, so it doesn't really matter what Jeremiah thinks anyway. Or how about this gem:

"But I say unto you, that ye resist not evil: but whosoever shall smite thee on thy right cheek, turn to him the other also." — Matthew 5:38–5:42 KJV

People use this to justify anarchy, demand for equality, or as an objection to (or evidence of) the vengeance philosophy of "an eye for an eye". The problem is nobody can agree on what the heck this book is saying, and people tend to come to opposite and sometimes horrifying conclusions that would be illegal in any modern society.

Another question, why is Jesus always depicted as being white? Mary, and all the angels as well, they're all snow-white vanilla. They lived in the desert for goodness sake! And what happens to all the colored folks when they die? According to every biblical work of art, when we die we turn into white folks. This would explain how Jesus could have been born and raised in the Middle East, because when he was resurrected he came back with less melanin. It must be that the shining rays of heaven irradiate all of the pigment from people's skin. Except, people who are exposed to more sunlight get darker skin. Unless you use chemicals, then you get orange skin. Perhaps the same thing happened to Jesus as Siddhartha Gautama (the Buddha). He went from a Hindu prince to a monk, and then the Hindu religion got ahold of him after his death and he went from mortal teacher to another Hindu deity. In a similar way, when Christianity came to Europe, Jesus and Yahweh became pale as bone. This similarity would make sense; Jesus and Buddha were great roommates in Japan (look it up). There is one last explanation; perhaps religions are based more on the people that made them than the divinity they wrote about. Unfortunately, that argument also has a serious drawback. Namely it upsets people, and disgruntled folks tend to stop listening to what you're saying. So we're not going to say that here.

Also, I never understood the whole wine and bread comparison with J.C.'s blood and flesh. I thought Yahweh hated paganism. Drinking blood is vampirism, and eating flesh is cannibalism! Christians even celebrate Easter like pagans. For Christ's sake, it's not even on the same day every year, so when on Earth did somebody steal his corps from that cave? Apparently after Passover, which is always a Sunday I guess? There's nothing wrong with celebrating an event, but that still doesn't explain the rabbits and chocolate eggs. Religion is legitimately unnerving in how basic and broad it is, like a brain without wrinkles. People take the whole, "knowledge is the ultimate sin" thing from the book of Genesis a bit too literally sometimes. I like knowing things, that's how we have clean water coming out of a metal tube in the wall, and a big box that keeps meat and vegetables from getting eaten by bacteria and fungus, and ice cream, and books, and agriculture, and airplanes, and trains, and rocket ships, and plasma lamps, and skyscrapers, and air conditioning, and custom mattresses, and cars, and antibiotics, and antivenin, and aspirin, and ships, and global trade, and television, and sushi in Colorado. Actually, screw it, all of Genesis is bogus anyway. None of the creation stories make any sense.

It isn't their fault or anything. How could you expect a person to know about the actual way the world works without a proper understanding of particle physics and spectroscopy? I don't see any mention of accelerators or telescopes in ancient holy texts. That's because they didn't have them, and lord knows nobody had the insight to dig into the Earth and carbon date all of the skeletons inside. It's no surprise, no one even knew what an atom, or carbon dating, or what the nuclear forces were anyway. It's no wonder absolutely no religion ever has any idea how the world came to be, none of the people who made up those stories had done any testable experiments. They just said, "This is how the world is, like it was. All hail Thy God, King of the Universe (except really only the king of a small patch of land on a single continent on Earth, because we don't know what a galaxy was)." It's quite admirable really, those people who wrote all of the ancient texts without knowing anything about nature. That's why they are all so blatantly general that they actually seem true. They are also arrogantly human centrist. But hey, humans are the apex predator on the planet, so why not build some self esteem through story telling? Plus they did get one thing right; the universe has been behaving exactly the same since the beginning of time. It just so happens that those laws require stars to explode, planets to form, water, atmosphere, magnetism, and volcanism to be present before all the smart hairless apes could show up and wreck the joint.

Some blasphemers try to reshape their religion to coincide with modern cosmology. People are so desperate to conform, and yet so scared to admit defeat. Seriously, how do you justify any creation story? Genesis doesn't even get the order right. The first thing Yahweh made was light. Sure, the universe was mostly light when there wasn't any space between particles and the universe was hotter than the Sun, but that only lasted for like three hundred thousand years, after that the universe was dark again for just as long before the first stars were born. You'd think Yahweh would have mentioned that to the Israelites, instead of lazily stating, "let there be light, and there was light". It even states that Yahweh made the Earth BEFORE the stars for goodness sake! The fish and birds also came before the land animals. Fish sure, but birds need to hop before they can fly. What about insects? They fly and creep, and they made it onto land before the fish diverged into amphibians, so where are they in all this? Yahweh also made two great lights, one for day and one for night, which I can only assume are the Sun and Moon. The only problem, the Sun formed before the Earth, and the Moon is up during the day, so maybe people didn't go outside very often back then. Also, man was made first, and a woman was the first sinner AND made from a man? That sounds like something a man who didn't respect his mother very much would write down. The Bible teaches men to hate women right out of the gate. Also, we aren't made of dust. Too much silica and not enough water. The plants and animals also don't need Adam to tend to them, they got by just fine for about a billion years before any people sprung up and started eating all of them. Most importantly, why on Earth is the tree of knowledge even in The Garden of Eden at all? Why is there a tree that kills people in the garden in the first place?? And why didn't Yahweh stop Adam and Eve from eating it if he cared so much, and why would He bother telling them not to eat it if he knew they were going to anyway??? I guess Yahweh can't control the talking animals.

And what of god's punishments for eating of the fruit? Adam just had to do some manual labor. Poor him, having to work for a living like a common animal. The serpent, which caused the whole mess in the first place, basically got off scot-free! Its punishment was to crawl on its belly and eat dirt. Clearly Yahweh wasn't aware of snakes that swim, climb trees, or eat crocodiles. Although I suppose not being able to talk anymore is rather unfortunate. The only person who was really punished was poor Eve, who was sentenced to serve her husband and experience horrible pain during childbirth. You know, until humans invented anesthesia. We still haven't grown out of the whole sexism thing yet either. Seriously, what was Yahweh thinking? He couldn't just kill Adam and Eve and make brand new people form scratch? None of this makes any sense. Whatever, He works in dark and mysterious ways, or some other snake dung. I guess we should give God a break, it's not like He's omniscient, omnipotent, and omnibenevolent or anything. The story of Genesis clearly demonstrates this.

Here's a cute excuse, what if a week in the life of Yahweh was equivalent to fourteen billion human years? What if those six days of creation were really fourteen billion years, and the seventh day is the modern era, when everyone had a camera on their phone and suddenly Yahweh stopped talking to people. Maybe we live in the seventh day! Fair point, but I don't understand why He would leave out so many interesting details in his books, nor why He would tell different prophets conflicting stories. Also, the seventh day was hallowed, which either means it really was just a Sunday and Yahweh made everything from Monday though Saturday, or everyone everywhere needs to hallow everyday equally, both of which are ridiculous. And why does Yahweh need to rest anyway? Apparently He wasn't omnipotent enough. He created a boulder too heavy to lift and microwaved a burrito too hot to eat with the whole universe thing. Yahweh needs to get in shape, maybe learn some Tai Chi and Kung Fu. He probably ate too much meat, drank too much, and smoked too much dank on the days He created all those things. Which makes sense; most young people use Sunday as their hangover day as well.

But seriously, the Old Testament God talked to people all the time. He also actively sent plagues and floods to smite his creations on a regular basis, conveniently when people were being slightly more sinful than they always are. Then He chilled out, sent down his son Christ, then had a conversation with Muhammad in a cave through the archangel Gabriel six hundred years after J.C.'s human sacrifice, and then the Muslims started taxing Christens and Jews. None of this involved the Hindus, Buddhists, and various Asians, Africans, and Americans. He just let them do their thing for a couple of millennia, until the Christian Europeans stole all of their resources, enslaved them, robbed them of their culture, killed six million of his first chosen people in concentration camps, and drew up borders with their guns and without approval, leading to many unhappy natives. Yet somehow, neither the New Testament nor the Quran have any useful information about avoiding plagues, preventing world war (they sort of venerate it actually), how to fix nitrogen from the air to increase crop yields, or how to build public restrooms with toilets. Instead, the divine texts are just filled with anecdotes about how to live life that are so vague that anyone could make up an interpretation. It's almost like The Bible and every other divine text were actually written by normal people who were just giving their best guess like the rest of us. But that would make people unhappy, so we won't say that here.

So, Buddha was Hindu, before he became enlightened. Similarly, Christians just tacked on a new testament about a completely different dude than the Jewish god Yahweh. That's exactly what Islam did with Muhammad! Christians just started calling Yahweh "God", and Muslims just started calling Yahweh "Allah". Then came reformation. Martin Luther divided Catholicism around the sixteenth century. Before that they had a lot of popes that Dante would eventually put in hell, and one King Henry VIII that wanted a new wife, and so invented the Church of England. After that some pasty folks sailed across the ocean and bickered about a bunch of their favorite version of the bible for some time. One guy named Joe Smith even said he saw Jeezy Creezy in America, but then he lost all of his books. Don't worry about all the dozens of native religions across the largest continents on the planet; the Europeans took care of them right quick. All they had to do was work 'em to death and not wash their hands. Monotheist god comes loaded with petulance, bad hygiene, and guns, like an action star that never bathed. Jesus Christ was such a Superstar that their black slaves picked it up too, and just like sports, totally did it better. And they were doing this all the while wondering why god would put natives on the Earth in the first place. It was clear that Yahweh totally bequeathed that land to them; because monotheist god is an industrial capitalist complete with railroads, and did I mention they also had guns? Jesus, on the other hand, who had died about seventy-six lifetimes ago, just told everyone to be groovy and drink some wine. Aha! THAT's why everyone loves J.C. so much, he liked to party! Also, guns, but mostly alcohol. That's also why the Russians picked Christianity.

But all religions have some good life rules as well. Buddhism has four noble truths: all life is suffering and people cling to useless crap, desire is the source of suffering and karma, one must rid themselves of desire to end suffering, and finally one may rid themselves of desire through discipline and by following the eightfold path. This path involves finding the right view, right intention, right speech, right action, right livelihood, right effort, right mindfulness, and right concentration. In short, understanding oneself and resisting worldly temptation leads to order in society, but everyone must find peace individually. This was very attractive with the oppressed members of the bottom castes of India, much in the same way the teachings of Jesus were most popular by the people being oppressed by the Romans a few centuries after Buddha died.

Similarly, Islam has five pillars to live by. First, there is only one god, same as the Jewish and Christian faith, just with a brand new prophet. Second you have to pray five times daily, but only after cleaning yourself. This is much more stringent than either Christianity or Judaism, where people can pray whenever they feel like it, including in the middle of a football game with crumbs all over their faces. Third, since Allah (Yahweh) owns everything, people should be charitable and work to decrease wealth inequality and help the needy. Most faiths profess charity, but it's only mandatory in the Islamic faith, and unlike other religions, good behavior doesn't increase your likelihood of getting into Heaven. For example, if you picked Christianity, then you get to hoard all your money and allow people to continue to suffer on Earth, since just believing in Christ absolves you of sin and guarantees both of you passage to Heaven. Fourth, followers of the final word of Allah (Yahweh) must participate in fasting, to increase respect and dependence on God. This would be rather unpopular to most westerners, who enjoy meat and being fat too much, although they do like dieting. Fifth, everyone of Islamic faith must make the pilgrimage to Mecca, which as previously noted, became rather difficult after multiple terrorist attacks incited mass prejudice against peaceful travellers.

Judaism, (ergo Christianity) has ten commandments. The Decalogue begins with, "I am the Lord thy God, which have brought thee out of the land of Egypt, out of the house of bondage." Besides beginning the exact same way as the first pillar of Islam, notice the exact specificity of Egypt. The Jewish (ergo Christian) god began through hardship. You might recall that it's not uncommon for religions to center around great hardship. By the way, for context, the enslavement of the Jewish people couldn't have occurred more than five thousand years ago, about another five thousand years after humans figured out agriculture. Anyway, moving on. The actual commandments can be abbreviated as Yahweh saying:

I. "I'm your only god."
II. "I am the only god"
III. "Seriously, worship anything else, and I will murder you. Until the afterlife that is. Wait, I haven't invented Hell yet. Just wait a few thousand years until my son comes along, then I'll make *him* damn you. "
IV. "Don't say my name (it's Allahweh, or GOD ALMIGHTY, KING OF THE UNIVERSE!!!)"
V. "Don't do anything on Saturdays OR ELSE"
VI. "Be nice to mommy and daddy (not me, your heavenly father. Your other ones, that actually went through the trouble of raising you and feeding your grubby little mouth)"
VII. "Don't kill anybody (unless it's the Amalekites, or the Canaanites)"
VIII. "If you got married, don't be a hoe. That goes for men, women, and anything in between. Thy lord God is not prejudiced, just homophobic."
IX. "Don't take other people's stuff"
X. "Be nice to Ned Flanders. Also, don't be jealous of Ned's stuff, and don't be jealous of Ned's wife (even though she is way more objectively beautiful that your significant other. The grass is always greener), and SERIOUSLY, DO NOT MESS WITH OTHER PEOPLE'S STUFF! (especially Ned's, he's literally your neighbor. But in general, do not covet other people's stuff. It'll just, like, bum you out man).

The commandments then close by acknowledging that the god of the Hebrews (ergo Christians) is a jealous god. But not envious, because that's a sin! Basically like your clingy boyfriend/girlfriend from high school, except with the power to CREATE THE UNIVERSE, and only talk to one person at a time apparently. And those are the commandments, given straight from The Lord, thy God, King of the Universe. Right?

Did anyone even go with Moses and Joshua when Moses was getting the ten commandments? They were gone for FORTY DAYS! Is anyone else starting to see a pattern with people being gone for long periods of time and them coming back like, "I talked to God, listen to me!" (I'm looking at you Muhammad and Joseph Smith). The way it's written it sounds a lot more like they just hiked up a mountain, carved out some rules on flat rock, then came back and said, "Don't be bad people, and also we can only have one religion, God says so". That's not divine, that's repressive, even if it was meant to unify a group of people for a certain cause. It's also very inconsiderate to all of the religions that popped up in Asia and the Americas thousands of years prior, not that Moses even knew about them. Also, I've always wondered how a bush could talk. And then I saw Jeff Dunham, and I leaned what ventriloquism is. Any Homo erectus could start a fire after all. It feels like poor Moses got conned into saving his people. Or maybe he just needed something to rally the Hebrew slaves behind. That particular story, if heard anytime after the fall of Rome, would be a testimony from a person, which is testimony from a book (The New Testament), which is testimony from another person, which is testimony from another book (The Old Testament), which is testimony (or testament) from another person, which is testimony from a voice in the clouds that only one person gets to talk to. The book this comes from also wasn't even allowed to be rewritten, because apparently it is so infallible that everything in it is always true and will be for all eternity. Until Martin Luther of course. I'd rather just listen to a person that actually saw a supernova through a telescope, predicts the locations of stars around black holes, or navigates the ocean using constellations, because I can do that too, given some practice.

Remember Cameron, the girl with half of her brain? One might attribute her survival to one of the hallmarks of religion, miracles. Indeed, people are constantly referencing miracles as evidence for divine intervention. In modern society, essentially all of them can be chalked up to good luck and a misunderstanding of how nature works. However, what of miracles that supposedly occurred before cell phones and neurosurgery were invented?

Here's a fun magic trick; turning water into wine and making bread loafs for five thousand people. Drunk starving peasants are definitely reliable sources of information. And let us not forget walking on water. It's ingenious rhetoric for describing somebody who was probably just standing on the shore. The walking on water scene plays out in the sea of Galilee, around 53 km in circumference, and 21 km long and 13 km wide (remember "Americans", one kilometer is sixty percent of a mile). It's a small patch of land close to where the big J.C likely lived, since he recruited his best friends to be his spiritual wingmen there, and performed a lot of magic shows there as well. One summer, I drove between 29 and 33 miles from my apartment to Harvard. It took around an hour and ten minutes in Boston traffic, and forty minutes during lunch. So if the roads were flat, one could drive around the lake in roughly a half hour, up to an hour. A boat would be far slower than highway; likely around a half to a third the speed with good winds, maybe lower if wind speed is down. We'll assume sailing rates between ten and fifty percent the speed of a modern vehicle on a freeway. Likely the size of the lake will have changed throughout history as weather shifts. However, it was just a short seventeen centuries before the industrial revolution, so the overall climate will be warmer overall, leaving the dessert lake likely with less water. We'll assume it could either have twice as much water volume or half. We have fairly high uncertainty, but it leaves the lake as being the same size on average for the last two thousand years. If it's smaller, it would be easy to navigate, and it would take longer if it were bigger.

Anyway, according to John, they were around six kilometers out to sail—so not too far, no more than four miles. Not a small distance, but people still run it for fun, and you could walk that amount several times in a day. Suffice it to say, Jesus wouldn't have had to sail very far to get to his prophets. The lake's maximum depth is only 43 meters, around the height of twenty-seven people. Near shore it's likely about half that. Peter tried to walk on water too, but fell into the lake, because the storm scared him, which isn't unusual for anyone who has been sailing during bad weather, or just been deluged by heavy rain. Jesus said to have faith, but why didn't he give him a second chance to try his faith? Also, the storm ended when Jesus got onto the boat, rather conveniently. Why didn't he walk back with them on the water when he could be seen? John also notes that they immediately reached shore, which would suggest that the storm blew them back to land, which would make the lake more shallow and easier to "walk" on, especially if there were hidden rocks that were only barely underwater. And what's with the shady behavior of Jesus, also noted by John, about not answering the people's questions of where he's been or what he's doing. Try to pull that after the invention of Twitter. Jesus would have never been elected president; he'd be ridiculed as a criminal who buys votes with lots of bread and wine. Also, with sick superpowers that nobody had ever done, you would think Jesus would recruit everyone for his cause and heavenly father whilst standing IN A LAKE! Sounds more like smart branding and propaganda than the will of god.

Probably the scariest thing in human history is that in the past churches had the power over government, and the people in power who controlled the church were just the ones who happened to be able to read the Bible and make up their own interpretation. That is excruciatingly terrifying, because people tend to get offended when they think they have the world figured out and you tell them they are wrong, and those people didn't even know what a quasar was! It makes my spine tingle, no lest because there are still people who think that it's morally acceptable to tell others how to think, and as a scientist I happen to enjoy thinking feely. Those people also tended to not listen to the women in their societies and looked down on a variety of people. Yet most people care about sports more than their religion. I can even prove how arbitrary faith is in a single declaration: "Yankees rule!" Just kidding, I spent time at Harvard. Go Sox! Alright, alright, you got me, I don't watch sports. Also, if Barats and Bareta can summarize the entirety of the bible in one minute, there isn't very much to think about on those pages is there?

One of my favorite anecdotes about a monotheist God comes from Neil deGrasse Tyson, speaking about yet another famous scientist, Isaac Newton. When writing his *Principia*, Newton was able to describe the motions of all the heavenly bodies. The calculus he developed shed light onto previously divine works of the gods, beautifully tracing the orbits of the moon, comets, and the planets. Only when his understanding waned, and his equations were no longer sufficient, did he evoke the hand of God. This has been the trend of human knowledge since our days on the savannah. What were previously the works of a divine overseeing force eventually became the predictable consequences of natural laws that occur independently of any intelligence or will. From this point of view, "god" is nothing more than the amalgam of all the things we have yet to understand. If that is the case, I hope that the idea of god never goes away.

So why is it that so many cultures around the world all have some notion of a higher power? Here's a pro-tip, just because one sees people come to similar conclusions doesn't mean they are true, it just means people react the same way to the same things, BECAUSE EVERYBODY IS

THE SAME! Well, not exactly, but on the scale of genes, atoms, and stars, every human is basically identical. Moreover, we all live on the same planet and therefore have similar mythology. Dragons are awesome and terrifying, and breathing fire is pretty badass. Death is scary, which is why nobody likes looking at spooky scary skeletons, which by the way are living inside of you right now! *~Cue scary ghost noises~* Your skeleton needs to be alive by the way. It works hard to give you structure and make your blood, which you kind of need for breathing and stuff. Speaking of which, everybody dies when they lose blood, which may make one think that blood has life force. Plus if you just watch a body decompose, you'll see the body dry and organs rupture. Hence why every culture in the world has some variation of vampires, from Dracula, to Edward, to the Australian Yara-ma-Yha-Who, and the Mesopotamian Lamashtu, y El Chupacabra. This is also why zombies are so popular. Death is just scary. So if someone started going around promising eternal life, or salvation, that would sell like hot cakes! You could get anyone to follow you if you promised that all the horrible things in this life were only temporary, and all the bad people would be punished and the righteous would be rewarded. If people think what they do in the present will affect them after they die, you can basically convince anybody to act any way you want. This is how Catholic kings were able to maintain power while their serfs toiled and how Indians could implement the caste system to keep people from aspiring to be anything more than a stall cleaner. Humans are clever like that; we can reinvent reality, which is essentially what every religion does. It's just trying to get people to shut up and be nice to each other, except without ever giving anyone a good reason to do so.

But no matter how you slice it, it's the same pile of excrement; "Do what I say, don't think for yourself, your suffering is only temporary. And don't kill each other. Well, unless you broke the law, or if you were born different, or if you think different, or if you cheat, or if you don't sink in water, or if you weigh the same as a duck, then we'll pay people to kill you in front of a crowd. There's no television yet, so the people need something to entertain them. We'll make a day of it! Praise the lord, thy God (or Gods), King (or kings and queens, and their kids) of the universe! For his (or their) word is law, and the text in their book(s) are divine, and only I get to interpret them, because you might have a different opinion, and then there goes the whole social order! Do you want that? I didn't think so. So just shut up and praise God (or gods), your heavenly father and savior (or the various gods that have specialty jobs, like weather and harvests, the things that actually keep people alive). For believing in God is the only path to salvation! Or his son, you can believe in him too. The poor hippy carpenter one, not the thousands of kings who only claimed to be the descendants of god, and whose people believed them for hundreds of years. They were all fake. Or you can believe in some dude who meditated in a cave and spoke to god, that did actually happen. Don't draw him though! Or some dude named Joseph Smith, who spoke to God AND Jesus! He was one of the few white folks to actually start a religion based on a pre-existing Middle Eastern one, and you know how white people are always right about everything. Except colonialism, slavery, monarchy, and generally how to treat other human beings as equals, but everyone makes mistakes. All of these people definitely spoke to god(s), and they are the only path(s) to righteousness. You can be a good person if you don't believe in God, or can't decide which religion scares you less, but no Heaven for you! You get to go to Hell, that place that was invented after everyone who wrote the Old Testament had been dead for a few centuries. You'll be spending the rest of eternity suffering with all the billions of people who lived their entire lives unaware of our one true religion, doomed to suffer while we rest in paradise, because a couple decades of blind faith are equivalent to an infinite number of lifetimes of either peace or suffering. We know that our books were written by random people we've never met, who lived long before the discovery of all the wonders of God's creation, and God only talked to certain people because he was too lazy to just straight up speak every human being at once like he could because he is your God, King of the Universe, and probably has a microphone lying around somewhere. Don't question his dark and mysterious ways, or down to Hell you go! We know that there isn't anything in our books about black holes, relativity, quantum mechanics, natural selection, plate tectonics, the actual order of the planets, mass extinctions, dinosaurs, cellular respiration, how to use electricity, how to go to outer space, or how delicious bacon is, but that doesn't matter. The word of God, as told by people who totally spoke to him (or them)—because god picks favorites—is about being a righteous person. Unless they disagree with you, or steal your land or food, then slaughter those fools! All hail God, King of the Universe, the source of prosperity and suffering, wealth and poverty, scientists and popes, conservatives and liberals, Jesus and Satan, believers and nonbelievers, and the free will that several animals have but is only important in humans. He rules them all with a divine fist, and sits on his shiny marble throne with popcorn watching from on high while getting high on power watching all his creations blow each other up, rape their fellow humans, kill his animals, and pollute the tiny pebble floating in space that He stuck them on. So don't go trying to solve all of these problems by working together and not fighting, for God has ordained that we all treat each other like shit, speak different languages, make up national borders, build walls, invent nicer gods, and not rock the boat by trying to learn about other cultures. Don't question his methods, and always obey his laws, lest some jackass might try to kill you and send you straight to God for a grade-A scolding. Unless your religion doesn't have a Heaven, or Hell, or if your views on them are different from your neighbor, then you're in the clear! Because the most benevolent thing God did was give you the free will to disagree with one another." Classic religion, just trying to promote peace, yet just can't help but breed intolerance and hate.

But none of the gods dreamt of by humans have had a single thing to do with any of it. It was a combination of all things, the stars in the sky, the microbes in the dirt, and the people who invented all of the world's religions who are the agents of change in the cosmos. You can blame god if you want, but if you just keep passing the ball and never take any accountability for your actions or acknowledge the actions of what you can observe, then your blind faith is just going to turn you into a useless human being.

So is there a god? I'll let you decide that for yourself. And you should let other people decide for themselves as well. Otherwise you are sort of interpreting the whole "life" thing incorrectly, and contributing to death, chaos, and world suffering, so shame on you. Stop telling people what to believe and just enjoy your short life that the universe is completely apathetic to. Then, at the end of days we can all stand in purgatory together, and as the angels gaze down upon us; we can all wonder why the one true faith was Elvis all along. Or Yahweh. Or The Big Lebowski. Or maybe we'll all get reincarnated as sheep in the next life, because clearly Yahweh's son thinks we need someone to shepherd us like little lambs all the time, because apparently we're all just toddlers to Him. Or maybe the Sun turns into a red giant and we engineer ways to integrate our consciousness into computers and voyage through space until the heat death of the cosmos as everything fades into blissful oblivion. Or maybe we'll hop universes and just start a new big bang. It's anyone's guess, and nobody can prove one way or the other which one will happen. Humans, and our mythology, are like a really good sitcom; you never know what's going to happen next! Unless the puritans were right, then everyone who came before 1558 and after 1692 will be in partying together in Hell. So it goes.

Tinkering with Brain Chemistry

Speaking of parties, how about that hundred billion neuron fiesta going on between your ears? I know we've already discussed the vast complexities of the neural network, but think for a moment what the consequences are for such complexity. With over a hundred billion individual pieces, it's quite remarkable how our brains are able to maintain order and balance, and it should come as no surprise how easy it is to offset that careful equilibrium. It's more accurate to say that the chemical balance of the brain is constantly shifting based on environmental conditions, after all adaptability is the key to survival. For most living creatures, their neural network isn't complex enough to change too much without putting those creatures in mortal danger, as the most primitive neurobiology found at the stem and base of animal brains mostly just regulates autonomic functions. However, add more nerves, and eventually you get a creature that can consciously reflect on its environment and say, "I like this, it must be good for my survival" and "I hate this, it might kill me". These natural responses manifest in our emotions of fear, love, pain, pleasure, happiness, sadness, anger, etc. Normally, the neurochemicals that balance these moods respond to things we experience that may be the difference between life and death, or at least what we perceive to be a matter of life or death, like a stressful job interview. However, since at the end of the day our entire existence is based on nothing more than chemical interactions stimulating the flow of ions, it's no surprise that the intake of certain substances can masquerade as our natural brain chemicals. In this way, drug use is one of the hallmarks of an intelligent animal such as ourselves.

Like most of the strange things we do, more complex animals have been observed trying to feel some sort of way by partaking in psychoactive substances. We all know that domesticated cats love catnip, but so do their big cousins. Dolphins seek out a buzz by rubbing their faces on puffer fish, so you could say they "huff"le puffs (and I'm not apologizing for bad wizard puns). Elephants freaking drink like sailors and do hallucinogens if given the opportunity. Birds crush up ants to get high. Horses and reindeer eat magic mushrooms too, which may explain the Santa Claus story. Bees drink booze and even get actively punished for it by their colonies, being reprimanded by getting kicked out of the hive. Drug use is just as common in the human animal, many of those drugs even inspire spiritual experiences central to many indigenous cultures and religions. Interestingly, a large number of psychoactive drugs come from fungi. Yeast of course is the source of everyone's favorite toxin, alcohol. In addition, ergot causes hallucinations and has lysergic acid (LSD) inside. Since it grows on grains, it has been widely speculated that eating grains with the fungus led to the hysteria that caused the Salem Witch Trials. Psilocybin mushrooms are yet another popular example of the strange hallucinogenic properties of fungus.

But there are plenty of plants that produce mind-altering chemicals as well. For instance, take the coco plant native to South America. As the derivative of concentrated cocaine, natives have chewed this plant for centuries as a natural source of energy, without addictive consequences I might add. The most dangerous aspects of the substance only became apparent when the active chemical was concentrated into the now infamous white powder. South America is also home to the powerful hallucinogen Ayahuasca, a combination of roots and leaves that allows the body to absorb the Dimethyltryptamin (DMT) within them, DMT being structurally similar to the neurochemical serotonin and something that may occur naturally in many living things. This substance is brewed for many spiritual ceremonies as a means of achieving enlightenment and spiritual peace. There is even a hallucinogenic honey found in the Himalayas, crafted along vertical cliffs by bees the size of one's thumb that pollinate a plant with psychoactive chemicals. Natives there have braved a countless number of extremely dangerous climbs to get to the sacred concoction. So despite the way they are so often demonized by snobby prudes, drugs are a natural byproduct of our complex neural programming. People have been actively seeking a way to get high since time immemorial, as have any other animals intelligent enough to want to have a good time. Obviously that doesn't mean that some drugs aren't extremely dangerous. Just keep in mind that the people who are afraid of hallucinogens are usually the same people that forget that aspirin, coffee, and margaritas are just as much psychoactive drugs, and that many illegal drugs are far less addictive and less dangerous than legal ones.

To understand what makes some drugs different than others, we must first understand what they are and how they affect our brains. Drugs work by either increasing or decreasing the intake of neurochemicals between neurons. These chemicals include dopamine (which affects alertness), norepinephrine (which affects concentration), serotonin (which helps us relax), and several others. All of these have effect on mood, memory, and overall behavior. Dopamine especially is a major player when discussing recreational drugs. It is the so-called "feel good" hormone that is released by the brain anytime you do something that is crucial for survival, such as eating sugar, salt, and fat, or when procreating. This is a major reason why doing drugs feels so good, because it has the same effect chemically as eating pizza or having an orgasm. But the brain is far too complex to be so strongly influenced by just one chemical. Every recreational drug has a unique effect on the brain.

Alcohol, despite being a party drug, is a type depressant. It inhibits GABA receptors, and also keeps excitatory glutamate from binding to synapses. It affects the frontal lobe of the brain, the area that makes us most human, and as a result hinders memory and impulse control. THC mimics anandamide, a neurotransmitter that hinders the release of dopamine. Unlike anandamide, THC takes time to break down, which is why the feeling of relaxation and poor short-term memory is perpetuated with the drug's influence. Lysergic acid diethylamide mimics serotonin and binds to areas that usually interact with serotonin, sometimes hindering and sometimes exciting receptors all throughout the brain. One main area of activity is in the visual cortex, hence the hallucinations. LSD has also been shown to help relieve cluster headaches and depression, even with the hallucinogenic effects removed. MDMA is also similar to serotonin. It gets absorbed by serotonin receptors, but also kicks out serotonin stored inside nerves, thereby trapping it between synapses. It's derivatives, including molly and ecstasy, alter mood and sleep, making them popular amongst those who like to rave. Heroin works in a way similar to THC by mimicking natural opiates that inhibit dopamine, but works in different parts of the brain. Most notably, it has effects of reducing pain, which is why morphine—a very pure form of heroin—is one of the most common painkillers used in hospitals. Unlike THC, which is almost completely non-addictive, heroin interacts strongly within the reward pathways in the brain, making it habit forming. Cocaine traps dopamine in the synaptic cleft, keeping it from being reabsorbed and causing a runaway increase of the feel good chemical. It also concentrates highly in pleasure centers of the brain and in areas that control movement, sometimes leading to fidgety behavior. Cocaine is traditionally used by snorting, but can also be smoked in the form of crack. Absorbing the substance as a gas through the lungs is far more efficient than breathing it as a crystal through the nose, thereby amplifying its effects and making it far more dangerous. Caffeine works in a similar way, but without the euphoric effects and to a drastically lessened degree. Nicotine mimics acetylcholine, binding to receptors and opening them, thereby sending signals that eventually release a temporary increase in dopamine. Methamphetamine works in nearly the same way as cocaine, but is far more active in the brain's reward pathway, making it among the most addictive substances. The key differences between meth and cocaine are that meth also causes the release of dopamine in addition to trapping it in the synaptic cleft, and it takes far longer to break down in the body.

As the drugs become more addictive, users run a greater risk of overdosing. This occurs because of our brain's remarkable pliability. Unfortunately, by adapting to altered levels of neurochemicals, the brains of drug users need to continually increase their doses in order to achieve the same high. Through extended use, an addict with a high tolerance runs ever-greater risk of taking in fatal levels of their drug of choice. But overdose doesn't just occur when people build up a tolerance to the point of the substance becoming toxic. More often it happens when people build up a tolerance, take a break, and then try to get high again using the same dosage. Their bodies are no longer used to the high concentrations of the drug, and so they O.D.

Addiction and dependence often arise as a response to hardship and sorrow, and there are few greater sources of sorrow than war. Some say that more U.S. servicemen in World War II that started smoking cigarettes during the war died from smoking than died from combat. This is debatable, but according to the world health organization and American cancer society, around six million people die annually from tobacco attributed to cigarette use. Even taking into consideration the fluctuation of use over time, this would still account for more total deaths than every war in the twentieth century combined. War also causes dependence in more severe substances. For instance, during the Vietnam War many U.S. soldiers began using heroin. Citizens were afraid that soldiers would return home as junkies. However, this didn't happen. After returning to a warm and loving home, the vast majority of soldiers who were addicted to heroin kicked the habit. Of the apparent 15% of total soldiers who became users, 95% didn't relapse when they returned to American soil. Not only do these examples illustrate just another one of the tragedies and horrors of war, but they also show us how complex the issue of addiction is.

When it comes to certain substances, such as cocaine, heroin, meth, and nicotine, there is no question that drugs can have addictive properties. However, just because some substances are more prone to addiction than others, blaming addiction on drugs alone is completely false. People can become addicted to many things, including social media, athletics, food, pain, drugs, sex, and rock 'n' roll. Okay, I was mostly kidding about the last one, but the point still stands. People can become addicted to anything; it just so happens that drugs are a consistent and simple way to fill the voids that plague the soul of an addictive personality. When people don't get the stimulation they need from "normal" social activity and loving relationships, drug dependence becomes far more likely. The dangers of drug addiction don't come from drugs themselves, they stem from deeper problems in society. When people feel hopeless and alone, that emptiness creates an abyss that those people will fill with anything they can. Everybody needs to get high somehow, but if one relinquishes themselves to substance abuse, they take themselves on a dangerous path to an early grave. As with any form of despair, perhaps the greatest way to fight the war on drugs is to create a society that doesn't ostracize people for falling into drug dependence, and instead encourages them to understand their desires and find purpose to their lives in a calm and understanding way.

When it comes to what makes drugs dangerous, developed nations tend to be quite backwards about which substances they deem suitable for legalization. More often that not, benign drugs are deemed dangerous, while the ones that truly bedevil society are perfectly legal. Specifically, I find it extremely strange that alcohol and nicotine have become the two most socially accepted drugs, despite them being among the most dangerous and harmful.

What does alcohol (a.k.a. ethanol/ethyl-alcohol) look like? It's an incredibly simple molecule, made of two carbons stuck together. One of them has three hydrogen atoms attached in a pyramid-like shape, and the other has two hydrogen atoms and one oxygen atom, which itself has to have a hydrogen to remain stable. It sort of looks like a no eared dog. Keep in mind also that ethanol is just a molecule. Compared to a single cell, it's like an ant on a boulder. It's no wonder this puny little molecule can have such influence on the brain. Relative to other neurotransmitters—which have several dozen atoms—alcohol can readily fit in and interact with transmitters in the brain. Not only that, but because oxygen is so greedy with electrons, ethanol can easily react with structures in the body by taking on or getting rid of hydrogen atoms. In other words, it can act as both an acid and a base. This is the reason why alcohol is essentially poison to the body, which is why it immediately tries to remove alcohol when ingested. But despite being one of the most dangerous drugs on the market, alcohol is so ingrained in human culture that removing it would be sacrilege. Wine is even used by Catholics as a sign of divinity. For god's sake, one of Jesus's so-called miracles involved getting people drunk. But drinking is a universal religious concept. According to historian Janet Martin's *Medieval Russia*, The Kievan Rus may have chose Byzantine Christianity because of the Islamic faith's prohibition on alcohol and not Judaism because the god of the Jews had permitted his chosen people to be deprived of their country in the 11th century. Drinking is so culturally significant, that the only amendment in the whole of the United States constitution to ever be repealed was the one prohibiting alcohol. It's literally the only time in history where the constitution changes its mind. Granted, the wives who called for prohibition did have fair grounds in doing so, as alcohol related domestic abuse was rampant throughout the eighteenth century and continued into the early nineteenth. People were abusing their alcohol privileges, so it's no surprise that they ended up being revoked. Nevertheless, prohibition only lasted from 1920 to 1933, ending with F.D.R.'s new deal, when Americans all needed a drink. If nothing else, prohibition should teach us that if people want to get high, they will find a way, and simply making something illegal is a poor substitute for teaching people to use drugs responsibly. This is where the movement against cigarette use was far more successful that prohibition, that and the fact that people don't like getting lung cancer.

Let's not forget that there are far more drugs sold legally that can be just as dangerous and addictive as their illegal counterparts. The National institute on drug abuse cites that drug related deaths from prescription drugs are higher than heroine or cocaine by a factor of ten or so. While this is certainly associated with strict regulation of dangerous substances, it is also striking in that it shows how people can still get high and die legally just as easily as they can illegally. This has begged the question of whether or not strong drug regulations are ever useful. Again, if people want to do drugs, they will find a way. So perhaps preventative measures are more important. Drug abuse and addiction has its roots in the complex neural networks that we are all born with. Understanding these problems require unbiased understandings of psychology, neurology, and chemistry, and it is imperative that researchers share their knowledge with the public to help everyone understand which drugs are dangerous, and which ones are not. If people are blindly afraid of drugs and the people who use them simply because they're illegal, then we can never hope to solve the problems related to those drugs.

It is clear that the public perception of drugs is heavily skewed from reality. This is best illustrated in data collected by Dr. Jack E. Henningfield of the National Institute on Drug Abuse and Dr. Neal L. Benowitz of the University of California at San Francisco, cited on *DrugWarFacts.org*. They corroborated national surveys and case studies from varied sources including the U.S. Department of Health and Human Services and the Surgeon General in order to compare the dangers of popular drugs. They related six drugs—nicotine, heroin, cocaine, alcohol, caffeine, and marijuana—based on five criteria: dependence, withdrawal, tolerance, reinforcement, and intoxication, rated on a scale from 1 to 6. Among the selected substances, marijuana was ranked as creating the least dependence, causing the least negative effects from withdrawal, and having the least effect on creating the tolerance that causes so many addicts to overdose. It was the only substance to receive the lowest rank of a 1 in those categories, with caffeine just slightly more dangerous with ranks of 2 in those same categories. As expected, the greatest dependence was caused by heroin at a 6, followed by nicotine as a 5. But unexpectedly, alcohol received a 6 as being the most intoxicating drug with the greatest symptoms of withdrawal, because like I said, it's literally poison. All this data and far more are available to the public, and with drug use being such a weighty topic I encourage everybody to conduct their own research, as with all cultural phenomena.

This brings us to the end. Not just to our conversation about all of humanity's weird cultural habits, but our discussion on all the major fields of academia. I hope I was able to demonstrate the importance of knowledge and how everything in the universe that we know is truly intertwined. Our cosmic and evolutionary history shapes our culture, our observations and logic reshape our understanding, and our language accelerates our learning ever further. But we still need one final thread to tie it all together. We must understand how it is that we can know anything at all.

9 | Cosmic Consciousness

"We are the cosmos made conscious and life is the means by which the universe understands itself."
— *Brian Cox*

"Independent inquiry is needed in your search for truth, not dependence on anyone else's view or a mere book"
— *Bruce Lee: Wisdom for the Way*

Let me pose a question to you. It is one that pervades my every thought, during every moment, and drives everything I do. How is it that we, an aggregate of matter and energy that took billions of years to develop, being made of pieces that have been inside stars and dinosaurs, are able to contemplate our own existence? How is that even possible? Many great thinkers before me have postulated that there must certainly be something more than just our physical bodies, and that our ability to think and reason can only be explained by our having some sort of ethereal soul, where our being truly lies. Some might contest that it is impossible to quantify the human experience in a logical way. I find this explanation to be wholly dissatisfying, for none of these great thinkers knew what a neuron was, nor did they know how supernova occur. They were doing the best they could; there is no doubt about that. However I can't help but feel that the likes of Plato and Descartes would have refined their worldview had they been given access to a microscope or space telescope. Nietzsche probably would have been mostly unchanged, but my point still stands. Though I know not how it happens, I postulate that everything we think, say, and do can be described using science and mathematics. This is what the sum total of my life experience thus far has led me to believe.

It's abundantly clear that there is something transcendent about the experience of being alive. Some people call it spirituality, and regardless of whether or not a person believes that there is such a thing as the supernatural, most if not all people can attest to having profound and deeply spiritual experiences. Some have it when they experience the loss of a loved one. Or perhaps they might find that feeling when they fall in love. Others experience it when they go through the miracle of childbirth. Some have powerful experiences when they witness the glory of nature. When one reaches the summit of a mountain, experiences a hurricane, a flood, a plague, or sees the Earth from the Moon, they will almost certainly describe the experience as being larger than life and existential beyond words. For many, this is the reason why we study astronomy as well. In fact, I would go so far as to say that seeing a patch of sky the size of a salt grain filled with dozens of galaxies tens of trillions of times larger than our entire planet Earth is the most profound and meaningful experience a human being could ever have. These are the moments we live for; significant not because of how many breathes we took while experiencing them, but because they took our breathe away. It is this feeling of a deep spiritual connection with the universe that spurs us to continue muddling through our otherwise torturous existence, to the point that we would actually risk our lives for it. It manifests itself in all of the ways we find meaning in our lives, be it through religion, thrill seeking, sports, family, friendship, work, or even war.

However, if I have proven anything to you during this conversation, I hope that I have demonstrated that everything we experience, from the mundane to the majestic, is nothing more that the natural laws of the universe playing out inside of your skull and beneath your skin. More importantly, this fact shouldn't devalue our lives or our existence. How remarkable is it that not only is our being traceable to the stars, but that very the laws that create chemicals, light, and everything in the universe are the same ones that lead to life and consciousness? Not only that, but despite the near universal pessimism of human beings who convince themselves that such matters of spirituality are beyond human understanding, some people have taken up this seemingly impossible task to understand why we exist. It is thanks to those brave few who had the courage to ask questions and persistently force nature to give up her secrets, that we have gotten out of the savannah and now soar over the Earth on metal wings powered by the very same carbon that plants and animals used to exist millions of years ago. Generation by generation, we have chiseled away at the world's mysteries, and with every new discovery, it becomes increasingly apparent that we are not special agents in the universe, but that humans and the universe are one and the same. You are the universe looking back at itself.

There are no ifs or buts about it. Our experiences are based on our neurons, cells that communicate with bundles of elements found throughout the cosmos. We all begin life as one of these cells, that single cell multiplies, and its replicons specialize into all of the unique organs and cells within our bodies, based on nothing more than the activation of specific genes that regulate when pieces are produced, how much is produced, and where those pieces go. We live out our lives as our outer cells die and replace themselves, and we must constantly eat and breathe to allow new ingredients to replace the ones we've lost. As we grow older we experience new things, change our minds, and if we're lucky even our convictions. We grow, and we change, until the person we were in our youth is nothing more than a prototype of our current selves. The body is always being replaced. They say that nerve cells last as long as we do, but the fact that water and electrolytes need to be replaced is proof that isn't exactly true. It may also be the case that neurons have special repair enzymes. It seems unreasonable that they don't need maintenance over the course of a lifetime, using a quarter of our energy and all. Just because individual cells don't die doesn't mean they are unchanged at a molecular level. But when we do die, bacteria and fungus recycle our elements and use them to live. Our existence is transferred throughout the living things on the planet; plants use them to develop new seeds, animals eat those plants and predatory creatures—including other hairless apes like us—may even eat those animals. Some of those ingredients may be integrated into the inanimate rock of the Earth, right up until the Sun engulfs our planet four billion years from now, and expunges those elements back into the vacuum, over and over again until the universe expands so far that all matter and energy is evenly distributed and nothing can live ever again.

I see that the universe is expanding, and as it continues to expand time is always moving forward and nothing is still. Your heart's constantly beating, your brain's constantly thinking. The Earth continues to spin and fall toward the Sun. The Sun does as well, but the solar system as a whole also falls with the Milky Way toward Andromeda, in between the clusters of billions of galaxies thinning apart as the universe grows dark and cold. We are young, and then we are old, and then others are born, and they grow old. The information of our generation lives on in the next, and over time we become more aware of what it means to be the universe. Things frail apart if they get too hot, but they can survive the cold as long as they can siphon heat from the environment, which ultimately derives from nuclear furnaces powered by gravity. Time goes forward, and that's why changing is synonymous with living. And so I wonder, what would happen if the universe were collapsing? If we are gaining time now, it would seem that in a collapsing universe, we might run out of it.

There is nothing good or bad about change, and no indication that we have anything resembling a "soul" that separates us from the material world. All things in existence are subject to change, even our spirit, and when we die, all of the complex processes that allow us to experience the world and know ourselves halt, and all of our pieces that allow those processes to occur are broken down and assimilated back into the universe in which they have been a part of since time zero. Once we understand that, then we can look clearly into the future and delve deeper into the past. To illustrate this point, take one of the most unique and breathtaking events a human can experience: a solar eclipse. Due to their rarity, and the fact that the Sun is the reason all complex life is possible on the Earth, seeing the Sun disappear behind the moon is one of the few experiences that can never be fully elucidated with words. But that doesn't mean we can't try. The best description of this deep spiritual feeling that I have found is in Pink Floyd's song *Eclipse*. To paraphrase (because copyright laws are a pain in the neck):

Everything that you touch, everything that you see, everything that you taste, all you feel, and everything that you love, and everything that you hate, all you distrust, all you save, and everything that you give, and everything that you deal, and everything that you buy, beg borrow or steal. All you create, and all you destroy, and everything that you do, and everything that you say, and everything that you eat, and everyone you meet, and everything that you slight, and everyone you fight, and everything that is now, and everything that is gone, and everything that 's to come, and everything under the Sun is in tune, while the Sun is eclipsed by the moon

The reason so few experiences are comparable to an eclipse is because of its totality. Everything you have ever said, done, or cared about, all of them take place on one planet. Like an eclipse, the universe is indiscriminate in its actions. The laws of nature apply equally to all things. Your worst enemies and the loves of your life, all of them only have value if you give them value. There is no more humbling experience than realizing that everything that has ever mattered to you and everything that has ever mattered to anyone all took place on a tiny speck of dust drifting around a medium sized star like any other, inside of one of many galaxies, in an infinite universe all made of the same thing. It's a very good reason to not take anything too seriously or get stressed about every little thing. Again, don't get the idea that you aren't unique. There's a difference between being unique and being special. There isn't a thing in the known universe as complex as that organ in between your ears, and you should be proud that a small piece of the universe is organized in such a way that octillions of tiny ingredients can come together to contemplate their own existence.

Consciousness is the most remarkable phenomena in the universe, and perhaps the greatest mystery. At this point, some people might begin to question their very existence. This has happed to many people throughout history. Some have even come up with the clever idea that all existence is a simulation. I quite like the idea, and I've heard that there are many compelling arguments to support it. However, when it comes to predicting anything outside of the universe, I withhold judgment until both theory and experiment are able to produce repeatable and falsifiable evidence for such an existence. Regardless of the origins of our universe or our existence, there is one thing we know for certain.

There are few expressions born from the minds of philosophers that make it to the mainstream, but René Descartes managed to do this, and it is likely because his idea so concisely describes reality. Descartes is most famous for his philosophy on skepticism. He realized that because our senses can very easily be mistaken, and often give us false information about reality, he could call into question everything that he thought was real. Not only that, but he realized that there was no reliable way for us to know whether every other human is really another sentient being or just a preprogrammed shadow. He stated that as far as we know, we could just be nothing more that brains in jars, or beings tricked by a supernatural deceiver into believing a false reality. This is more or less the plot synopsis of *The Matrix* film trilogy. It could very well be that our physical bodies are nothing more than an illusion. Descartes thought experiment, then, was to assume that everything he knew was false in an effort to understand fundamentally what it means to "be", what was undeniably true. What he realized was that even if everything he knew was wrong, the one thing he could know for sure was that he existed, because of the simple fact that he was able to think and ask the question of whether he existed in the first place. His conclusion is summarized in one of the most famous phrases in metaphysics: *Cogito ergo sum*: I think, therefore I am.

So we know that we exist, and we are fairly safe in assuming that most of what we experience is real as well, insomuch as we need to assume that if we're going to eat and make money. But how do we do that? How do we do that thing were we know we exist? Put another way; how does thinking? Here's one method that I have thought of. I call it The God

Experiment, because a lot of people like to give god credit for everything, and the essence of knowing that we exist is in understanding why we are able to know that we exist. First, define what the word "god" means to you, and then think in isolation. Whether you refer to it as meditation or praying makes no difference. Just find a place where you don't get any other sensory input so that the only thing you experience comes from your brain. Then close you eyes and cover your ears. Allow your imagination to go free for a few moments. Think about someone you know, a place you'd like to go, or dinosaurs, anything you like. Think of your favorite song. Notice how the thought of a melody or a lyric causes a cascade of other thoughts, of feelings and people and places, and how each of those places and people and feelings remind you of other thoughts, until you see the connections between everything you've ever experienced. Finally, say a sentence without making a sound. Speak without speaking. How are you doing that? There is nobody to talk to you; there is only your brain. Everything that defines "you" is just sitting there, inside your skull, doing everything that you do. You aren't your face, or your deformities, or anything else that you can see in a mirror or an x-ray. You are a bundle of nerves attached to a non-thinking body comprised of trillions of individual cells almost completely under your command. You are an organ, a clump of fat and salty water, made of billions of neurons firing between each other, sending messages that result in your ability to think. Everything that you know, everything that you feel, it's all there, floating in your skull and wondering what the hell is going on. Well, besides the 500 million neurons of the enteric nerves in your digestive system and the nerves lining your outer body. In a sense your entire body is thinking. I even hesitate to definitively claim that your non-nerve cells and the trillions of bacteria lining your body don't also think considering how they also communicate via chemicals. Try this experiment out to see where your thoughts take you. Use it to refine your resolve to live, and define what it means to be you.

The main difference between humans and the rest of the universe is that humans care illogically about trivial things for no real reason other than to survive while the universe cares about nothing for reasons that can be described coherently and logically. The universe is gray, its very foundation being based on quantum uncertainty and relativity. The Earth is titanic to a mountain, but miniscule to a star. You are ancient to a peach, but infant to a tree. In many ways relativity is a philosophical truth about reality as much as a statement about time and space. It is quite exhilarating to be a part of such complexity, necessarily intertwined with all of existence, something monks might refer to as the Tao. In many ways this isn't a new idea, but it is something that is useful to remind ourselves every once in a while. Remembering that all things are both unique and one is a wonderful way to bring fulfillment to any life experience. When you look long into the universe, the universe also looks into you.

I'm sure that even this isn't enough to convince some people that there isn't some supernatural component to our lives, and that's perfectly fine. My aim is only to put a little more science in conscience. But these ideas go to the very core of our being, and so they may be difficult for some to grapple with. Thinking about thinking is much more difficult than just thinking, and it's also terrifying. In that regard, conscience makes cowards of us all. But are mediums, ghosts, the force, or ESP actually real things? Can people channel the unity of nature to perform miraculous feats? Some might find answers in the strange phenomena of entanglement and other quantum effects. Can the brain entangle itself with the environment and with other people? While this is certainly an area that could provide interesting research, it still seems quite farfetched. These types of questions are about as useful as asking how many angels can fit on a pinhead or who would win in a fight between a horse sized duck and a hundred duck sized horses. For any matter, one must be careful applying the laws of quantum mechanics to the macroscopic world. The famous thought experiment of Schrodinger's cat illustrates this idea perfectly. Nobody would expect a real cat to be both alive and dead, like a particle that is both spin up and spin down. The problem with applying the strange reality of the very small is the sheer number of small objects that make up the world we experience. While subatomic particles may become entangled, it seems ridiculous to imagine that an object the size of the brain, made up of octillions of subatomic particles, could entangle with a completely separate object separated by octillions of particles of air and the trillions of microscopic organisms between the brain and another object. Quantum effects simply become too diluted on the large scale to justify any kinds of supernatural event, especially with other brains. Ghosts and the like are far more easily explained by fungus making your house creak.

Or how about this question: do we have free will? Well, kind of. It seems obvious that we do on a daily basis, but plenty of people would disagree. You can manipulate the inanimate matter around you, like your clothes and your body. But there are a lot of independent bundles of matter around you, a countless number in fact. The living ones can steal resources from you, and weather or meteors could kill you randomly. Even your brain can't be aware of all the microorganisms buzzing around your body, nor genetic mutations in the DNA of a single cell. Meteors may have been put in place at the birth of the solar system and gamma ray bursts may occur anytime due to a quantum quirk in the heart of a star. But in the twenty-first century we can make atomic guitars and hypothetically deter meteors, which is a far greater command over nature than the dinosaurs had. The whole history of humanity seems to revolve around our desire for greater control of the universe, in a bid to gain ever-greater freedom and autonomy. In truth, this has more or less been the story of life as whole. The perception of free will is truly fascinating, because it begs the question of whether the entire universe itself has a will of its own. Again, some might draw insight from hypothetical ideas rooted in science. Quantum superposition, inflation, and string theory all independently suggest the existence of a multiverse, where every outcome occurs and where entire universes play out under different laws. It may very well be that Schrodinger's cat really winds up alive in one universe and dead in another. In the same way, we may not have free will if we are tethered to one universe with a set outcome, but at this time such a claim could be nothing more than speculation. In any case, I would conjecture that the reason people so often feel a connection to a higher power in the cosmos is simple. Of course you feel connected to the universe, you're inside of it. Not only that, but you're literally made of ingredients found throughout the cosmos. I would hope the universe would be connected to itself. That doesn't mean there's anything supernatural going on that at least somebody couldn't comprehend, and it certainly doesn't mean the universe as a whole cares about you or your well being.

Nevertheless, it is quite remarkable how rare life is in the universe, especially intelligent life. The Drake equation is famous for approximating these odds, but we're going to make our own estimate here, based only on the sizes of objects, the volume of space, and the percentage of stuff within it. On the Earth, life has managed to creep its way into every corner of the atmosphere, miles beneath the surface, at the bottom of the ocean, and even in the saltiest, hottest, and coldest corners of the planet. That's why life will continue just fine even if humans kill themselves off themselves with climate change or nuclear holocaust. However, despite it's pervasiveness throughout the outer fringes of the planet, the lithosphere is still just a thin layer. It is often said that if one could hold the Earth in their hand, with all the mountains and canyons, it would still be smoother than a bowling ball. Our atmosphere too is deceptively thin. On average, the combined thickness of the outer crust and atmosphere is a mere 50 kilometers, about 128 times thinner than the radius of the planet.

Assuming every single bit of atmosphere and crust has something living in it, then the amount of life on the Earth is equivalent to the total volume of the Earth's crust, about 25 billion cubic kilometers. Now, there may very well be life elsewhere in the universe, and in truth it's quite likely. Hydrogen is the most abundant element, and oxygen is the third, meaning that water must be one of the most common chemicals in the universe. We observe evidence of liquid water even on the likes of moons such as Europa. Considering the abundance of planets around stars, it doesn't seem the least bit of an exaggeration to assume that many of those planets also have moons. But despite the number of potential worlds with liquid water, one must also consider the percentage of heavy elements needed for life to exist. Not only is less than one percent of the universe composed of elements heavier than helium, but much of those elements don't even seem to go into forming planets. For instance, the Sun has about 466 and a quarter Earth's worth of iron inside of it, and iron is one of the least common heavy elements, relative to an already miniscule percentage of atoms heavier that helium. Coupled with the fact that most observed planets are either too gaseous, too hot, or too cold to have liquid water, I think it's safe to assume that even if millions of worlds have life, it would take up so little space and would be rare enough to use the volume of Earth's life as an approximation for life in the observable universe (especially since we don't have evidence for life anywhere else). The radius of the observable universe is about 46 billion light years, or 4.4 x 10^23 kilometer, which means the volume is 3.6 x 10^71 cubic kilometers. So, in terms of volumetric space in the observable universe, all of life is about .000000000000000000000000000000000000000000000000000000001% of everything that exists. Soooo yeah, life is pretty rare, even if my math is off by a factor of a million. Essentially everything that exists is cold (or hot), dead matter (or energy). But, remember that while basically 0% of the universe is alive, 100% of living things are made of the universe!

On to the question of sentient life, something that is apparent in many creatures, but none remotely close to that of humans. To understand why, we can just look at the case study of life on our little grain of dust. As we know, sentience occurs when an organism has a sufficiently high number of nerve cells with sufficiently high numbers of connections. Animals with smaller brains are slightly less aware of themselves than animals with larger brains. But that's not all. Intellect is also based on the number of connections between nerves, and the number of connections between organisms with brains through language. So how many neurons are in a thought? Are different thoughts different amounts of nerve cells? Under what circumstances (e.g. nutrition, genes, education, and experience) are bundles of nerves able to form with enough dendrites to form a self-aware organism? While I can't answer those questions directly, we can get hints by observing the rest of the animal kingdom.

What do whales do all day? Not only do many of them have bodies far larger than humans, but they also have brains two or more times larger than we do. With so little of the ocean explored, whales could have hypothetically started whole whale civilizations underwater, and simply turned off the lights every time humans stuck their heads in the ocean. Unfortunately (or fortunately?), this is quite ridiculous, for exactly the reasons we've already laid out. It doesn't matter if their brains are bigger, because they don't have hands to make tools or a language complex enough to share ideas. They certainly have a language, and even seem to demonstrate local dialects, but nothing to the extent of *Hamlet*. They might be smart enough to memorize the whole ocean landscape, but I doubt they could internalize Riemann sums. Nobody will deny that Ahab was certainly crazy for chasing after a hyper-intelligent sea-hippo, but the number of whales hunted by real humans is proof of how crucial our subtle advances are, beyond just our adaptive brains. A similar argument can be made for animals with different neurology, most notably invertebrates. Some of them have the ability to regenerate their neurons. Researchers at The University of Illinois study slug brains, and have shown that they grow throughout their lives. This may be because of genetics, and because they have fewer neurons to work with, having only tens of thousands as opposed to our 80 billion or so. We also know that octopi are intelligent, despite how their 500 million neurons are distributed al throughout their bodies. In many ways, an octopus's head is less clever than the sum of its legs. But again, the fact that these animals tend to communicate less differentiates them from humans. Instead, what if we look at the animals humans have decided to interact with the most, via domestication?

Dogs lend themselves well to human interaction, which may be why they were domesticated before agriculture. For one, humans and canines are both hunters with a strategy of making social packs, and so could naturally form a symbiotic relationship, something that is common amongst all animals, even frogs and spiders. In that way humans and wolves could have fed each other. Modern dogs have also evolved most closely with humans, based on how they seem to be one of the few pets that are able to eat the same things people do. They also respond most readily to human commands, which means they comprehend human language. This is unlike cats, which just hung out with people because we had excess food being eaten by mice, and further unlike dolphins, which we have to imprison in theme parks to do our bidding. Although it isn't all bad, the dolphins do say thanks for all the fish. Furthermore, cat's have still become quite close to humans, despite their aloof nature. After all, cats and dogs are the only non-primate animals that seek eye contact, a sure indication of their symbiosis with humans.

Naturally, herd animals came after agriculture, when people had abundant crop supplies that herds likely gathered around. This would include cats, which are also useful to people because they hunt diseased pests that could have threatened villages. All the other animals—that could still kill us, didn't want to be eaten by us, or are just generally uninterested in our antics—stayed in the so-called, "wild". The poor ignorant animals that weren't terrified of humans became today's burgers, breasts, and bratwursts. Their only crime was being delicious. Of course, by the twenty-first century a person could conceivably eat any organism without much resistance. We could even genetically alter them to eliminate venom and parasites. Or, even better, we could just leave those suckers to eat each other on Earth while we eat 3D printed steaks in space, with our cats and dogs. It seems that none of the other animals quite need complex intelligence to survive, but it also doesn't seem like the traits they did evolve with lend themselves to all the combinations of traits that allowed us to develop our brains, namely collaboration, language, free appendages and the ability to digest plants, animals, and fungi. So despite the fact that we don't have the largest or most unique brains, the combination of all our physical traits makes us the ideal animal to adapt using critical thinking. Still, we can further differentiate our intelligence by looking at the way we spend our time.

All animals also need to sleep regularly. Some whales sleep vertically, and dolphins might have half of their brains asleep at any given time. Octopus need to sleep, and they even change colors during REM cycles. Herd animals that live in groups sleep on their feet most of the time and only lay down when they have protection. Lots of animals have to come up with new and exciting ways to be unconscious and safe at the same time. That's the thing though, all animals sleep, and the amount of needed sleep is often tied to their intelligence. While most animals rest to conserve energy, many of them also need to be aware of their surroundings to look for danger, and as a result may only need to sleep in short bursts. It would seem that only the most intelligent creatures need enough beauty sleep for their brains to process information and allow their bodies to recover. Or perhaps, only the animals whose environments allowed for extended rest were able to grow more intelligent. That's the other thing, if an animal isn't eating or sleeping, only then can it think in depth. That's true for us as well, we just so happen to be really good at hoarding food. Most animals don't have time to ponder existential questions, because they have to spend their waking hours eating or hiding from predators. Humans are weird in how we end up sleeping too much or two little and need different amounts of sleep based on our mood and work schedules. But that's why smart hairless apes can have cars and Internet, and stay up all night lost online on a work night, then be groggy all day and have to run on coffee, and then be jealous of the animals who have it way easier without their jobs in place of the constant threat of being eaten.

Here's another thought experiment; if humans leave the Earth, the remnants of our society would crumble and erode away. Animals would retake the land humans stole, and refill the niches. Would Earth be like a giant modern Zoo, where people could hunt exotic animals as their populations flourish without pollution? Would there be people left behind who lived in hunter-gatherer societies? Would the uncontacted societies living in the jungles radiate out of heir homes to re-colonize the planet? Would they learn how to use our old technology, fly into space to catch up to us? Would they find a new society, or the wreckage of the first humans to travel to space? What if a new, more intelligent species develops, just like the mammals did after the reptiles, who dominated after the fish. Would they develop agriculture and weapons of mass destruction far faster than the humans did? Do we want that competition, or could we work together? That's essentially the main question behind every alien invasion movie as well as the experience of every human being who has ever come into contact with another group of people they previously didn't know existed.

Unfortunately, the gut instinct of most people is to assume that everyone else is different than they are. While this is true superficially, that line of thinking can easily lead people into thinking that they are superior to others. If this dehumanization is taken too far, people might end up seeing each other as being entirely alien from one another. It's more accurate to look at other people as different versions of yourself, displaying a life you could have lived, better or worse. That should also be exciting though, because you can pick and choose between attributes you enjoy from other lifestyles, if you have the opportunity. All you need is to do is learn about as many different people as possible. All people are knowledgeable in different things, even though that doesn't necessarily make them intelligent. In reality, intelligence is a multifaceted concept, and it is something anyone can attain. As Mr. Rogers would put it, you can plant ideas in the garden of you mind. All you have to do is think, and they'll grow.

There has never been a better time to do this. Thanks to the Internet, more people than ever can expand their knowledge through the experiences of tens of billions of people. Having an Internet connection and access to the whole of human knowledge is crucial for anyone who wants to maximize their intelligence. The Internet was even crucial for my completion of homework—not just in high school, but throughout college as well—and it had no negative consequence on my ability to take tests. Many people criticize the web for causing a whole generation to lose their ability to retain information, thereby decreasing their attention spans, the so-called "Google Effect". However, this was the same argument Socrates made about written language at a time when all knowledge was passed on orally. Just as storing information in text has freed people to focus their brain power on critical thinking and memorizing ideas in a larger frame, storing information in our pockets and laptops will allow humans to spend even more time thinking critically about the world they live in. Using a search engine is also far more efficient than finding a book on a shelf. In a book you have to laboriously sift through pages to find the information you want, even with a index. Using the Internet all you need is at least a few keywords and at most a well-formed question. MRI's even show that people's brains are more active when surfing the web than reading a book, given that they are used to finding information in this way. It allows people to think more deeply, and allows them to have access to more data, thus enabling them to make better informed decisions based on the trends they see, so long as they check multiple sources. In the same way that people who only get their news from one channel will inherently be less informed, people who live their entire lives based on one book or make assumptions about a topic from reading one online article will simply not be as informed or intelligent as a person who checks multiple sources. Of course, one must also think of the potential drawbacks of the Internet. One common criticism is that the freedom to choose the websites we visit places people into bubbles, filled with people that think exactly the way they do. While it is indeed wonderful how everybody's Internet is tailored to each individual, that freedom can also easily become a trap. Hopefully real-world interactions will be enough to counteract this effect. Of course, there may even be a time when humans will be able to directly plug into the Internet. If that happens, people could communicate telepathically, but they could also be potentially brainwashed or hacked. It's a curious dilemma. However, it seems that if people are clever enough to wire their brains into a global network, they would also be clever enough to design protection measures to defend their freedom of privacy. It would birth a whole new branch of Internet security based on neuroscience! Regardless of what the future holds, it is clear that our history points us in a direction of greater information, connectedness, and hopefully prosperity. We must only continue to get smarter.

So do you want to know how to become smarter? The key to intelligence is simple; never assume you know anything for certain, only what is most probable. The moment you think you know something with absolute certainty, you give up trying to collect more data to reinforce your view on how the world works. If you assume you're right, but you're not, then you will only shut out any information that goes against your views, and you'll be left in the dark until the day you die. You must never stop trying to prove yourself wrong. That's the essence of scientific thought as well. Our worldviews are based on the things we've experienced. It expands with new observations and fills with repeated experiences. If you stop experiencing new things, your worldview will never grow, only fill to capacity. An easy way to do this is to listen to what other people have to say, especially if they come from a completely different part of the world than you. They have a lifetime of experiences that you've never had, and they probably processed them the same way you would have under the same circumstances. This of course is no easy task. All of us have a confirmation bias; we want to hear things that jive with our current points of view. Hearing things contradictory to what we think or believe can often times be uncomfortable or even painful. Hearing something we don't believe might even reinforce our current beliefs. But acknowledging this bias, both within ourselves and in others, can help us communicate in a way that is more sympathetic and kind. You can't persuade someone by being brash or rude, because it's hard to accept statements made by callous rude people. If you want to become smarter, you need empathy, kindness, and patience with other people. Never assume that others will agree with you easily, because you're both probably thinking the same thing, "I am right, they are wrong", when the truth is probably closer to, "I'm partly right, and so are they". I don't know if thinking this way would lead to world peace, but I can say it has allowed me make countless amazing friends.

In order to understand yourself, you need to understand others. In order to understand others, you need to learn about the lives of as many different people throughout history as possible. To understand history, you need to understand human psychology. To understand psychology, you need to understand the brain. In order to fully understand the brain, you need to understand biology and ecology. To fully grasp biology, you need to understand chemistry. To fully understand chemistry, you need to know the physics of the large and small, and to do that you need a basic understanding of how numbers and logic work. In order to learn about so many complex things, you need to be able to think critically, to ask questions, be curious, and embrace ignorance. One can do all these things with nothing more than an open mind. Only after all that can people truly, completely, understand themselves, each other, and the universe in which we exist.

Part III: Disciplines for Livelihood

10 | Working to Survive versus Working to Live

"To suspect your own mortality is to know the beginning of terror. To learn irrefutably that you are mortal is to know the end of terror."
— Frank Herbert

As we've just discussed, the focal trait that differentiates humans from other animals and allowed us to grow more intelligent over time was our ability to acquire enough food to not spend every waking hour hunting and eating in small groups. Food is the key to our success. Humans were foraging and hunting for a hundred and ninety thousand years before we developed agriculture. Granted, fossil remains suggest that many of them probably had higher levels of health and more casual time than agriculturalists, but the main difference is that people couldn't specialize in tasks without excess food within the community. Instead, everybody would have had to essentially live off the land, and everyone would have had to be survivalists. After the agricultural revolution less people had to work to feed entire populations, and as a result human progress exploded. Eventually progress plateaued, until the industrial revolution, and now people in developed countries have so much food that eating has become a second hand inconvenience. There are fewer people actually producing food than ever before in human history, freeing people to pursue all of the various careers that people now associate with "work". Think about how remarkable that is. The human species is so successful that they can sustain a global population of seven billion with the hard work of only a few million. Less than one percent of living people feed the planet, and the rest of us have the great privilege of advancing society.

It's important to remember that the majority of what we know comes from the minds of other people, whether it's our parents, the people who write our school textbooks, our friends, or just random people we run across in our lives. But it's also important to recognize that the situation is no different when it comes to the material things we own. I don't know about you, but virtually nothing I own was made by me. I didn't build the place I live in, nor any of my furniture (with one exception), or any of my appliances. I have a cell phone in my pocket, food, the dishes and utensils I use to eat them, a lava lamp and a plasma ball, everything in my bathroom, and I made absolutely none of it. I'm very thankful for that fact, because if I had to learn how to make all of those things, I wouldn't have the time or resources to learn how nature works. Don't get me wrong, I'm all for self-sufficiency. However, the fact of the matter is that for a civilized society to run and develop, we must work with and put our trust in one another, period. More heads are better than few, not because they can all work together doing one thing, but because they can focus on more things in greater detail. Now the human hydra is seven billion heads strong, and we have the potential to accomplish feats our ancestors couldn't even fathom.

But there's a problem; millions of those heads are either in poverty and unable to contribute to society, criminals who are unwilling to contribute, or terrorists and religious zealots who were never taught the full history of the world and therefore spread fear and death as a result of their ignorance. That is the topic for the remainder of our conversation. We've reviewed what we know, how we know it, and now it's finally time to discuss how we can use that knowledge to keep doing the whole living thing that we have been so successful at these last few short millennia. Now that many people no longer need to work just to feed themselves, more and more people have begun trying to find fulfillment in their work. These days finding a career is more about finding a way to make enough money to do the things you want. This leads us to one of the major pillars of capitalism. Many criticize western society for its corruption and greed of money, and this is certainly true for many. However, many people also acknowledge that money can't buy happiness, but on the same token there is nothing wrong with producing something that people want to buy. No matter what good or service you provide, if somebody is willing to pay you, it's probably useful in some way. There are grey areas, especially if you're getting paid to break the law. Then you might want to think twice. Although as we've discussed there was a time in American history when slavery was legal and another when alcohol was illegal, so clearly you shouldn't base your decisions solely on whether things are "legal" or not.

It's probably more important that you don't cause anyone harm.

One of the most difficult things a person can do is marry their passion with success. Instead of working by doing something they love, many people have to work just to survive and feed their families. The goal of every workplace should be to inspire people to contribute to society and encourage the people who work there to always strive to better themselves whilst giving them the opportunity to do so. Since we will be talking about so many professional disciplines, I may not be able to portray all of them as accurately as someone who lives and breathes the lifestyle associated with those jobs. I don't have much expertise in nearly anything written here, but as a citizen of Earth I can speak generally about how different professions impact the human community. Regardless of the means, the general rule to success is the same: find something you enjoy doing that people will pay you to do. In order for people to want to give you money, you need to be able to sell yourself. But that's only a piece of the puzzle. One of the most important lessons I've learned about working came from one of my fellow instructors, and my chiropractor. He told me that you should never feel guilty charging people for what your time is worth. I think of these words when my humility and self-consciousness hinder my ability to self-improve. The difficulty is deciding what your time is worth, and then convincing other people to agree with you.

People are irrational, especially when money is concerned. You probably didn't need me to tell you that, but it's honestly quite shocking how ridiculous we can be, especially around money. For instance, there's a popular ultimatum game discussed in many economics and psychology classes. I myself first heard it in a global development class I took as an elective. The game takes place between two people. One person gets a sum of money (say $100) and they get to make an offer to the other person. The caveat is that they both have to agree to the deal, and if one of them refuses, neither one gets anything. What's the best deal to make? Well, technically the best offer the first person could make to the second one is a penny, keeping the rest for themself. Since they both started off with nothing, it's still a net gain for each one. Logically this makes sense, but I'm sure anyone can see a problem with somebody offering us dirt while keeping the gold. Many folks would probably rather get nothing than get swindled. This is where empathy comes into economics, making it far more complex than basic statistics.

This irrationality also manifests itself in advertising. Language is crucial to selling your product. In many ways, consumer companies are more skilled at producing propaganda than dictatorships. Think of it this way; would you rather earn $1 everyday or earn $365 a year? Now flip that around, would it be easier for you to spend $1 everyday or $365 once a year on a subscription? These are the decisions people have to make everyday based on their individual income and the bills they have to pay. Multiply that by a few billion people with several hundred of those decisions per person and you get the global economy. If it wasn't already complicated enough, think of all the different types of jobs one can have. To name a few, doctors, lawyers, bankers, chefs, architects, diplomats, educators, and entertainers all contribute to society in some way. Of course, the amount of money each profession makes also vary wildly.

Despite my bias, I have to say that scientists get bar none the coolest jobs, and the coolest toys. They get to smash particles together to learn about the early universe with countrywide accelerators, they get to do magic with chemicals, and they get to interact with and study amazing animals like tigers, elephants, and even cobras. Not to mention they are some of the first people to work in space. There is countless random research that is tremendously important. Take the example of bat researchers studying white nose fungus. Most of us probably don' think of them much, but bats are critical pollinators, and they also help maintain low insect populations. Most importantly, they eat mosquitos, the single most lethal animal to human beings on Earth. When they get sick and die out, every part of the animal kingdom pays a price, as crops suffer and insects multiply and make the problem worse. Nobody can say that researchers aren't dedicated to their jobs. I once even saw someone brushing a funnel-web spider with a tiny brush. They were holding it down with an index finger; the poor thing couldn't do anything but squirm while this hairless ape was cleaning it. Of course, that person had to be some sort of professional spider handler. Those critters don't play games, and they're built to kill. The same is true for any lethal animal, whether it's jellies, snakes, frogs, apes, or big cats, you need to respect and understand creatures that could end your life on a whim. But there's no reason you can't also bond with them, and in fact it's the only way to study nature properly. Plus, how cool is it that with enough knowledge, human beings can take something lethal and turn it into a pastime? Honestly I think that's one of the best parts about being human, telling nature, "Hey, behave yourself. I think I'd rather not die today thank you very much."

There is but one professional discipline I find truly insufferable, that of politics. Some politicians, it seems, live in a constant state of absolutes. "Things must be this way or that, and this thing is always wrong and this person is always evil". Those people should spend some time reading the work of Simone de Beauvoir, famous for her philosophy on ambiguity, who argued that freedom only comes by making decisions in the moment while simultaneously recognizing that the world is unpredictable. People who don't recognize that the world can't be described simply as absolutely right or wrong are some of the most blatantly ignorant and unkind people I have ever seen. Which is tragic because I'm sure under any informal circumstance they would be very nice people. Politicians themselves really aren't so bad. After all, their whole job description is to make government policies that reflect the views of at least some of their citizens. If they don't keep the people happy, they don't get to keep their job. That's the idea at least. Politicians are sort of like police in that their job is to protect and serve, but some of them abuse their power and end up killing people, albeit less directly. The problem arises when politicians either exploit the system to stay in power, like in many developing nations, or when the people who support them are so ignorant that the elected official is forced to butt heads with others who actually understand what's going on. Politicians should make the most informed decisions, but they can't do that if the people who put them in power aren't informed themselves.

But no matter the profession, I have the highest exalted respect for the time and effort people put into their jobs. I too have made a decision for what path my life will take, that of scientific research and education. That being said, there is nothing I could ever say about being a farmer, senator, law enforcer, factory worker, journalist, musician, actor, software engineer, etc. that a person who earns their livelihood doing those jobs wouldn't say better. From this point on, much of what I discuss will be no better than second hand experience. I'm just a humble astrophysicist. However, as a human being who depends on other human beings to do their jobs well, I can discuss how those jobs affect my life. I have learned a decent amount just through observation alone.

11 | Combatting Nature and Ourselves

"A great many people think they are thinking when they are merely rearranging their prejudices."
— William James

There are two paths to survival, and countless means to follow them. One can either eliminate things that cause them to die, or promote the things that help keep them alive. You might call this a sort of "living utilitarianism". It should also be noted that not only does the instinct to live shape morality, but it can also override it. Of course this isn't always the case, many people have strong enough willpower to enforce their morals over their survival instinct. In that sense living utilitarianism may or may not additionally be altruistic. If you and another person are in danger, mortal or not, you may or may not save them depending on the circumstances. But as a general rule, we're wired the same way as all other life, insomuch as we seem to not want to die. Life, uh, finds a way.

The problem is, sometimes we are fooled into perceiving things as being dangerous when that isn't the case. This can lead us to wasted time, like when people decide to pray away problems instead of looking for a solution, or it can lead to the wanton destruction of resources, like factory farming or murder. Again, it should be made clear that there are times when all you can do is pray and cross your fingers, just as there are times when murder is a choice of life or death. For instance, when one is in the middle of a raging storm, or if you found yourself in a spaceship next to a supernova, after you've secured yourself there's nothing you can do but sit back and hope you don't die. Likewise, if somebody is running at you with a knife over a loaf of bread in the middle of a famine, or if a bacterial colony is eating you alive, or if you are being hunted by a pack of wolves, few would contest to you killing those organisms as a means of self-preservation. Of course, the case of killing another human is clearly more extreme than murdering bacteria or wolves. Although all three instances are conflicts for survival—where if you end up dying it will at least be in order to keep something else alive—the fate of another human being should take special consideration.

This begs the question, when is it ever justified to kill another organism, and how should we take sentience into account? In broader language, when should we fight against the world we live in, and when should we collaborate with the world we live in? The heart of the next two chapters is to answer these questions. For instance, if a boulder were going to fall down and kill you, you would probably like to zap it into pieces with a phaser. But what if the boulder was an asteroid, and blowing it up would put people's lives in jeopardy? What's the difference between killing something to defend your family and going to war with everyone who lives with them? Is it possible for all living things to only kill each other if they plan to eat them? Most importantly, if the whole universe is made of the same ingredients as all living things, and the universe is going to expand until all matter and energy are inanimate, then what does it mean to be alive? Why does all life seem so desperate to not die? These are the fundamental questions that shape us as professional adults in a global community of plants, animals, fungus, and single cells, so it would behoove us to try and answer them.

Engineering

As a general rule, there are three main types of schools that one can consider applying to at Universities in the United States. A majority of students enter some sort of school for arts and sciences, which includes a whole spectrum of degrees from physics and astronomy, to language studies and political science. Alternatively, one may enter business school, and while I didn't actually converse with very many during my time in college; I assume they were all learning how to make more money than everyone else. But, for a select few learned individuals that enjoy constant stress, there are the schools of engineering, which has perhaps the most rigorous curriculum amongst all university degrees. This is no coincidence, as engineers are arguably the most important human beings walking the face of the planet.

Because degrees in astrophysics are technically considered to be liberal arts in general, most of us actually don't spend very much time working with engineers. Although, we do have a very similar first year curriculum, as engineers need to learn basic physics and calculus before beginning to specialize into their particular fields, and the first year is often the last time we get to see our engineering friends before they are left to have their souls crushed within the eldritch halls of their own personal buildings. I however had the distinct privilege of living with engineers, rather than other science or liberal arts students. It was completely coincidental as well; it just so happened that one of the few freshman residence halls that had personal bathrooms in every room (as opposed to the dreaded and often disgusting communal ones) was also an engineering dorm. They were the people I ate with, studied with, lived in several homes with, and the people I shared my adolescent antics with. I even took part of my math courses through the engineering department, as opposed to going through the mathematics department, which is more common amongst astrophysics students at my school. There was less of an emphasis on theory, and more of an emphasis on rigid and precise application of theory. Those math classes are a major reason why I have so much respect for engineers. Their curriculum is absolutely brutal, to the point that the department really seems like it's trying to weed out inept students.

By the way, despite what you might hear in the news about fraternity shenanigans and couches being set ablaze, nobody I ever met in school partied harder than the engineers. It's a common misconception, which mostly arises due to the fact that engineers are subject to vastly higher levels of stress than the common frat, and also because most people in fraternities aren't as cognitively inclined, and as a result tend to make more foolish decisions when they drink than engineers. Remember, drinking makes a fool of everyone, but stupid people are still stupid no matter what mental state they are in. Engineering fraternities can be a mixed bag, because most of them regulate drinking, although only for formal events. Regardless, never underestimate an engineer's ability to have a good time.

So why is engineering so mind-numbingly strenuous, and why is it so important for humanity? Well, in essence, it's because engineers are the people that not only learn about the complex ideas and figures in people's imaginations, but then turn those ideas into real, tangible objects. They turn fantasy into reality, harness the laws of nature, and use them to build all of the things society needs to function. In some cases, the lives of hundreds of thousands of people rest on their shoulders. In the context of whether we fight or work with nature, engineers are the people that keep nature from killing us. Nature is scary, as anyone who has been stranded on a deserted island, at sea, in the dessert, on a mountain, or in a rainforest can attest to. Folks who grow up in cities often never have to learn about how treacherous living in nature can be. Unless they go camping that is, but even then a lot of them just drive honking RV's into the woods and call it "camping". But you can't really blame people for not wanting to go back into nature. Not only is it lethal, but living without refrigerators, ovens, or clean drinking water is just frankly inconvenient. It's the same reason people don't like to visit developing nations, and why the fact that people still have to live that way is a grievous crime against humanity committed by all developed nations. That's why we construct huge buildings with air conditioning, electric lighting, central heating, luxurious furniture, and plumbing that takes our waste to a magical land far far away instead of a ditch on the lawn; because we go outside and say to ourselves, "This is pretty, but I think I'd rather not live here".

One needs look no farther than every historical record before the industrial revolution to understand how deadly it is to live. Take the famous *Wave of Kanagawa,* which depicts Japanese boatmen caught in a raging storm clinging to their rowboats as a tidal wave looms overhead. It's beautifully drawn, and if one takes the image to scale, it is clear that the wave is several stories tall. Can you imagine? Picture a wall of water as high as a building threatening to sink your existence into the vast depths below. And even those waves are nothing but ripples on the face of the planet. A planet, I'll remind you, that is nothing more than a raison in the face of the Sun. A tiny solar flare could consume our planet several times over. But even that pales in the face of a supernova that could engulf several solar systems. It's no wonder the old gods of Mesopotamia, Greece, Mexico, and the Old Testament are such vindictive and destructive deities. Even before people had the knowledge to explain natural disasters, they were still well aware of the fact that nature wouldn't hesitate to kill them. Only after we harnessed electricity and fossil fuels were we able to live cushy lives.

When I'm bored at home or driving on a nicely paved road while stuck in traffic, I often try to imagine what the world looked like before hairless apes started replacing nature with concrete. I did this especially during long Colorado winters. In Boulder, one lives where long rolling planes lay at the base of Flatiron Mountains. During the winter, I would imagine frosted grasslands with pine trees distributed loosely along the landscape, providing little coverage as the land ices over tens of degrees below the freezing point. Then I would look outside and instead see roads, markets, and rows of houses ordered nicely in the tundra, each one nearly as warm as a fresh spring day. It's at that point I realize that humans really don't belong on most of the planet. We're the most successful invasive species, reshaping the global landscape to match the African Savannah. But again, the fact that our brains give us the ability to collaborate and build structures that reshape the environment and harness the laws of nature is nothing short of remarkable. However, the fact that we essentially need to kill native species in order to build infrastructure, while remarkable, poses a serious threat to biodiversity. Sure some animals like raccoons and pigeons have adapted to city life, but the fact remains that humans don't like to share their space with other animals unless they're cute. As a result, most native plants and animals face extinction due to human expansion. Ironically, humans have begun to reverse the stronghold that nature once held over us, and now risk destroying ourselves by killing off the rest of Earth's inhabitants. But where clever engineering got us into trouble, clever engineering can get us out. Cleary, one of the greatest tasks we face revolves around harnessing energy that won't run out or overheat the planet. For that task, engineers of the future will need to understand how to utilize the energy already flowing through the planet in order to power our now high-tech lives and transport us between destinations quickly and efficiently. As that mostly involves harnessing nature, we will reserve those topics one more chapter ahead. Nearly every problem can be solved by some sort of engineer, from making better electronics to devising advanced robots. As such, we won't spend too much time discussing all the problems we need engineers to solve. Here we will instead focus a little on just one major problem with human overpopulation, the problem of where we're going to put everyone.

One of the most brilliant aspects of visiting a city is seeing forests of skyscrapers. They are one of the most impressive feats of engineering, mostly because they defy gravity. But the brilliance of these towers of glass and iron is in their special efficiency. It simply makes more sense to counteract overpopulation by expanding upward instead of outward. For one, there are fewer things living above ground than on ground, and so creating taller buildings allows more people to live in a city without increasing the amount of square miles the city takes up. In addition, skyscrapers are efficient for businesses because they keep large numbers of employees close together, and I'm sure CEO's like being on top both literally as well as figuratively. However, the trait that makes tall buildings impressive is also what hinders them.

In order to build structures hundreds of meters tall, people have to lift the materials from the ground, something that requires more and more energy the taller the building is. Some have devised clever mechanisms that utilize vacuum tubes to transport materials, otherwise settling for cranes or in extreme cases helicopters. This is just one logistical challenge architects face. Taller buildings also have to deal with increased wind speeds, although hypothetically placing several tall buildings together as windbreakers could mitigate this effect. It might also be possible for taller buildings to utilize high-speed winds for electricity, and there are in fact some buildings that already practice this. Lastly, there is the elevator problem. A minor issue is that tall buildings require a new elevator every several stories, since a single elevator can only go so far upward. The more important issue is with the air. The thing about the atmosphere is that it's thicker at the bottom, which means that if a building is too high, people at the top floor will need more red blood cells than those at the bottom just to breathe properly. Granted this would provide a new way for athletes to get altitude training without traveling to Colorado, but for casual business and living it is rather inconvenient. Some might say that these issues call into question the idea of using skyscrapers at all, but solving them could revolutionize the way we live. If we could accomplish building structures all the way to space, we would be one step closer to developing space elevators to the moon and buildings that could allow people to live in the upper atmosphere of Venus, where temperatures are more similar to home. Another potential solution to building infrastructure lies with 3D printing. It may one day be the case that humans are able to remotely build entire facilities simply by printing them. Not only could buildings be printed until they reach the sky, but if we figure out how to repurpose materials already present in the environment, we could even build cities on distant worlds, including The Moon!

Engineering takes a certain combination of creativity and rigid work ethic. It's something that certainly isn't for the faint of heart or the weak minded, and I think that's why so many of my closest friends, the people I respect the most, are engineers. In fact, my personal hero is an engineer. His name is William Kamkwamba. He grew up in Malawi, the poorest country in the world, and accomplished more by the age of fifteen than the richest person in the world (Bill Gates founded Microsoft in 1975, when he was 20). Growing up, his parents didn't have enough money to put him through school, and he eventually had to drop out to help his family. But he had an insatiable thirst for knowledge, and with a few trips to the library and a bicycle light, he figured out how to build a windmill to bring electricity to not just his home, but his entire village. After that they were able to use a few tiny bulbs and a radio. I first heard about him through his memoir *The Boy Who Harnessed the Wind*, a book that we had to read for our summer assignment before my senior year of high school (thanks IB), and even today is one of the few books for school that I have read cover to cover. To me his story is fundamental proof that being born with wealth can sometimes be more of a handicap than a benefit. When people have nothing to fight for, no cause to serve, they don't do anything useful for society. Moreover, people all around the world have the capacity to accomplish great things. This is why charities are important. You shouldn't look down on poor societies, you should lament the fact that they have to waste time struggling to get an education and feed themselves instead of developing technologies and innovations that could benefit you and the entire planet. You're shooting yourself in the foot when you don't allow people to contribute to society by keeping them poor, sick, and ignorant.

So if engineering allows us to adapt and counter the relentless and soulless forces of nature, what keeps us from being killed by other living things? For the last few millennia, humans have been successful enough to mostly claim sovereignty over all other creatures. But those who claim that we are outside of the food chain often forget about the one thing lying under our nose that still devours us; the microscopic organisms that we are made of. People often say that bigger is always better, but even the largest predator still gets sick. This is true for us as well. The last thing that persistently causes us to die isn't the environment or any large predator; it's disease. Luckily, medicine has been advancing just as rapidly as our ability to harness electricity.

Medicine

Even though I already gave a disclaimer, for the sake of posterity I must make a special specification: **I am not a medical professional, nor am I qualified to give medical advice.** I'm just a curious scientist who marvels at all the ways humans have come up with to live past twenty. That's important, because people spend as many (usually more) years learning to become a medical professional as they might to become a doctor of philosophy. The problem of course is that anybody with an opinion and in good health might try to tell other people what to put into their bodies to make them feel better, even without any knowledge of chemistry or biology. This is most obviously seen in medical professionals before the advent of cell theory. Shamans feed herbs and seeds with magical properties to patients, like rhinoceros horn, which as we're already discussed is tantamount to feeding people hair. Clergymen during the Black Death tried to use leaches and prayer to cure the plague, instead of replacing their straw roofs infested with flea-covered rats or washing their hands. That or burn hundreds of Jewish people alive on Valentine's Day. What I'm saying is, without an adequate knowledge of biology, anatomy, and physiology, and several years of dedicated practice, you're more likely to kill somebody than cure any disease. People meant well, but they simply didn't understand how the world really worked, they just bent over to their gods and hoped they didn't get screwed over. On the same token, I have dedicated my life to general science trivia and philosophy, not medicine. So take everything I say with a grain of salt. But hey, at least I know what a cell is!

What is bar none the worst human pastime? My guess is getting sick. There's nothing as inconvenient as being in the middle of a successful life, when out of nowhere you come down with a sudden case of plague and die. People don't make plans to break their arm or catch the flu, and since nobody knew what a cell was for ninety-nine percent of human history, there are very few accounts of how much disease has impacted human society. Instead, people replaced reality with fantastic myths glorifying human agency or divine will, when in fact people were just getting sick and not having access to any good pills. In reality, despite what most textbooks might emphasize, quite a few critical moments in history were likely decided behind the scenes by some sort of sickness.

The nice thing about disease, if anything, is that it is indiscriminate. Famous people, regardless of how history may paint them, are no less likely to get sick than anyone else. For instance, Alexander the Great likely died of a mere fever. Tutankhamen (who is only famous because previous pharaoh had their graves robbed by ancient people instead of twentieth century British people) probably died of infection and/or malaria. Disease has also killed way more people than any human-caused event. South Americans of the Inca and Aztec died of disease before European conquistadores even arrived in the early sixteenth century, resulting in civil wars amongst the native empires, making it easier for the Spanish to take over. Even though it may have contributed to better infrastructure and economic growth, The Black Death killed 75 to 200 million people (remember World War II fatalities are estimated to have been 60 million). The Spanish Flu killed some 50 million (as opposed to World War I's maximum of 15 million fatalities). But even all of those pale in comparison to the death and disease brought upon the native people of North America. After the arrival of Europeans, some 80-90% of North

American Indians died of smallpox and syphilis. Coupled with the success of the colonists, their subsequent need for expansion, and a healthy dose of racism and ignorance, the native population eventually became overrun by pale immigrants. The petulance was by no means anyone's fault, but the fact that vaccines had yet to be invented ended up becoming the secret lynchpin for creating the world's most powerful nation.

Disease has been notoriously dreadful because of how difficult it is to deal with. Nature is lethal, but at least volcanoes and hurricanes aren't alive. Pathogens on the other hand evolve, not to mention they're ridiculously small. Generally speaking, the smaller a pathogen is, the more difficult it is to treat. For instance, bacteria—which cause everything from bubonic plague to ear infections—require nothing more than antibiotics to treat. As it turns out, fungus also hate to get infected by bacteria, which is why humans can use the chemicals they produce to create such drugs. Viruses are obviously more difficult to deal with, being one one-hundredth the size of bacterium, coupled with the fact that they actually infiltrate other cells. The only real way to deal with them is to prepare our immune systems before infection, because as of right now, our natural immunity is the most powerful tool for our health. This is why vaccines were such a stroke of genius, and why viruses that target the immune system, like HIV, are especially difficult to treat. These types of diseases are so terrifying and adaptive that many people think the world will end from a global pandemic. This fear manifests itself in every zombie apocalypse film, even though zombies themselves are quite ridiculous on the grounds that humans without their brains are tantamount to squishy rag dolls when placed anywhere outside of a city. In any case, humans seem to have a good enough grasp of genetics and evolution to quarantine sick individuals and whip up cures before anything spreads across the planet. Moreover, people tend to underestimate the complex sophistication of the human immune system (and human paranoia). After all, we're the descendants of four hundred million years of multicellular organisms showing single celled organisms who's boss.

In my opinion, the most terrifying diseases are the ones that don't get transmitted, because you have little warning before you get them. Humans have been fairly successful at eliminating cell-based diseases from viruses and bacteria, and this has led us to living far longer than our ancestors. For the majority of human history, people have died too early from being eaten by pathogens or carnivores to even consider old age. For the first time, a large number of humans are actually growing old and wrinkled, and we're only now seeing the consequences of living a long life.

Dementia, for instance, is something that will almost certainly only happen to you if you're lucky enough to grow old. In many ways, it is a completely new occurrence, and a terrifying one at that. As someone who quite enjoys their brain, the idea of losing your humanity as your neurons slowly die is a fate worse than death. It is also quite expensive. People spend hundreds of billions of dollars caring for dementia patients. This brings to light many ethical dilemmas on how to deal with patients who are on their deathbed and no longer even cognizant. Is euthanasia ethical if a person is in a vegetative state with no hope of recovery? If everyone dies anyway, wouldn't it be better if we just did our best to give people a happy send off? This is the weighty question many medical professionals must face on a daily basis.

But of all the diseases that plague humanity, none has proved more persistently lethal than cancer. Bacteria and fungus will kill your cells from the inside and viruses will infiltrate your DNA, but even in a vacuum, you can still get cancer. It is a genetic disorder; a random product of the same genetic shuffling that allows life to survive in the first place. Cancer is nothing more your own cells going rogue, and multiplying indiscriminately. Cancer is an anatomical civil war. It can strike anyone, anytime, anywhere, regardless of what foreign toxins enter your body. They spread throughout the body, and disrupt vital organ functions by impeding flow and sapping resources. This problem has become the forefront disease of the twentieth and early twenty first century. We can fight crime, kill bacteria, and prepare our bodies for viruses, but there is no law, no pill, and no vaccine that can cure cancer, at least not yet. Its cause is molecular and native, and this is why cancer has only become prevalent after sufficient medical advancements have been made. But cancer has always been with us. Breast cancer was even recorded by Imhotep as far back as ancient Egypt. But that doesn't mean it needs to stay with us.

I can't give you a cure for cancer, but natural preventions are already hard at work within your body. Your immune system is among the most highly advanced in the animal kingdom, and your guardian cells will stop at nothing to keep you alive. But they can't do their job without your help. Regardless of the blight, your mind and body are always the first and foremost defense against your death. Your white blood cells can sniff out and eliminate cancerous cells before they clump into tumors and spread throughout the body. As with all bodily functions, a healthy immune system is only as good as the nutrients you feed it. Cells break down and die, and in order to replace and replicate, they need a wide variety of complex chemicals. One must therefore eat a wide variety of foods, and as we tell the kids, different colored ones at that. A varied diet will also give you more energy to take on the day, enough that you might feel motivated to exercise. A cancer cell can appear anywhere in the body, and there is a type of cancer for every organ in the body. However, I've noticed that many of the most common types of cancers occur in parts of the body that are commonly inert. Breasts and prostates don't get exercise like other parts of the body, they don't move, and it must surely be easier for cancerous cells to multiply in areas where they won't be jostled around. Lungs of course move constantly, but life is stressful and a lot of people living in developed nations find solace in smoking. Regardless, all current and potential victims of cancer can find alleviation in exercise and movement. Without the need to hunt our food, daily exercise has increasingly become more of a chore. But without some form of activity, your body simply cannot function to its fullest. While it might seem counterintuitive, you need to increase circulation in your body, because even though it circulates cancerous cells, it also allows your immune system to eliminate diseases more easily. Cancer is only a problem when miniscule cells multiply into visible limps. It would seem to me that dividing out cancerous cells throughout your body would make it easier for your body to conquer those rebels.

Despite all of the scary ways we could end up six feet under, humans have done a pretty good job figuring out ways to give the finger to the reaper. Irregular heartbeat? Have a pacemaker! Body doesn't produce enough insulin? There's a pump for that! We can even counteract the effects of venom, arguably the most ingenious means evolution has developed to help animals eat. Potentially the most hilarious (and expensive) medical practice, the process of creating antivenom is actually quite similar to the process of making a vaccine. The only difference is that instead of injecting inert viruses into humans so that they produce antibodies, venom is injected into large animals so that they produce the antivenom. This is similar to how researchers test vaccines and drugs by first introducing them into the blue blood of horseshoe crabs, in order to see how their amebocyte-based immune system responds, since it reacts tens of times faster than vertebrate immune systems. After that the serum is purified and frozen until it is needed, which is quite inconvenient since most of the places with venomous animals don't have enough electricity for refrigerators. I say that this is comical because in order to attain the venom, the animals in question have to be milked. Don't get me wrong, that is a dangerous job that isn't for the faint of heart. But seriously, just think about it from an evolutionary perspective. Here are these ferocious spiders and snakes who have been developing fatal toxins for millions of years to avoid putting themselves in danger while they hunt or to deter predators, and then all of a sudden these goofy hairless apes just start putting them in cages and forcing them to spit into jars. All because they keep biting us out of paranoia instead of leaving us alone or visa versa. But to be fair, most venomous animals don't like to waste venom on things they can't eat, and humans do have a tendency to murder every other living organism that doesn't look like them.

But antivenom only scratches the surface of human cleverness. Most notably, the technological revolution has offered numerous breakthroughs in medical development. For instance, robot prosthetics and exoskeletons are being developed for paraplegics, making them more than just handicapable. Nanotechnologies are also being researched as ways of enhancing the body's natural immune response, helping us cure illnesses, and even reducing clogged arteries. We are even beginning to learn how to enhance our brains. One day, people may be able to get neural implants, giving people who are paralyzed technology to turn their thoughts into reality. If one wants to speed up learning, they could try a trans-cranial direct current stimulation thinking cap, a simple loop of wire that supposedly enhances neuroplasticity. For those with damaged brains from injury or dementia, stem cells might be able to provide future brain replacements. There will even be options for prosthetic surgery for both superficial and major body replacements. Let's not forget the wonders of 3D printing. Whether it's organs, prosthetics, or even drugs like ibuprofen, we're not far from the day when medical supplies and body parts will be as readily available as a book report. The blind will be able to see, the deaf will hear, and the lame will perform circus tricks! We've come a long way from the days when a paper cut could lead to a long and excruciating demise.

But perhaps the most exciting advancement in medicine has got to be the discovery of clustered regularly interspaced short palindromic repeats (CRISPR). CRISPR-Cas9, as it's full name is written, is a big deal. It's the BIGGEST deal since the smallpox vaccine. If I could, I'd spend a good book writing about all of its implications. It all starts with the age-old struggle between prokaryotes and viruses. As you know, viruses inject their genetic information into their hosts. What primitive single cells have developed to counter this onslaught is a protein called Cas9. Its job is to match a strand of RNA from a virus with a sequence that has been inserted into the cell by another virus, and cut the sequence out. This is a sort of immune response to viruses that keeps them from replicating inside of the cell. Here's the magic: bioengineers can use this system to directly modify genes far more quickly, far more accurately, and for far less money than ever before. This protein acts like a genetic GPS system that can precisely add or remove genes from an organism within a generation. Again, it's hard to emphasize how monumental this is. With this technology, we may one day be able to eliminate all diseases, including immune viruses, cancer, and birth defects. Researchers have already demonstrated the ability to remove HIV from lab mice. This technology could open the doors for designer babies free of disease, born with enhanced metabolism, better intelligence, greater athleticism, and even longer lifespans. This might even be our key to surviving space travel. Of course, modifying humans comes with countless ethical issues that will undoubtedly be squawked over by politicians. There is no doubt that modifying humans might seem terrifying in the hands of the narrow minded. What if dictators decide to breed armies of genetically enhanced soldiers? But that's exactly why we need to keep open lines of communication and transparency like we have for all new technologies. When push comes to shove, humans are quite adamant about decreasing the amount of suffering in the world. But this does lead us to the controversial nature of medicine and politics.

Obviously it is both lucrative and mandatory for a society to have people who can keep everyone else alive and able to do their jobs. Unfortunately, the combination of high demand and huge pressure makes getting into medical school a very cutthroat ordeal. Medical schools have become overly competitive as a result of this, with strenuous curriculum that creates a bottleneck for students, and too few high paying positions to justify the backbreaking effort it takes to get a degree. There isn't necessarily a problem having more medical professionals. The human body is radically complex, and when you have multiple people you get greater specialization, and those people begin to rely more on one another. The difficulty is when there isn't enough funding or demand (i.e. jobs) for medical professionals. To that I would reply: why not send some of them to the Moon? First-Aid training would obviously be mandatory for anyone going into the vacuum, but clearly people will need medical professionals in the long term as well. Or people could go in shifts like in the Peace Corps. It would be hugely prestigious, and if anyone wants a shot at being an astronaut, they need to be in peak human condition, ready and adaptable. Surely with such high standards, it may be that the elite medical professionals who keep their positions so long that nobody can replace them could be filtered to a new branch of medicine. With those experts gone it would free up work on the ground for experienced young graduates who could fill the niche of providing the best medical care available for ninety-nine percent of human history.

Another byproduct of the competitive nature of medicine is the increased number of individuals who give medical advice without a degree or experience practicing medicine. Even if students don't make the cut to become doctors and nurses—or they don't want to accept low paying work in clinics or developing nations—they will still likely try to use their expertise to make some sort of money. This may be how several doctors have ended up becoming television show hosts. The Internet only exacerbates the issue of increased medical sources, since anybody can anonymously post healthcare information without giving their credentials. Hence, it falls on the responsibility of the patient to think critically about where they are getting their medical information. Just because you hear someone make a claim justified by the statement "one study showed that" or "studies show" doesn't make that claim true. When it comes to your health especially, it's wiser to take the word of a professional over your closest friends. Moreover, everybody has unique physiology, which means even medical professionals might not know the perfect treatment for you. When it comes to your health, just ask yourself, "Do I feel better?" and also remember the power of your brain, as evidenced by the placebo effect. In minor cases, just thinking about healing your body is enough to get the job done.

Probably the best way to mitigate the dangers of inaccurate medical information is to become well informed. That's not to insinuate that everyone should memorize every intricate piece and interaction of every organ and tissue in the body. However, a primary understanding of human anatomy and physiology should be included as basic human knowledge. After all, this type of information is useful for basic first aid, as well as preventative health, diet, and overall well-being. Understanding one's own body is no different than understanding the intricacies of one's own house. You may not be able to repair the electricity (nerves) or the plumbing (digestive system), but one can at least have the basic knowledge to keep those systems from breaking down. The only difference I could think of would be the fact that looking inside of the human body is disgusting. For many centuries, conducting human autopsies was deemed taboo, and the ancient world was far more prone to bad medical practices as a result. It's not anyone's fault, it only makes sense that seeing our insides would make people uncomfortable. Blood and gore are usually signs of death, and people usually don't like dying. Why else would skeletons and ghosts be symbols of holidays like Halloween and horror movie tropes? But everybody has a spooky scary skeleton inside of them, it's not a big deal. If one can learn to become comfortable with their own anatomy, they would become better doctors than the majority of all shamans, clerics, and priests in human history. But sometimes there's no getting around the fact that you need help. What then?

With healthcare being so imperative to society, on would presume that access to it should be considered a universal right to all citizens. Of course, there is no such thing as a free lunch, and medicine is one of the most expensive human endeavors. People need to pay for their healthcare somehow, and the two main ways of doing so are either through taxes, or by directly paying for personal health insurance. Money complicates everything, and so many different nations have devised different ways of keeping their citizens healthy. For instance, a government can have a single payer system, like Canada. Those governments fund healthcare through taxes (like education), but doctor's offices are often privately owned, with income from the government. The United States has a mixed system, where the government funds healthcare for a few people who are either over 65, military servicemen, or in poverty, and where everyone else gets their healthcare from employers or individual sources. While this system is often criticized for enabling poor people to parasitize tax dollars, it often results in many people simply not having healthcare. For instance, the census bureau cites one in ten Americans not having health insurance in 2014, although that was less than previous years. Many of these individuals don't have healthcare because they work part time, and so aren't impoverished, but their employers don't offer health insurance for part-time work. Above all, the biggest downfall of the mixed system in the United States is the fact that neither local nor state governments negotiate medical providers for the costs of drugs, hip replacements, etc. As a result, there is little incentive to lower costs because these providers can charge whatever they want since their products keep people from dying, and you can't put a price on life very easily. All of these nuances, coupled with the higher wages for medical professionals, lead to Americans paying more for healthcare (even in taxes) than every other developed nation.

While it's clear that every person needs to be healthy to contribute to society, it's not clear whether or not people should have to already be working in order to receive medical insurance. This is only further complicated by the fact that there are so many private insurance companies to choose from. On one hand, competition brings down insurance prices, but on the other hand having a complicated system has resulted in thousands of people just not being taken care of when their lives are in jeopardy. Furthermore, people who do have access to healthcare tend to behave like hypochondriacs and waste resources getting checked for illnesses they don't have. The difficulty is deciding who should be taken care of, and who's going to pay. From a large-scale point of view, it would seem that the best solution to this problem is an educated population with a culture of exercise and good health, coupled with a tax ensured healthcare system that takes care of people during unseen emergencies. It's clear that sick people are basically useless to society, which means it's in everyone's best interest for all humans to be as healthy as possible. If people know how to keep themselves healthy, they won't need to waste tax dollars and medical resources unless they absolutely need them. This will only become easier as the world becomes more connected, people become more knowledgeable, and medical technology increases. But as with all things that involve humans, reaching these goals will be complex and difficult. Keeping ourselves alive isn't a black or white issue, and it doesn't help that there is an equally complicated issue that keeps people from working together. I'm of course referring to the difficult task of treating each other as human beings.

War & Conflict

The time has come, I have to say, to talk of many things. Of shoes and ships and sealing wax, and that strange thing we do involving mass slaughter, genocide, and the destruction of our own humanity. That thing we call war. Goo Goo G'joob.

Now, when one imagines war, many things may come to mind. Most likely, you might imagine vast herds of armed troops, of ships and planes, guns and tanks, smoke and fire. Indeed, this type of war is frightful beyond imagine. Fighting over resources is not only deeply intertwined with the history of Homo Sapiens, but arguably all living organisms. Humans, however, have the power to cause destruction unparalleled in the kingdom of life. Power to bring war built with iron and lightning and fire. War that sweeps in like a tempest, with no discernment between soldier and civilian. War that makes devils tremble and angels weep. War that leaves not even ravens to scavenge from the Earth. But war comes in many faces beside the often romanticized scenes focused on so commonly in history courses. Some of these might be benign, while others have the potential to erode the bedrock of society. What's more, some kinds of war can find themselves in our everyday lives in a way armed combat doesn't. Class war, drug war, flame war, race war, gender war, even mere bullying are all sprung by fear and hatred. But while any of these things may incite terror or violence independently, at their heart it would seem that they all stem from the same root cause: our inability or refusal to see each other as being equally complex as ourselves, a failure to see others as ourselves.

Paradoxically, humans are so good at discerning differences in things that we are actually very bad at recognizing other people as being the same as we are. What's more, if the mentality of difference is pervasive enough in a society, people can actually acquiesce to their own marginalization. Peasants in middle age monarchies lived their entire lives believing that some people are simply born to be poor and serve the crown, just as some women genuinely believed that their jobs were confined to the home, or that people with dark skin believed they were inherently inferior to people with pale skin. Obviously (or apparently not) this is rubbish, nonsense, and the only possible justification for causing other people to suffer. These misconceptions have resulted in the most harrowing, disgraceful, and downright pathetic behaviors of the human animal. Moreover, hate only breeds more hate. When you marginalize someone else, they will pay you back in kind, until everyone feels justified in hating other people because everyone feels like a victim. The pain that we have felt through our lives and over generations have left deep scars in society, and you may feel uncomfortable discussing them. However, if we want to heal our wounds, we must take a moment to stop our hatred, do away with our sorrow, and try to see the world through the eyes of the people we hate. The problem is never far from the solution, and so we must not be afraid to speak rationally and calmly about these issues. Only by understanding the problem can one dissolve the problem. If you are too uncomfortable, don't feel ashamed to skip on to the next section.

Let's get the big one out of the way first. Large-scale wars have been a part of history about as long as city states. They are endlessly romanticized as the builders of great empires and seen as glorious sacrifices symbolic of national patriotism. But make no mistake, war is hell. There's nothing glamorous about dying for land that doesn't belong to you. The only people who chronicle the glory of war are the one's who were on the winning side, many of whom never set foot on the battlefield. Perhaps there was a time in history when defending your kin with swords and cavalry merited some sort of pride, but those days were gone the second humans began using gunpowder. To truly understand something so horrific and beyond imagination, we can only turn to the voices of individuals who lived through wars. Here are just a few famous accounts of some of the brave soldiers who put their experiences to paper, and shed light on the consequences of our stupidity.

Karl Marlantes, served in Vietnam, author of *What it is like to Go to War*:

> "Choosing sides is the fundamental first choice that a warrior makes ... The second fundamental choice of the warrior is to be willing to use violence to protect someone against intended or implied violence."

Erich Maria Remarque, served in World War I, author of *All Quiet on the Western Front*:

> "We are not youth any longer. We don't want to take the world by storm. We are fleeing. We fly from ourselves. From our life. We were eighteen and had begun to love life and the world; and we had to shoot it to pieces. The first bomb, the first explosion, burst in our hearts. We are cut off from activity, from striving, from progress. We believe in such things no longer, we believe in war."

Kurt Vonnegut, served in World War II during the bombing of Dresden, author of *Slaughterhouse-Five*:

> "It is so short and jumbled and jangled, Sam, because there is nothing intelligent to say about a massacre. Everybody is supposed to be dead, to never say anything or want anything ever again. Everything is supposed to be very quiet after a massacre, and it always is, except for the birds. And what do the birds say? All there is to say about a massacre, things like 'Poo-tee-weet?' "

War is senseless. Perhaps that's part of the appeal. It's not surprising that it was seen as glorious until the twentieth century, when film and photography could truly capture the scenes. Until then, war led to the expansion of territory and greater wealth. Then technology spread—as it always does—and old tactics became outdated against machine guns and chemicals. Soldiers dug trenches that would become their graves, living in fear of bombardment and being buried alive. After that the carnage only increased. Pilots flew their planes into enemy ships, prisoners were worked to death, frozen alive, fumigated like bugs and cooked like cattle. Civilians were battered and raped, left for dead with their heads on steaks and their bare bodies laid in piles with bayonets left in their genitals. Cities were bombed to dust, with human silhouettes drawn in ash on the walls and the stairs. Homes, schools, hospitals, churches, restaurants, and libraries were sacked, burned, and demolished by the force a thousand burning Suns. Burning napalm fell from the skies, melting away cloth and skin alike. Countless soldiers looked each other dead in the eye, saw their humanity, and then ripped it to bloody shreds until all that remained was oblivion.

And to what means did all this suffering lead? For what grand purpose did a hundred million souls take up arms and toss their bodies onto piles of corpses? Was it for some great divinity that either doesn't exist or doesn't care enough to reveal itself? Was it for some glimmering rocks or wads of cash? Or maybe it was for a fancy throne and a fancy hat? Did they die for queen and country, for the glory of the fatherland and the banner of liberty? Was it to keep bloodlines pure and untainted by race mixing? Was it for strategic borders and regulated food supplies? Was it because a foreign power halfway across the world lived a different lifestyle than they? Were they evil for thinking we were demons? Was it for power, utopia, and control? Was it to claim dominion over a tiny speck of dust drifting in an empty vacuum, just so someone could be remembered before the last grain of their hourglass fell? Why? Tell me why so many men, women, and children were mercilessly beaten, penetrated, hanged, shot, stabbed, starved, burned, frozen, drowned, and blown into tiny chunks by their brothers and sisters? Why do people, just like you and I, find it in their hearts so necessary to kill their humanity and cast the ongoing wrath of the cosmos onto their fellow human beings? Why would we waste our time?

They were children, every last one of them. Every single hero and tyrant of war, every criminal and victim, none had the wisdom of where they were going or where they had been. They had next to no information about the history of the world they lived in, nor the direction it was headed. All they had were false promises and a singular purpose to eliminate the enemy. They stopped at nothing to preserve their home or their genes, and killed themselves in the process. Nations imprisoned their own citizens in internment camps and gulags in case they were enemies of the state. They drew each other to fear and to hatred, in return for rallying their allies against a common enemy. If nothing else, that is the singular redeeming quality of war. The feeling of serving a purpose greater than oneself, galvanized by one's duty to their comrades and their homeland, something that gives soldiers a sense of purpose that most of us can only dream of.

Perhaps the most terrifying thing about war is that at the end of the day, there isn't a singular concrete object for us to place the blame. One certainly can't blame the heroes who laid down their lives for their countries. How can you blame someone for wanting to defend their home, for wanting to perpetuate the freedom of their people? Even the Nazi schutzstaffel put their lives on the line for what they thought was a just cause. On the battlefield, it doesn't matter who you are, you're life is on the line, and even bitter enemies have to acknowledge that in the end, you're both fighting for survival. This is why veterans who survived going through hell should be venerated to the highest degree, as some of the only people willing to die for the sake of others. So then should we blame war on the people who forced those kids to become murderers for the state? Again, no. One can't even wholly place the blame on the generals and politicians who ordered these atrocities to occur. They may not have seen combat, but sometimes living with your crimes is more torturous than being put to rest. The leaders of nations act with much of the same conviction as their troops, and they too were just trying to survive. No, none of these causes are sufficient. But deep down, I think everyone knows who's to blame for war. We already know why genocide and mass murder are so hard to fathom, and why we find those actions so vastly horrific. It's because, at its heart, war is a human endeavor. The blame doesn't fall on individuals, it falls on all of us. You and me. We're terrified by the idea that placed in the same situation, under the exact same circumstances, with the same lack of knowledge, it could have been us shooting foreign soldiers, bombing their families and impaling their children. The scariest thing is the fact that it could have been us. The fact that in some sick way, we actually get pleasure from destruction and take solace in eliminating what we deem to be a threat to our way of life. War is the epitome of our hatred, to ourselves and one another.

However, when faced with our demons, we can either cower in fear or learn from past mistakes. This has indeed been the case throughout human history. As the scale of war became larger and larger, so too did the size of the forces facing off against one another. Where once tiny villages saw each other as rivals, eventually cities and eventually kingdoms and eventually aligned countries went at each other's throats, until the age of globalization. Only now after two hundred thousand years are we beginning to see how insignificant our differences are in light of our genetic and cosmic history. Now in the twenty-first century, pacifism is at an all time high. The vast majority of conflicts have shrunk from wars between states to civil wars within borders. War is now not only expensive, but widely regarded as depraved and primitive. We are beginning to see how universal the desire to live is and how persistent human beings are to maintain peace. The last refuges for people who actually want to incite violence and conflict are now limited to the most secluded areas where people don't know that the universe is billions of years old and unfathomably big. These of course include North Korea and sections of the Middle East, the most conservative, destitute parts of the planet where people can't think freely, lest they be put to death.

As we discussed back in Chapter 4, resources are one of the primary motivators for conflict, especially if the resource in question is food. Interestingly, lands that were once rich in resources have become some of the poorest countries on Earth, as a result of colonization. Since then, wealth inequality has only continued to grow throughout the world and even within nations. The rich get richer, the majority of wealth is held in the hands of the few, and when they die they still don't give up their money. This leads us to the most intangible form of discrimination, classism.

Like any other form of discrimination, class warfare takes place between two groups of people because of the way they were born, something nobody has any control over. The poor resent the rich for never having to do any hard work and because money is power, and the rich look down on the poor as if they don't work hard and because the poor resent them. Often times people of affluence aren't even aware of the burdens held by the less well off because they never come into contact with each other. For generations and generations, the rich passed on their wealth to their descendants while the poor were left in the dirt. As kingdoms grew the gap did as well. One could maintain such a system only if the people were convinced that god had ordained it, but when people started working for wages things changed. Two opposing ideologies sprung forth in response to the disparity in wealth that had been accumulating. One was capitalism, which allowed poor people to earn a seat at the rich table if they were lucky and clever enough to create something people wanted to buy. The second was socialism, which promises fiscal equality at the cost of banning people from having any opportunity to amass wealth.

Both of these competing ideologies have strengths and shortcoming, but like all ideals they share the common weakness of not taking into account human nature. Sure, if you live in a capitalist society you could get rich overnight, but you're far more likely to make it big if you were born well off because your parents worked hard and became successful. So capitalism perpetuates a cycle of people being born with handicaps. Then when people make money all they can think about is how to make more until they start hoarding wealth that serves no practical purpose besides status. People don't like to be worse off, and so they are always unlikely to give away the large sums of money they don't need, and then you just have poor people who hate rich people all over again. On the other hand, if you live in a socialist state, you don't have any incentive to innovate because all of your hard work will just be given away to people who didn't do anything. What's more, people are dreadful at maintaining equality. Nowhere is there a better example of this than in George Orwell's fictional tale *Animal Farm*, which replaces the proletariats with farm animals and the bourgeoisie with a single farmer. Tired of being used by their two legged oppressor, a group of animals takes over their farm, establishes a new regime, and even fashions seven commandments; dubbing all animals equal and abolishing all human behaviors like drinking and sleeping in beds. This lasts until the smartest animals, the pigs, start calling the shots, manipulating their fellow animals, and conforming to human lifestyle. By the end, their seven commandments are whittled down to one simple statement, "All animals are equal, but some animals are more equal than others". This tale serves as an allegory for every communist nation in history, from Soviet Russia to communist China and North Korea. As much as these regimes tried to spout equality, there was no restraint on the people in power from giving in to their bestial need for power. As a result, the U.S.S.R collapsed, both Russia and China slowly began adopting capitalist values, and North Korea remains the most isolated and juvenile state on Earth. This had less to do with capitalist or communist ideals, and more to do with how Western capitalists had stolen wealth and built their power on slave labor while Eastern communists gave in to their hubris and failed to uphold the Marxist ideology. Some claim that the basis of capitalism is human nature, while socialism is based on human choice and planning. The problem is that our choices are almost always dictated by our instinct, and our nature is malleable and adaptive. If that wasn't the case, we would have gone extinct already. But it is thanks to our adaptability that we can mend the global wealth gap, so long as we acknowledge the fact that the poor are no less capable that the rich and at a certain point money is just a number.

What makes poverty so difficult to rectify is a combination of the sheer number of impoverished people and the fact that many of them live in completely separate continents as wealthy people. Peter Singer best articulated this dilemma in his thought experiment about a drowning child. Imagine yourself rushing to work with clean pressed clothes. You're running late for a big meeting that could lead you to a promotion and your boss has a posterior that could turn coal into diamonds. Your job is within walking distance, and your route takes you through a park. That day, you see a young child drowning in the middle of a pond. It's early in the morning, and you don't see any other person nearby. I'm going to assume that you aren't a capitalist pig or a Wall Street wolf, and say that you would probably be willing to muddle up your clothes and arrive to work late if it meant saving a life. But this example is highly specific. My biggest criticism is that it doesn't make sense why a child would wind up in the middle of a pond if they couldn't swim, but I guess that's beside the point. What if there were more people around? Several psychological experiments suggest that people have a tendency toward inaction when witnessing tragedies in large groups. This is known as the bystander effect, and it affects poverty on the large scale. Why would you donate a dollar to a charity, when a billion other people could do the same, or when some billionaire could donate a hundred dollars to a dozen charities with less of a dent to their bank account? Or what if instead of a pond it was a lake, and instead of one child it was a thousand? Would you even bother swimming all the way out to try and rescue one of them, or would it be wiser to ask someone with a boat to save more of them? This is analogous to having one's government provide poverty relief. Finally, consider if the drowning children were half a world away, and you had no way of knowing if they were being saved or not. Now we can see why poverty is so difficult to defeat. People see it as too daunting, and they have very few means of knowing whether the contributions they make are actually going to those in need, or just to the corrupt governments that are too inept to feed their people. But while these obstacles might seem daunting, if we remember that the people who live in poverty are also the ones most likely to want to get out of it, we can focus our efforts on giving them the means to prosper, rather than tossing out pocket change in the hopes that it will reach them. This is where education comes into play, and that is a matter of such gravitas that we will save it for Chapter 14. That just leaves one final question about the drowning child; what if they were black?

Racism can be summarized in one interaction, and it's not elaborate or complex, but horrifyingly simple. You go to a gas station to use the restroom. Someone stands outside and tells you that it's broken. You say "thank you", and move along to the next station. This was what happened to the late Muhammad Ali riding in a limousine through Mississippi. The toilet wasn't broken, and the clerk referred to him and his friend as "boy". A similar thing happened after he won his first gold metal in 1960. He proudly wore his American pride around his neck, and was still denied service to restaurants in his own hometown. He was so taken aback that he chucked that hunk of metal into the river. He was only eighteen. This man was the heavy weight champion of the world, but was still wisely scared to initiate conflict with white folks in a backwater country, because you never knew where a bigot with a shotgun or a badge might be lurking. It's entirely possible that nothing would have happened if he had stood his ground, but he wasn't worried about one person. Anyone can fight one person, and a fighter like him could hold a half dozen. But when a quarter of the people on a continent were born and raised being told they were better than you, then you'd better believe they wouldn't hesitate to make that fictitious nonsense into malignant truth. That's the scary thing about racism, just as much with sexism, and even nationalism. It's always the same, no matter who it happens to. "It doesn't make a difference what you've done. You don't matter. You're not as good as I am. You can't accomplish anything worthwhile. Why? Because you're different than me. Everyone agrees. That's the way it's been, and that's the way its going to stay. You were born trash, and you'll always be trash."

What balderdash. I'm happy to say that as time has progressed, people have become more tolerant and accepting of one another than any time in history. But the scars of racism are still far from healed. Many wealthy and light skinned people born in the late twentieth century probably have a hard time understanding the resentment many poor minorities might have towards them. After all, many of them have been getting reparations such as affirmative action. Slavery has been abolished in the United States for 150 years, and the country is being overwhelmed with dark skinned immigrants. It's not like poverty is an issue for only people of color either, there are plenty of poor people of European descent. All of these things are true, and I would like to make it clear that the majority of colored folks don't hate white folks, and they acknowledge that the majority of white folks aren't bigoted nor should they be blamed for colonialism and slavery. That was in the past, and there is nothing anyone can do to change the actions of their ancestors.

That being said, people of European descent have zero justification to claim that they are being oppressed, because they are born with privileges that people of color don't have, ESPECIALLY if they're male. The reason is simple: while Europeans were experiencing the revels of the industrial revolution and accumulating wealth either through trade or by stealing resources from their colonies, every other part of the planet was being forcibly held under their boot and kept from accumulating wealth for fear of being shot. Natives in North and South America were decimated and Africans were forced to work without pay and denied the right to read and learn. Asia managed to do fairly well by isolating itself from the West while blending some of its culture with their own, but other than that, Europeans were allowed centuries of time to accumulate wealth while dark skinned people were treated as less than human and denied the right to make money. That so-called, "white-privilege" manifests itself in wealthy white neighborhoods, gang-invested ghettos, poor schools populated mostly by minorities, and the low representation of non-

Europeans in film, television, politics, news, academics, and generally most positions of influence. Again, all of the people who are to blame are long dead. It's just important to remember that everyone in the world who isn't descended from someone born on that tiny peninsula off the tip of Asia is still playing economic catch up.

Lastly, for anyone who questions whether slavery still affects us in the twenty first century, let's set the record straight. Slavery is as old as civilization, because forcing people to work without pay and minimal food is a good punishment for losing a war, and a better way to grow the economy. The chattel slavery that would become infamous in American history developed from the Romans, who the barbarians and Jews knew as crazy white folks. Many of the African slaves were even traded to the Europeans by fellow Africans. Slave labor was also used in the silver mines of South America during Spanish conquest. Parents even dismembered their children so that they wouldn't have to work in the mines for fear of mercury poisoning. The United States owes a large amount of its land to a slave revolt in Haiti. The uprising made Napoleon, who was fighting the British at the time, give up on American colonization and eventually sell Thomas Jefferson the Louisiana territory in 1803. That same 828,000 square miles of land would of course have to be cleared of the natives and negotiated for slave states before it could become part of the country, and we all know how well that turned out (for the confederacy). People defended slavery to the last possible moment. Some justified it by saying that the slaves were fed and clothed by whites, or that the economy required it. People actually claimed that they had "freedom to own slaves", which just demonstrates how arbitrary the idea of freedom is. Others even went so far as to invoke the Bible and The Curse of Ham to justify slavery, which just demonstrates how arbitrary the word of Yahweh is. They said all this despite the fact that treating people like mules and separating their families reduces the number of innovators in their population who could work on inventing electricity, cure diseases, or create the Internet a few centuries early. But they weren't thinking about the Internet, they were too busy growing cotton and tobacco. Again, the African slaves were forcibly kept poor, which means that when they were finally freed, they had zero money, little education, and a homeland full of bigots. They essentially had to start from scratch whilst every other American at least had the right to earn a wage.

But as bad as people of color have had it for the last few centuries, there is one demographic that has been handed the short end of the societal stick since the days of Adam. You might even say they comprise half of the human population. If you want proof of how ingrained patriarchy is in society, look no further than the Abrahamic religions and the English language. How many times do you hear the word "man" as a reference to all of humanity? Have you ever used the terms *mankind, man-made,* or *modern man*? People speak as if all of humanity was one gender. What's worse, a large number of quotes about women are demonizing. Hell hath no fury like a scared patriarchy burning witches, literally or figuratively. We even see it in the United States, the "bedrock of democracy", in that after its founding in 1776, it took 141 years for the first woman to join congress, 205 years for the first woman to join the supreme court, and in 240 years we have had exactly zero female presidents. To top it off, the first time one was nominated by a major party, she gets a conservative opponent who's primary criticism just so happens to be the blatant display of racism and misogyny that has been building up for several millennia. This bias is also seen internationally. Between 1901 and 2014, there have been 822 male winners of the Nobel Prize, and just 47 female winners. This again is baffling, considering how much more useful societies would be if they utilized the intellect of every person. That's not to say that there aren't numerous matriarchal societies in the world, they just don't happen to have much global influence.

As someone who identifies as a male, I can only speak second hand about how patriarchy marginalizes women based on the experiences of those affected by it. Even though this is technically true about nearly everything discussed in this book, due to the widespread minimization of the issue, I especially can't hope to completely articulate its full ramifications. But I'll do my best to make the case that inequalities have no place in modern society.

Women historically have faced a unique type of discrimination regarding sex. Beyond the inferior wages and apathy that other groups face, women also have to deal with the sexual predation that arises when their coworkers (especially supervisors) objectify them. So, as one would expect, sex is the source of sexism. Or in other words, the fact that we are animals who have to procreate in order for our species to survive is the source of sexism. At the end of the day, it is a battle between our primal urge to pass on our genes, and our complex reasoning that dictates us to not sexually harass each other at work. This is probably why, I would argue, sexism is the most difficult form of discrimination to fight against. Race is a made up concept, and classism only exists because of money, which doesn't buy happiness (even though money has its roots in food, something we also need). If we don't have sex, then all of humanity ends.

What's more, we are biologically predisposed to place sex over everything else in our life, which sort of explains why some people desperately try to keep boys and girls separated when they're teenagers. One can see how this mentality erodes society when taken to the extreme in Margaret Atwood's *The Handmaid's Tale*, which chronicles how a religious society that vilifies sex represses all members of society and only leads to unrest. So while my main focus is the marginalization of human females in society, it's important to remember that like men, women enjoy sex, and their hormones can make them just as crazy and prejudiced. Also like men, they don't like being taken advantage of. The real problem is when one gender objectifies the other and sees them as being something other than human, be it angelic, demonic, or simply a tool for procreation. And with gays, lesbians, bisexuals, asexuals, transsexuals, and everything in-between, we can quickly see why so many people find it easier to just be intolerant bigots than imagine other human beings complexly. Humans are just so complicated! But at the end of the day, we are all equally human, all made of stardust, and should be treated as such.

The easiest way I have found to grasp this issue in the most unbiased manner is to compare humans to other animals, and attempt to derive how marginalization can occur as a result of our heightened intellect (or lack thereof). First we can look at societies that may be considered by some to be patriarchal in nature. Take for example lions. Males of the species have the primary task of finding a pride, asserting dominance over the patriarch, and fathering cubs with all of the females, killing off the cubs of the previous ruler, who's genes could be considered inferior (if not a waste of resources). The male then serves to protect territory from the likes of hyenas and rival males, in order to best insure the survival of his genes, which is why they have their symbolic manes as protection. As a result of this competition, male cubs are far less likely to reach adulthood than females, but if they survive long enough to become king, they get to sit at home and be first to dinner. On the other hand, lionesses famously act as the huntresses, acquiring food, training cubs, and raising young. They sustain the pride, and as a result are smaller and sleeker, with greater agility to capture prey, and they also work collaboratively. They may migrate as well, but unlike male lions that often roam with another male or alone, the females may potentially have the extra burden of feeding cubs. Similarly to males, lionesses also fight to ensure the passage of their genes, but they do so more subtly by excluding females from outside of their pride. This is actually a relatively common tactic amongst female animals. Even in humans, it often makes more sense for females to be passive aggressive, because they are the only ones who can nourish offspring. In any case, we can see that although they serve different roles, both sexes rely on one another and have a mutually beneficial relationship. As a final example of patriarchy, one may look at lowland gorillas or chimpanzees, in which a dominant male gets first choice of mates and leads negotiations and migrations. It should be noted, however, that females also have a hierarchy, where the dominant female gets first access to food and has significant influence over migrations based on where the food is. Generally speaking, these roles are common amongst animals; males compete for females by demonstrating favorable genes that maintain the species, while females acquire the resources necessary to keep everyone alive and nourish their young, and many times get to choose amongst male suitors. But nature is creative, and there are plenty of instances where females make all of the decisions.

Now let's look at a reversal of roles. Nature is abundant with matriarchal societies, some even to great extremes. The differences between the male and female dominated species seem to be surprisingly minimal overall. Females still require more resources logistically, but instead of being chosen by dominant males, females get to choose from prospective males (which isn't dissimilar to what I just explained). In less complex species, males are more or less just around to provide gametes. Bees and ants for instance only have a few males to mate with a queen, with the majority of the population being comprised of females. Similar demographics can be seen in some fish species. The female anglerfish is notoriously tens of times larger than the teensy males that end up fusing into the bodies of females in order to mate. Similar kamikaze style mating is observed in numerous arthropods like praying mantises and black widows, which actually get eaten should their mate need nutrition for their offspring. Animals with higher complexity exhibit far more inclusive matriarchies. Elephants and orcas, some of the most intelligent vertebrates, both live in large social groups run by females. Orcas are slightly more extreme than most species, with both male and female offspring living with their mothers for life, and the matriarchs living up to ninety years, decades past menopause. Orca offspring are even observed to be several times more likely to die if their mother dies, speaking to the influence that they hold over their societies. Elephants also spend several years raising their young, but their males will leave the family in search for new groups after fifteen years or so. Elephants resemble lions in this regard, which is why some actually argue that lion prides are matriarchal in nature. Finally, one of our closest relatives, the bonobos, also live in female-run societies. Being nearly identical to chimpanzees, bonobos behavior is remarkably fascinating to observe. The main social difference between the two seems to be the more liberal nature of bonobo sexuality. While chimps may kill the offspring of rival males, female bonobos tend to have so many partners that males can't distinguish the lineage of any of their young. As a result, bonobo families tend to have less familial conflict, and more universal child care.

Finally to humans. By observing the behavior of our fellow animals through a scientific lens, we have been able to see all of nature's complexity, and how every species is unique in the way they form bonds and find mates. There are so many creatures with so many radically unique lifestyles, that it can be difficult to distinguish the causes of certain traits. More than anything, the observations in the wild seem to indicate that relationships between opposite sexes are wide and diverse. Regardless of their behavior, they simply depend on what is most beneficial to the survival of each species as a whole. When it comes to social structure, tying together causes and effects only becomes more difficult with more complex organisms. Whether an animal is complex or not, they all share the need to pass on their genes, and all animals, regardless of sex, take the genetic success of their partners very seriously. So, without looking at pay gaps or human culture throughout history, one might be able to find the heart of patriarchy through the lens of natural selection. As I've stated, the best way I can think to go about this issue is by bringing to light what I believe best represents the argument for my demographic, and to work from there to find a compromise. It's unfair, but this is my book, so if it bothers you write your own. Ironically, the strongest case I have heard for why women have been marginalized since the agricultural revolution comes from a *Cracked* article, in which writer David Wong lays out an argument for why society trains men to be misogynistic pigs and wolves on the hunt. In his conclusion he writes:

> "It's all about you. All of it. All of civilization. So where you see a world in which males dominate the boards of the Fortune 500, and own Congress, and sit at the head of all but a handful of the world's nations, men see themselves as utterly helpless. Because all of those powerful people only became powerful because they heard that women like power.
> This is really the heart of it, right here. This is why no amount of male domination will ever be enough, why no level of control or privilege or female submission will ever satisfy us. We can put you under a burqa, we can force you out of the workplace – it won't matter. You're still all we think about, and that gives you power over us. And we resent you for it."

Now, I am aware that I cannot speak for all men, and I know for a fact that hearing any justification for millennia of poor behavior is like a white person apologizing for slavery. However, I personally find this line of thinking to be more or less true, considering our evolution as a species, and long history of not treating each other like human beings, regardless of how we look. It echoes a central theme in all social struggles: both parties feel like the victim, and neither one wants to acknowledge the complaints of the other side. It's classic hairless ape behavior. But make no mistake; regardless of the circumstance, there is no excuse for bad behavior, and no redemption that can be given through words alone. Only action can overturn action. But good behavior begins with good ideas.

Some may complain that sexual attraction makes it impossible for men to fully treat women as equals. Women have the power over life, and just as in many animal species the females have sexual power over men. In many ways males are far more disposable, as evidenced by the fact that our haploid cells never stop being produced and we don't require the resources needed for pregnancy. Therefore, it may have been justified to keep women at home where they could raise the next generation while men put their lives at risk in the outside world. This would have only been compounded by how inconvenient the mental cycle can be without modern amenities. Perhaps this line of thinking would have been true back when people didn't have an understanding of their own anatomy and hormones, and couldn't control population growth without separating the sexes. However, this mindset has no place in modern society. The universe is much bigger now, and you can't live on a small continent anymore. Not only does it not take into account people who don't identify as being between male and female, or the people who aren't even attracted to the opposite sex, but it also doesn't take into account advances in birth control, medical discoveries that mitigate the effects of periods, or a full understanding of how life even arose in the first place. What's more, when people are able to fully look past external features, and instead focus on the brains inside those shells, they will be able to achieve ever-greater prosperity. Hormones will always make this difficult, but they are critical to the survival of not just humans, but nearly all eukaryotic organisms. We just have to understand that sometimes those same neurochemicals make people act in ridiculous ways. After all, that's just another thing we all have in common.

With all the hate, fear, and bigotry in the world, its no wonder people have trouble escaping mitigation. When you hear so many voices telling you "no", it becomes increasingly difficult to tell yourself "yes". Humans are gullible in every way. We are also daftly subservient to authority. As the Milgram experiment showed, people will even go so far as to kill someone if some quack in a lab coat tells them it's for research. Tell someone they're descended from some god, they believe you, tell them they're descended from feces, and they believe you. Yet with all these assumptions throughout history, nobody ever bothered to look down with an electron microscope or up at the sky with a spectrograph and say, "Oh, I guess we are all literally made of the same thing, and when we die we just get eaten by microscopic bubbles. I don't know why I got so insulted earlier." Although in defense of past humans, it took us around a hundred and ninety thousand years to figure out agriculture. They did their best; and people are growing more compassionate every day. That's why it's so important for us to expect greatness from ourselves and accept nothing less from each other. What you do matters, and you matter as a person. You can contribute a helping hand. We're all the same. That's the way it's been, and that's the way it has to stay, for it is written into the laws of nature.

So how do we continue progressing in our tolerance? A major step would be to stop referencing people's physical traits in the first place. Asking a person their race or gender is both racist and sexist, as well as prejudiced. Just by asking, you are insinuating that the race or gender of the person actually matters, when it doesn't. It prevents equality from ever coming to fruition. How many ways can I say this? It's just a bad idea. It puts people in boxes, and it's deplorable. I don't care where you come from or what you look like, you deserve to go to this school, and you deserve to get this promotion. I only care about your past actions, and you had better darn well be able to justify them with honesty and sincerity. That's the inherent problem with the conservative argument. They mean well, but they keep putting people in boxes, like when an "American" tries to apologize to a Native America or when they call them "red". Just stop bringing it up, and you help solve the problem. Redeem yourself and your tainted history through action, not though empty words. After all, Europeans used kind words when they first met the natives. But now they've almost gone the way of the not so tasty Dodo Bird. So, with the utmost tact; PLEASE JUST STOP ASKING WHERE PEOPLE'S ANCESTORS WERE BORN, WHAT THEIR ORIENTATION IS, OR WHAT THEIR INCOME CURRENTLY IS. Only ask what their experience is, if they've hurt somebody in the past, and if they've tried to redeem their mistakes. If everybody tries to keep balance and die in the black, then please just be nice to each other. Otherwise you are sort of interpreting the whole "life" thing incorrectly, and contributing to death, chaos, and world suffering, so shame on you.

The saddest thing, I think, is the waste of talent that results in conflict, the needless disposal of valuable resources. Think about every war, genocide, massacre, or homicide in history. We took those people, potential innovators that could research cancer or space travel. People that could have made food more delicious or accessible. People that could change the world, and we blew them up to smithereens, along with historic art, diverse phyla, and maybe dinosaur bones. A lot of narrow-minded people in power have murdered others. They didn't trust their fellow humans, and even today we still don't trust each other. And that's why we're all still stuck here on Earth, like chicks too afraid to leave the nest, because we forget to treat each other like Earthlings, and we kill each other over problems we could solve simply by working together and communicating our thoughts. If people need an enemy so badly, then why not declare war on death? Because that is certainly the one thing we are all not, dead. Let us make it our duty to preserve the life that has fought so hard to combat the cruel apathy of the cosmos. Of course in the end, there is no defeating death, but conceding would have meant that we would never have been born. History is wrought with underdogs fighting valiantly against certain defeat. There's a kind of reverence for that sort of perseverance.

Despite all of our shortcomings, the twenty-first century has heralded the most peaceful era in human history. People are living longer, less nations are going to war, and we have the Internet, where people can conduct all of their business at home and not deal with the people they hate if they're clever about it. The point to be made here is simple. As Charles Kenny would say, things are getting better. People are living longer and hating less. But there's still more work to do, and it's easy to find out how much work needs to be done in your own particular community. Look at your neighbors, your coworkers, the people who shop at your local market, do you see people with ancestry from all around the world? Do you see an equal gender distribution? Does everyone dress differently and have their own personal style? Did they all have access to the same quality of education you did? If the answer is no, there's still more work to be done. We can't help how we are born, but if we are smart we can decide how we die.

I still don't really understand why people dislike each other. I've watched two people become friends, befriended both of them, and then watched as they became enemies, despite them both still wanting to be friends with me. And during that drama all I want to do is save time and just have everyone be friends again. I'm sure this isn't unusual and probably happens to most people. It's hard enough to write down all of the equations of a single quantized hydrogen atom, especially during an exam. Human beings are far more complicated, and lazy. In fact, we're mostly just lazy. We are in fact so complicated that we would rather just shoot each other than try to imagine another person with the same complexity as ourselves. The thing about that though, is that it's kind of justified. To fully understand every person you've ever met, you would have to sit down with every single one, and exchange stories from every moment in your lives. Clearly, this is ridiculous. People are very sensitive about their personal information, and more importantly nobody would get anything productive done. Ergo, the most pragmatic solution is to simply assume that everybody is just the same person living under different circumstances, and to trust each other until they do something wrong. Before we can trust each other, we need to all shut up, put our guns down, our hands up, and ask, "why are we killing each other, instead of figuring out how to get off of the Earth before the Sun dies?" Nobody has any idea what's going on, but people do the best they can. If we figure out the whole saving humanity thing, we can all sit back in our space couches while watching *Simpsons* reruns and *Buzzfeed* videos using quantum computers.

Hate breeds hate, and I think people only hate each other if they legitimately feel like another group is attacking them. In other words, they have to think so highly of themselves, that they are too blind to see that someone else also feels oppressed and marginalized, exactly like they do. And if someone does hate you, it's probably because of an idea that they think you stand for, and not for the person you are. All in all, I've found that like all things in the universe, people are different arrangements of all the same ingredients. There are no boxes to put people in, just attributes that any given person may or may not have. You have no idea what sorts of messed up things other people have seen, and you should be thankful you haven't. There are too many people, and all of them are too unique, and hating people is too complicated to do properly, because you need to know EVERYTHING they stand for. That's too much work, being rude is too complicated, and I just don't care anymore. I've decided to give up and just be as nice to everybody as possible. I've got more interesting things to worry about than your gender, ethnicity, sexual preference, nationality, wealth, and in certain circumstances even your history or intelligence. I'll treat you with equal respect to what you show me, or probably more just to be safe. So kindly get over yourself.

Peace & Life

To be, or not to be? There is no question. Whether the atoms in your body are in space or six feet under, you're still a part of the cosmos. But that doesn't quite answer the question of what it means to be alive. What's the difference between life and, well, everything else? To answer this question we need to understand what happens to us when we die. Death occurs when an organism no longer undergoes the chemical reactions that allow it to grow or replicate. For animals cells, death occurs when oxygen and glucose are no longer taken into the cells, cellular respiration stops, and the cell breaks apart or gets devoured by another organism. Some say the cause of cell death lies in DNA, specifically in the ends of chromosomes known as telomerase. If this were the case, then telomerase would be the candle on the wick of a cell's life. After every cell division, the amount of telomerase decreases slightly. When it runs out, chromosomes fray apart and cells appear to no longer be able to replicate. The key is that those ingredients don't get transported outside time and space. Once an organism dies, it is nothing more than a salty bag of water without anything coming in or going out, essentially like an very complex rock.

The foremost difference, I believe, between the inert and the living, is that the former is subject to the laws of nature, while the latter can harness them. One of the defining traits of a living organism is its ability to siphon the energy from the environment and use it to grow and multiply. Furthermore, living things adapt to their environment, and change their environment in turn. Life adapts through natural selection, as the organisms that can survive reproduce, and the ones that can't die off. On the flip side, life shaped the Earth by giving it free oxygen and through the hairless apes that reintroduced carbon into the atmosphere and made the dark side of the planet visible for the first time as a result. In that sense, rigidity is death, and pliability is life.

More than anything, we should never vilify the notion of death. There's no sense in fearing the inevitable. To demonstrate, consider the following situation. It's the dead of winter, and a wolf needs to eat a rabbit to survive. Both the rabbit and the wolf are mothers, and the wolf has already fed the father rabbit to her cubs, and now she risks starvation if she doesn't catch the other rabbit, and she has no way of following the mother if she escapes into her den. Either one of two things has to happen. Either the rabbit dies so that the wolf can eat, or the rabbit gets away, leaving the wolf to starve. Which outcome is more favorable? (I apologize to any ecologists if this scenario isn't realistically compelling). It should be clear that no matter who survives, their luck should be celebrated, and the death of the other should be mourned. But one can't really favor one or the other with anything more than opinion, because both parties are just trying to survive. Regardless of who lives that day, death is still inevitable for both eventually. But at least they managed to have offspring. Living is tumultuous like that, which is why surviving to the end of every day should be celebrated as a triumph over death.

One can define death as a natural consequence of entropy, or chaos. In the same way time always moves forward, and the universe keeps expanding and cooling, all things that once lived eventually die. Life is the antithesis of entropy, just as death causes the undoing of complexity. That's the inherent difference between a rock and a bacterium; some level of complexity. Life funnels the second law of thermodynamics, and multiplies as a result. And because all life is ultimately powered by the Sun, we technically only exist because of gravity, a result of whatever causes energy to warp spacetime. The universe strives toward equilibrium and total homogeneity, but living things and gravity counteract that diffusion, and replace it with structure, and preserve that structure through diversity and despite adversity. Essentially, the universe is trying to kill every living thing in it, but life is stubbornly resilient. It's important to note, however, that the law of the universe is a tendency toward entropy. There are mathematically just fewer organized states any system could be in as opposed to disordered ones. The universe is expanding bigger and bigger as energy is dispersed evenly throughout space, which is the source of that law. You need to eat something that used to be alive in order for you to live, and something else is going to eat you when you are no longer able to put out enough energy to counteract entropic chaos. Most of the food you eat is just lost as hot infrared photons anyway. Entropy can only be counteracted in one universe for so long, eventually everything dies. But this isn't really new information. Beings have known this long before hairless apes were roaming around. My favorite version of this life lesson is best portrayed in the movie *Stranger than Fiction*, where Will Ferrell plays a robotic tax collector, cold and calculating. Then, one day he begins to hear a voice, and it's narrating his life. Eventually he finds out that his author is a real person, and she is famous for always killing her protagonists. While at first understandably distraught, with some help he realizes that if he knows he's going to die, he might as well just live his life, and make it the one he's always wanted. The simplistic beauty that unfolds addresses the question of how we should face our mortality, and how we would react with the exact knowledge of when and how we're going to die. This has been done by countless other sources of media throughout time. Sometimes everyone just needs a little reminder to hurry up and seize the day.

A great number of minds have pondered the future of life. Long from now, in the deep future, millennia after the Sun fades away and The Milky Way and Andromeda merge, where will life be? I'm going to assume foremost that some sort of intelligent beings will have made it out of the solar system by that point. At that time, they may or may not be what we would consider "human", but if they managed to become space faring, I would think that they would have retained their humanity enough to collaborate with each other, even if their genes are altered form Homo Sapiens. In order to survive, they will have had to develop some way to harness the energy output of the universe. Whether they use stellar photons, hawking radiation, or somehow find a way to harness the vacuum energy pervading space, they will need some way to feed themselves. Another possibility is that they may be able to assimilate the elements found in planets, moons, and asteroids, mining them for natural resources. Furthermore, despite what many alien invasion films would suggest, I find it highly unlikely that sentient beings would tamper with alien planets inhabited by life. For one thing, they will almost certainly have pathogens that the other species isn't adapted to, although they could probably develop medicines to counter them. More than anything, I'm convinced that the only way for any living thing to survive in space for any extended period of time will only be able to do so by understanding the history of the universe and the scarcity of life within it. I'm sure they could harvest resources without murdering an entire planet covered in organisms.

I also find the whole notion of alien civilizations finding less advanced civilizations highly suspect. First of all, I don't think most people who believe that aliens have visited Earth understand just how big space is. I'll remind you that traveling at light speed, it still takes four and a half YEARS to get to the closest star. Not only that, but creating wormholes seems entirely theoretical at this point, requiring stars of energy to be concentrated into an infinitesimally small space. That's not to say this can't be done, but it doesn't make any sense for beings with complex neurology to evolve and figure out how to do such a thing centuries or millennia before we ever could. I don't just say this because it took four billion years for cells to group into animals, and for animals to develop brains complex enough to begin fathoming the universe. Rather, I say it because the universe is uniform. One has to remember that the whole (potentially infinite) cosmos is the same age at every point, and has an almost perfectly even distribution of matter. That means that every point in the universe where life managed to take hold had the exact same amount of time to evolve and develop. Barring all of the worlds that didn't have conditions suitable to intelligent beings with opposable thumbs and voice boxes, every world that could possibly have life like us has had the exact same amount of time to get into space, likely after dealing with genocides and mass murders of their own. So I find it much more probable that all life in the universe is at the same point in development, and certainly not so far beyond our technology that they were able to traverse thousands of light years to happen to bump into our tiny spec of dust and water. But I can only speculate. Maybe aliens could arrive from a separate universe, if those exist.

One major concern, that even I find deeply troubling, is the idea of life developing into a singular mass consciousness. Some might refer to this as a hive mind, a famous example being the Borg from the *Star Trek* series. This is mostly disturbing because it deals with the notion of free will, something countless of philosophers have questioned since we had time to think. A hive minded civilization could potentially terrorize the galaxy by assimilating all matter into itself, including all sentient life, reducing all individuals to nothing more that the unified extensions of a higher being. It's not altogether clear whether this would be a bad thing or not. Sure, we enjoy what we think is our ability to make decisions, but we also squander that ability to do some quite idiotic things. This is best illustrated in an episode from the series *Rick and Morty*, an animated science fiction show that chronicles the adventures of an alcoholic super genius freedom fighter named Rick and his grandson Morty through an apathetic and wacky multiverse. Every episode tackles deep metaphysical and existential questions about our place in the universe, and one episode explores the notion of freewill by using the trope of a hive mind. In it, Rick runs into his ex-romance, a hive mind named Unity who has just recently assimilated an entire planet of sentient beings. They revive their passionate relationship through partying and copious fornication, and Rick's grandkids are left in the care of Unity, while the rest of the hive mind parties with Rick. One of the grandchildren spends her time trying to force the assimilated beings back into their original selves. However, to her chagrin, the hive mind has brought world peace to the planet, turning pedophiles and drug addicts into contributive members of society. Eventually, Rick and Unity party so hard that Unity actually loses control of some of the planet's inhabitants. Renewed with their free will, the aliens promptly start a race war riot based on the types of nipples they have. While this is admittedly a comical and somewhat juvenile depiction of humanity, it's not far from the truth. Real human beings have gotten into countless skirmishes over superficial things like which god they worship or the color of their skin. It begs the question, do we deserve our free will if it allows us to act like idiots? Is peace and harmony a worthy enough cause for the exchange of our free will? I would argue that individuality is worthwhile, so long as we can learn to respect our differences. No matter what, people can't seem to agree on anything besides the well being of their kin. However, a sufficiently advanced civilization should understand that all living things are to be considered kin.

But while a productive life should always be encouraged, it is just as important to remember not to demonize death, nor to think that just being alive is always better than being dead. After all, we animals need to kill other life forms in order to eat, and that has benefit to the ecosystem by keeping prey animals from overpopulating and overusing their resources. However, this also extends to predators—especially humans—because we are the planet's apex predator. We too can easily overuse our resources until there is nothing left. Something isn't valuable just because it's alive, that's nothing more than a slothful oversimplification. The value of anything, alive or not, is determined case by case. Like all things, the value of life is circumstantial. Plants have value alive because they produce the air we need to live, but they are also valuable dead because they feed the animals we eat and act as the foundations of our homes.

Human beings have near infinite potential, but only if they grow up under the same circumstances. If an egg is fertilized by someone who only wants to assert their power, but has no intention of raising a child, or if a couple are psychologically or financially too inept to raise a human being, then that child may grow up dysfunctional when they could have been more fulfilled had they been born with loving parents. People who claim that life is always precious never seem to be aware of tyrannical dictators that initiate genocides, and who may have been better off just not being born. Nor do they think of mindless bacteria that multiply by devouring the body's resources, cause plagues, but also allow humans to digest their food, act as a basis for global food chains, and of course made all aerobic life possible. Something as complex and elegant as life could never be boiled down to mere good or bad. Without Hitler, there would be sixty million more potential innovators who could have contributed to society, or there could have just been another tyrannical megalomaniac that would have done the same thing, because they hadn't experienced the horrors of a second world war. Similarly, we need to kill pathogens for public health, but without certain kinds, we would either be dead, or have never been born. The point then, is that life is neither good nor bad, it just is. But we can work together to decide under what circumstances it is best to keep something alive, and optimize those conditions, while deterring life that could cause unneeded suffering.

After looking back on all of human history, the ecology of other life forms, and the origins of the planets and stars, I have found no good reason for people to kill each other. Animals eat the plants that get their nutrients from the fungus that break down the other two, but people are smart. We can cultivate the land, develop artificial nutrition, and people have to usually be quite desperate to eat each other. Armies and states go to war for freedom or food, but people are smart. We have refined language that no other creature can match, we can learn about each other's lives, our hopes and fears, and we can compromise. Individuals have declared themselves rulers of people and nations, they herd them like sheep, work them to death in camps, make them toil on farms or in factories, they steal people's humanity and christen themselves as divine. Yet none of them were aware that what they were ruling over was but a small patch of land atop a moat of dust suspended in a sunbeam, nor how trivial their decades of juvenile grandeur were to the vast empty space around them. That's all murder really is, childish. Like everyone else, they too eventually turned to the dust that fed the next years crops. That is, of course, unless people built them shrines with shiny rocks. In that case bandits or archeologists would raid their tombs to make a profit, after the whole civilization changed by the generations.

Whenever I hear about some buffoon shooting other people, I always think about small neighborhoods, where parents and grandparents take their children to the park so the parents can rest and the other two generations can socialize. I think about birthday parties I've hosted for work, all the curious minds celebrating life. I figure those types of things have to pop up anywhere people are and anywhere in time. People are very resilient to hate, they just need to trust one another, and earn each other's trust in kind. Then anybody is liable to party. When people feel justified in harming others, it often indicates that they see themselves as being wronged by somebody or something, and they see violence as the only way to rectify that wrong. I have faith that when people understand how rare life actually is, they will be much more inclined to preserve it. Regarding people who have already been incarcerated and expunged from society, if they can earn back the trust from the people they've wronged, and share a drink together, I'm sure they too can find redemption. And if the criminal can forgive the society that led to their crime, then everyone will have forgiven each other. If they refuse to stop being counterproductive, then I would leave the matter to a professional judge and jury to decide.

Back to the topic of the metaphysics of life and death. It seems particularly illuminating to define both words in terms of the history of the universe. Picture a ball drifting through empty space. Newton's first law states that an object in motion stays in motion and an object at rest stays at rest. This isn't observed on the Earth because we have an atmosphere—which we need to breathe—and air molecules exert a pressure on moving objects that slows them down. But if the object was moving in space, without any light or matter to interact with, then it will neither accelerate nor decelerate. However, this may be called into question when considering that the object is also a blackbody. If it were a star, it would be glowing in the visible, but even if it were a billiard ball, it would still be glowing in the far infrared and microwave. It's not in thermal equilibrium, so it would cool off due to the second law of thermodynamics, and radiate away energy in the form of light, or in other words it would cool off. Since it's losing energy through light, that also means that it's losing mass, if just slightly. That energy is lost into the vacuum, which expands as a result of diffuse energy from the thermal output of the object. Now, with less mass, the object is putting a smaller dent into the fabric of spacetime. Does it feel less inertia? If momentum is conserved, then a smaller mass might imply that the object speeds up. If so, it would continue to speed up, distributing its energy throughout the vacuum, stripping off layer after layer, until it is only a proton, then a soup of fundamental particles. As it speeds up, time passes more slowly for the object, until it becomes like a massless boson of pure energy, first a gamma ray, highly wound up. Then, like redshifted light falling into a black hole, the photon would become unwound, go dark, and be assimilated into the vacuum. The certainty in the location goes from near certainty with the massive ball, to complete uncertainty as the energy decreases while the object speeds up, and since there is energy throughout the vacuum, those basic particles can pop up like foam bubbles anywhere throughout the universe, giving a complete uncertainty about where any one may be in the universe, but a complete certainty in their energy, because the universe is everything, and it is now pure diffuse energy. The universe has no dents from gravity, because energy is evenly distributed, and so from a sufficient scale, the universe appears flat. Particles are just dense blips of wavy energy that can occur anytime, anywhere, with exact and precise amounts of energy determining the type and properties of a particle. What is this thing that denatures matter back into energy and throughout space as time goes forward? Gravity. It increases space and distributes time, because a moving object escapes time by losing mass. The more massive an object is, the less time there is in space, which would explain why everything was together at the big bang. Whatever gravity is, at its core it interacts energy with space and time.

This would also imply why the universe tends toward entropy, and why the universe is expanding. As the universe reaches equilibrium, all energy is converted into an "empty" vacuum, which is allowed since the total energy density of the universe would remain the same, not violating the first law of thermodynamics, which states that energy is neither created nor destroyed. It might be that because we live in a dense part of the universe, the vacuum energy we measure is inherently greater than the "true" vacuum that is observed as cosmic expansion. That would also explain why space is expanding equally between galaxy clusters (areas with high density), but never within highly energy dense areas, like planets or the human body. If we could learn to harness the energy within a vacuum, we may be able to stop the universe from expanding forever in a big chill, or at the very least slow it down. What do you think? I have heard of no experiment to verify this, but it seems sound. One possible point of contention might be the fact that the expansion of the universe is speeding up every day, but this would simply mean the average vacuum energy density is simply not linearly proportional to mean particle energy density. This is also the result Stephen Hawking calculated for the evaporation of black holes, that smaller ones evaporate more quickly than larger ones. Plus, as we know, life tends to be quite obstinate when it comes to obeying nature, especially entropy. We just can't help but build structures and maintain them. Perhaps existence isn't a hopeless battle. We as living organisms simply need to restore balance to the forces of nature, by giving them order again.

That's what life is. It restores order, and harnesses energy to grow. It gobbles up the energy in the universe, microbes from rocks and gas, plants from the Sun, predators from prey, and the human mind from everything. When they die, they get assimilated into the organic and inorganic alike, but they are always part of the universe. You may not be the reincarnation of Jesus Christ, but some of the atoms that were inside of him will certainly pass through you at some point, and in the rocks, and eventually back into space when the Sun and Earth die. By the same token, you may be part tyrannosaurus rex. But you might also be part Hitler, so don't get too excited. In accordance to the first law of thermodynamics, everything in the universe is the same, alive or dead. The only difference is how complex and sentient it is.

With life being so rare, you can imagine how someone like me—who has flown to the edge of the universe and back a stupid number of times, lived vicariously through the eyes of many writers, poets, photographers, painters, filmmakers, and living people, and seen the suffering of thousands of innocent men, women, and infants—would consider it childish for someone to want to kill someone else for any reason, ever, at all, period. Seriously, you're going to die, so is your worst enemy, and as far as we can tell so is the universe. So why would anyone waste their time causing something to die prematurely? I wouldn't go so far as to suggest that nobody should ever swat a fly or eat meat. After all many animals can kill you, and bacon is delicious. One should just be cognizant of what they need to do in order to survive, and not wastefully murder something just for cheap thrills.

As technology and medicine continue to advance, and with altruism and knowledge on the rise, one could conceivably envision a future in which death is optional, and only taxes are mandatory. We may even one day break free of our flimsy biology, instead living through technology able to survive anywhere in space for trillions of years. Perhaps humanity will outlive even the dimmest red dwarves. These sorts of questions have answers that lay far beyond my lifetime, and so I'll leave it to future generations to decide how long they want to live. But if I were to place a wager, I think that after living for a few centuries in the bodies of twenty year olds or computers, most people would accept death's sweet embrace. After all, life, by definition, is nothing but hard work. Life is but a walking shadow, a poor player that struts and frets its hour upon the stage and then is heard no more. It is a tale told by lots of idiots, full of sound and fury, signifying nothing. But it sure is fascinating, don't you think?

12 | Collaborating with Nature and Each Other

"At present I do not know, but one day you will be able to tax them."
— Michael faraday, when asked by a politician what good his electrical discoveries were

We all know that actions are easier said than done. But I believe that thinking deeply is more difficult than either one. I would argue that this is the reason why innovations in society occur in sudden leaps interwoven with slow, gradual steps. This is also why conflict and ignorance are preferable over tolerance and peace. Conflict is easy. In fact it falls in line perfectly with the entropic nature of the universe. Of course it feels good to destroy things; putting effort into building something is so much more difficult than tearing it down. Bridges and buildings would crumble away if they were simply left alone, and if they are to last someone needs to maintain them. But that's what life is about. Life is hard because it goes against the nature of the cosmos, and that is possibly the singular greatest truth in this entire text. But the fact that living takes effort is what makes it rewarding, and the same is true for fostering relationships with other people and understanding nature. It's one thing to just let the world toss you around; it's a completely separate matter to direct your life in response to the world.

Government

Imagine an Earth in which only one person resides. That person also has all the resources and know-how to build a home and keep themselves fed. There are a few animals, but for the most part they are in complete solitude. Sort of like Minecraft, except far smaller. In this hypothetical world, that individual would essentially have unlimited freedom. They could go anywhere they wanted, and use any resource they needed. Essentially, they could say and do whatever they wanted without consequence.

Now imagine that they encountered another person, the only other human being on the planet. Furthermore, it just so happens that this other person also has a complete set of knowledge on how to survive, and they don't find the other person particularly attractive. In fact, they think the other person is kind of a smug prick. They even eat their bread butter-side down! So for all practical purposes, these people are completely useless to one another. If anything it would be more accurate to say that they are actually a hindrance to one another, taking half of the other's planet, eating the other's food, and using their resources. Now, the two might come to an agreement to not trespass on each other's territory, but who's going to punish either of them for stealing each other's cattle, or more simply just killing the other and taking their stuff? And before you say anything, they don't have the fear of hellfire to motivate them, they're both atheists. Multiply this by several million people, most of who actually think some divine power commands them to do something, and you can begin to see why human beings invented government (and why many governments invoke religion).

The question then becomes, "How do we organize our government in a way that reflects our society?" The answer this question depends on who you ask, and on what role they think the government plays in their lives. Much like many ancient traditions in human history, the origin of government is marred by time. The way a person views government speaks volumes on how they view humanity as a whole, and it essentially boils down to how much influence the government should have over the lives of citizens.

The role of government is often viewed as a debate between three major philosophies. The champions of these views are Hobbes, Rousseau, and Locke, and they argued their case for the role of government based on how they imagined life was like for human beings before the advent of government, the so-called, "state of nature". First there is the view credited to Thomas Hobbes, who perceived the role of government as a necessary sanction to protect us from our animal nature. According to Hobbes, the state of nature for humanity was dreadful, and without laws humans would indulge in wanton destruction. This is the view largely held by those who see anarchy as analogous to chaos. For Hobbes, the best form of government would be total monarchy with no citizen representation.

Then, there is the counter argument presented to us from Jean Jacques Rousseau. According to him, the state of nature was peaceful, and human beings are better off left to their own devices. This line of thinking would suggest that government is a corrupting force that snatches away the freedom of citizens. As such, Rousseau would contest that pure democracy is the most favorable form of governance, where the government only needs to be involved in matters that concern everybody. The bridge between these two extremes comes from John Locke. For him, it was clear that all humans are born with natural rights to life, liberty, and the pursuit of happiness. However, we also have freedoms beyond our natural rights that must be subject to change and even suspension for the good of society. Locke's ideal form of government would be a representative republic of officials elected by and servant to their citizens. One can see the history of the world's governments as lying along the spectrum outlined by these three viewpoints. As it happens, Locke's balanced view seems to be the most accurate and successful in modern history. The problem with both Hobbes and Rousseau is in their extreme outlooks, and one can see evidence of this in the evolution of human society worldwide. For one thing, Rousseau's view of the state of nature unwinds in the light of archaeology and evolution. Not only were the first humans just one animal among many, but even today we are still subject to the same laws of nature, despite no longer being menu items for large predators. It may very well be that our ability to organize ourselves through the medium of government is the sole reason we are so successful as a species. However, this does not defend the position of Hobbes either. It is rather common knowledge that the very worst regimes in history were the ones in which all the power was held by a singular monarch or dictator. It is clear that when too much influence is given to either the masses or to the government, the potential of a society is limited. At its core, a government is designed to serve and protect the citizens of its regime. A good government depends on its citizens just as its citizens depend on it.

This question of whether the government should be strong or weak is at the heart of politics today, most prominently in the United States. American politics is so often constipated with nuances and ads that it often seems too convoluted to understand. However, beneath all the rhetoric and campaigning lies the same fundamental issue: should the government be large or small, and when should it be involved in the lives of its citizens? This is the very ground upon which the country was founded, and this eternal question is what drives policy. Every aspect of American life is shaped by the principles of its founders, who had just escaped from a monarchy that didn't allow them representation. Their task was to both protect the people from the government, while also protecting the government from the people.

Although the values have changed with the times, the two party system that the country is known for has been the staple since day one. Today's Republicans and Democrats descend from yesterday's anti-federalists and Federalists respectively, and in fact this is why the United States is a Democratic-Republic. It is a democracy in that citizens vote on laws, but also a republic in that the citizens elect officials to represent them in the government. It makes sense that this system is so diverse in terms of structure, considering how young the United Sates is relative to almost every other nation, and one can see how the founders used their knowledge of history to construct what has so far been one of the most successful and stable governments in the modern history. Even though democrats and republicans so often despise one another, I think the debate between them is what makes America so great. They balance one another out, and truly depend on one another, strengthening the government when the people need it, and reeling it back in when it grows too large. That conflict between parties might keep congress from getting anything done, but it also keeps the government from becoming overly corrupt, because there are so many voices crying for different issues. In that sense the United States government is sort of like a schizophrenic eagle, supported by hundreds of thousands of people who don't just bleed red, but white and blue too.

That's not to say the American form of government is perfect, by any means. While trying to balance the power between the people and government has resulted in many successes, it has also birthed some failures. But I'd contest that most of the successes outshine the failures. For one thing, the Bill of Rights and all subsequent amendments have allowed the constitution to stand the tests of time. Not only that, but the checks and balances insured by the creation of the executive, legislative, and judicial branches of government have also promoted a strong, stable system. Even within the legislative branch, balance is ensured by the fact that there are two houses of congress, the house of representatives which is based on the population of each state, and the senate which grants each state equal representation regardless of its size. The various terms in office are also finely crafted to keep the government in line with the citizens. Representatives in the house serve short two-year terms in order to keep the house responsive to the needs of citizens, while Supreme Court justices serve life terms for the exact opposite reason, to ensure that they are not swayed by the need to barter for a vote. The president, a sort of pseudo monarch for the nation, of course serves a middle length term, lasting four years with a maximum of one re-election.

But while the system is well crafted overall, it is also weakened by the same philosophies that make it great. This is most plainly seen in the way officials are elected. The fact that some terms are so short makes it difficult for many officials to actually focus on the issues that matter to citizens, because they need to campaign for re-election, which is also quite expensive. Furthermore, the voting system has grown archaic. More money is spent on election campaigns (hundreds of millions of dollars) than the actual voting system, making it difficult for citizens to actually vote, and frankly wasting time. Although, this isn't as much of an issue considering that politics is treated by many Americans much like a sporting event, in which people vote for their team regardless of how a particular politician stands on any given issue. And let us not forget perhaps the most grievous hole in the governmental structure of my country, the Electoral College. In the spirit of protecting the government from the ignorant masses, the founders created the College as the means by which the president is elected, not through popular vote within the country as a whole, but based on elected officials who actually cast their votes state by state, officials who may or may not listen to the will of the people. As a result, the leader of the world's most powerful country can be elected even if the majority of citizens didn't actually vote for them. The only consolidation for this flaw is that the number of delegates sort of reflects the population of each state, although that is sort of nullified by the fact that everyone only votes for their party most of the time. Understandably, this is why we don't promote our form of democracy when spreading it around the world.

Regardless of your political stance, it's nearly impossible to deny the benefit and necessity of the services the government provides. What we must decide collectively are which services the government should provide, and whether they should be provided locally, federally, or privately. As with all things in the universe, I find that balance is the key. While it is true that people shouldn't become dependent on their government and individuals should strive to better themselves, governments have the power to do things no individual person can, so long as its' people trust them enough to give them their money in the form of taxes. I know, I know, everybody hates taxes. Except the problem is that they shouldn't. Taxes are the cost we pay for living in a civilized society. Although I should clarify, people don't necessarily hate paying taxes. In fact, many educated citizens understand the patriotic duty for everyone to pay for the governmental services that keep us happy and healthy. Taxes force you to put a stake in not just your government, but also in your fellow citizens, so long as the government is doing its job. In a country with trust, citizens can combine their power to do the impossible. And the more people you have working together, the more they can accomplish. This may be part of the reason developing nations don't have taxes, and therefore have more volatile governments. The people don't trust their government and their government doesn't trust its people, like a bad marriage ready to crumble. This isn't a problem if both parties know the other is fallible and human, and they depend on one another. Taxes are similar to banks in that people pool their money. They have the potential to galvanize growth. Just one example of this can be seen in the Silk Road, which reinvigorated thanks to the Mongols because they wanted to tax goods. Thanks to them, the West and the East were able to connect, spreading not only bubonic plague, but also economic wealth.

Taxes aren't much different from interactions one may have with a money holder. You put your money in a bank or credit union, which then gives out your money to people that need loans for cars, homes, or businesses. The only difference between this and taxes is that other citizens get access to your money as opposed to the government, even though the government is made up of citizens, and they should be spending tax dollars on their citizens in turn. They are crucially similar in how they both require people to trust each other with their money. People without trust and communication, therefore, could never support a national bank or government, as they would always become volatile. People need to know that other people aren't going to cheat them out of their livelihood. But, if people can have confidence that nobody is going to steal from them, then the people can coordinate their resources to make long lasting, impactful change to their lives. Just remember to treat each other as equals, and don't be horrible to people even though you can't see them, it's that simple.

You can tax any object, good or service as well. It doesn't solely need to be for revenue either; sometimes it can just be a means of influencing citizens. Many are familiar with the British taxes on tea and stamps that irritated the colonists in early American history, but some taxes are useful for hindering bad behavior as well. Countries tax cigarettes, alcohol, gasoline, soda, and even fat (in the form of fast food) to deter people from overconsumption and to promote a healthy population. Now, while they might have altruistic motivations, this may also seem overly paternalistic of a government, and citizens might protest what they get taxed. Taxes can also be distributed differently by different societies.

People can all be taxed the same amount (a regressive tax), taxed proportionally by their income (a progressive tax, with marginal income tax brackets), or taxed an equal percentage of their income (a proportional tax). All have their merits, and with the complexities of government services, tax deductions, and general confusion, they might all have separate uses depending on a nation's specific circumstances. It's complicated. Regardless, taxes are essentially the sole means for a government to make money, and so it is important to understand how they work. It should be mentioned that governments can and often need to borrow money if they spend more than they make in taxes. This is done when people or foreign nations buy bonds and lend the borrowing government money with interest. Again, this can only be done if a government is trustworthy enough that others think it will pay off its debts, like the United States is. But the primary means of gaining revenue remains through some sort of income taxes. They can be good or bad depending on what you tax, but they also give the citizens leverage over their government. When the government needs money from the citizens to run, it has to listen to them, and it is important to remember that government workers are also taxpayers, and more importantly citizens.

With so much at stake worldwide and at home on the United States economy, it pays to understand where our money is already going before we can decide where it should be going. The United States budget is broken into two main sections, mandatory and discretionary spending. Mandatory spending revolves around what keeps people alive, and so it includes funding for health related issues including Medicare for retired and elderly citizens, Social Security, food assistance, and unemployment. The percentage of the budget for mandatory spending is essentially set in stone, unlike the discretionary budget, which is what people generally refer to when discussing tax reform. The main portions of discretionary spending include everything from military spending, to education, housing, foreign aid, energy and development, transportation, and miscellaneous spending.

According to National Priorities Project, in 2015:

- The total budget spent was $3.8 trillion, $3.18 trillion (84%) of which were collected via taxes
- 65% spending was Mandatory
- 29% spending was discretionary
- 6% was used to pay off debts

Notably, military spending eclipses the amount of resources used on education and space exploration through NASA. In 2015, education spending was 6.3% of the discretionary (1.8% total) budget, defense spending was 54% of the discretionary (17% total), and the NASA budget was ~.3% of the military budget (.5% total). The United States spends more money—$601 billion—on defense than the next nine countries on Credit Suisse's index combined. Ironically, this high cost both encourages and hinders immigration. In defense of defensive spending, it makes the country extremely secure, which gives citizens peace of mind. On the other hand, nearly every person I have met from another country cites how frightening it is that our citizens are so adamant about owning tools made for the sole purpose of dealing death. That's not to say that private defense is tantamount to state defense, but it is ingrained in the culture. People are actually afraid to live in the country because average citizens are given so much excessive firepower. Not only that, but American law enforcement personnel are also seen as strange to many foreigners because they too carry firearms. Of course, because the United States also focuses on economic growth almost as much as security, many people still immigrate in search of safety and opportunity for their families. I would imagine immigrating into the States is like holding up inside of a bunker where every square mile has a cannon, and it all seems to stem from a combination of blatant paranoia and legitimate fear of being attacked. It seems very much like a system starved of trust and compassion. But it might also be healthy, who knows?

Despite what many say, foreign aid spending is also smaller than many of these other budgets, at less than .1% of the total budget, not even taking into account that it gets distributed to multiple countries. Foreign aid—as with any type of poverty relief—is tricky, because people may misuse handouts. On the other hand, as Economist William Easterly, author of *The White Man's Burden*, would argue: the system of blind loans is useless, but aid isn't. The people who know the problems of the developing world best are the people that live there. Human beings are smart, and if you give them the right tools, they can solve their own problems. If instead you just keep sending them money without doing anything to address their broken infrastructure, you would be no different than an overbearing parent.

If you still retain an unfavorable view of taxation, just remember that the United States was born largely for tax representation. They didn't mind paying taxes, they just wanted to have a say in what the crown was taxing. They even said that England was treating them like slaves, despite the fact that they simultaneously owned human beings as property. But that still undermines the fact that many people in colonial America didn't want to completely revolt against Britain. Taxes are so important to people, they will literally overlook social injustice just because the government didn't ask nicely for their money. If you don't like what the government is spending your money on, that's okay, just make sure you can articulate specifically what you think should change, why it should, and maybe how it could happen. If enough people care enough to ask for something, the government will pretty quickly bow down, and then promptly continue to bicker amongst itself whilst keeping the people safe until a majority of people want change again.

On the matter of how large the government should be, I think it is important to make a distinction between the size of the government and the influence of government. It should be clear to anyone who has read this far that the growing trend in globalization and connectedness we have seen throughout human history has been a force for peace and decreased violence. Along with a more globalized world, governments too will have to expand. I actually tend to have a more extreme view personally. From space there are no borders, and no matter where you go there are organisms (and people) just fighting for survival. I'd like to see our politics reflect this. Centuries into the future, I dream of a planet where all borders have been dissolved, and the entire planet lives under one government with representatives from smaller states all around the globe, much like the one in the *Futurama* cartoon series, except without Richard Nixon. Working together is the only way we can peacefully and successfully colonize the galaxy, but we need to learn how to respect one another first, and that will be far easier said than done. Regardless, a bigger government is inevitable, but that doesn't mean it needs to get involved in the everyday lives of citizens. An Orwellian Big Brother nation seen in *1984* terrifies me as much as anyone else.

In many ways governments seem distant from the normal citizen. Whether it's a monarch, parliament, dictator, or church, people who have historically been born in power, keep that power. But that isn't so in the United States, despite how many politicians often resort to stupidity and lying just to get re-elected. But that's not always a bad thing, so long as people eventually catch on. Here people cycle in and out of the government, and at the end of the workday, they are normal citizens. Whether it's a president or sanitation worker, everyone follows the same laws, and dies a citizen. Not only that, but the government also doesn't have complete power. Corporations also have a fair share of money. Of course, unlike Walmart, nations have trained armies and quickly pick up potentially dangerous technologies. Just another reason governments need to be shaped and supported by the people they serve.

So we are left with a decision on which specific aspects of our lives the government should be involved in, and by how much. For instance, a major area of debate can be found in the economy. We are all subject to money in civil society, and so the decisions we make about how the government interacts with the flow of our money are crucial. How much should the government get involved in the economy? Furthermore, when it does, how should it work to mitigate the damaging effects of a stagnant economy or one where the value of money is increasing or decreasing at a radical rate? Should the government change taxes to regulate how much money is in the economy, or should it alter it's spending to direct the flow of money in a way that would increase economic growth? As I've stated, with so many irrational agents and flawed humans contributing to society, economics is one of the most difficult professions one can participate in, and I deeply respect the work economists do. The level of involvement that the government should play in the economy has been a constantly evolving and a perpetually active area of research and debate. One extreme argument is that the government should keep its hands out of the market, as it can self regulate and therefore should be left to private agents. This was the common view until the Great Depression hit in the 1930's, at which point it became clear that an economy without the aid of the government was still susceptible to collapse. The opposite extreme can be described through Keynesian economics, which involves heavy regulation on businesses and having the government actively participate in the economy. This too is dangerous because it has the potential to harm citizens and use their money against their will, which is especially dangerous for the government to do considering it has a larger pool of money than most businesses. Many economies today now struggle with finding a balance between government participation and market independence.

For one thing, governments in general have a difficult time decreasing their spending, like many of us. One of the major criticisms of the Reagan administration, for example, was that even though it did a fabulous job lowering taxes for citizens—something that many would argue promotes citizens to spend money and bolster the economy—the government continued it's spending, most notably in defense, keeping the arms race alive with the Soviet Union. Many contest that Reagan bankrupted the Soviet state by forcing them to match spending, and that this resulted in the fall of the repressive regime. While this was almost certainly a contributing factor, the fact is that continued spending while lowering revenue also bankrupted the United States. Not only is this romantic view of the Reagan administration an oversimplified perspective that doesn't take into account the millions of citizens living behind the iron curtain calling for change, it also ignores the critical politics going on within the Kremlin, especially with Mikhail Gorbachev's new liberal policies of *Glasnost* and *Perestroika* in the eighties, which gave citizens greater freedom and opened up trade with western countries. Moreover, bolstering weapons is a sophomoric way to promote peace.

But the same criticism can be made for raising taxes and lowering spending. After the recession of 2008, Europe and the United States diverged in policy to fix their economies. The European Union took the stance of using high taxes and low spending, unlike the United States, which increased its deficit by having the government stimulate the economy. While this is still a relatively recent event, many now see that Europe has had slower growth and development than the United States. There was a price to pay for this, namely that under the Obama administration the national debt increased by $7.4 trillion dollars between 2008 and 2015. That being said, this also is an oversimplification. The national debt has been increasing for the last several decades, under every administration. This is to be expected considering the size of the economy has also been growing, meaning the total GDP of the country has been prospering on the whole. Not to get too partisan, but under the Reagan and both Bush administrations—which cut taxes (especially for the rich) to make citizens happy—added more to the national debt than any democratic president since Carter. I would be careful to mention that this is in regard to the percentage increase of the debt, not necessarily the dollar increase in debt, which is less useful considering that a healthy economy should experience minor inflation.

It's clear that if the government is wise in the way it spends money, it can help the economy, so long as it promotes its citizens to do so also. The government can also help the market system by preventing fraud and theft, although this too is controversial (like everything in this section). More than anything, the government can simply do things that even the wealthiest citizens can't, and it can enforce policies to make sure people don't cheat one another. How else would large coal companies reduce their emissions? The issue of global warming specifically echoes the need for government involvement in the market system, precisely because climate change is an issue that affects every human being.

As the financial system becomes more global, governments will naturally rely and become indebted to one another. If governments can't pay their bills, then the people they serve will also lose money, as will corporations, banks, and the governments of other nations in that chain. This would only compound as people lose their jobs and begin to request more money from their government, which neither one has. People would stop selling, which means they would stop buying, and economies everywhere would freeze. Again, as long as you have people with ingenuity, they will find a way to make money and get it circulating, by finding problems people will pay to get solved, again harkening to the importance of an educated populous. However, those innovators won't be benefitted if wealthy people are allowed to hold on to their money. It's just in our nature, we don't give unless we are told to explicitly, and that's what taxes do. This is the primary justification for the claim that the rich should be taxed more than the poor. It's important to remember that progressive taxes are based on incremental amounts in salary increase, but they are not absolute increases. To clarify, if there was a tax bracket at one dollar, a person making two dollars would get the same tax rate for the first dollar as a person who makes fifty cents, but the additional dollar they make would get taxed more. No matter how you slice it, the person with higher income still makes more money at the end of the day, which is why people who complain about entrepreneurs becoming demotivated by getting taxed more as they become wealthy really just sound like they are whining. It would be good for governments to invest in businesses so long as anyone can earn a grant by demonstrating the potential to help everybody. This lends to the idea that government should be more reactive than preventative, only helping when no individual can.

Many of the criticisms of big government in the United States stem from Franklin Roosevelt's New Deal policies that came out of the Great Depression. Indeed, F.D.R. did more to expand the government than possibly any president in our country's history, but that wasn't all bad. One of the few examples of the New Deal big government that came out of the Great Depression that few people complain about is social security and Medicare. This is why it is considered to be part of the mandatory budget, and why it is paid for using a regressive tax where everyone contributes an equal portion of what they have for something everyone wants, even though it is only for the elderly and impoverished to use. But that's becoming an issue as people live longer and have fewer children as a result of prosperity. There are several ways one might consider tackling this problem, all of which I'm sure would be controversial. The government could decrease aid by lowering funds, or it could raise taxes, or better yet raise the age of recipients. Another important outcome of government expansion came in the form of The Rural Electrification Act, enacted on May 20, 1936, which provided federal loans for the installation of electrical systems to serve isolated rural areas of the United States. Again, if a cause is good for everyone, nobody complains about paying taxes for those things.

In order to grapple with such an aged and complex topic as this, naturally I had to listen to multiple arguments, on both ends of the spectrum. I stumbled upon a video from Dennis Prager, who runs a very well designed YouTube channel, who published a video titled, *The Progressive Income Tax: A Tale of Three Brothers*, which made a case against unequal taxing. The story goes that three brothers, all born under the exact same conditions with the same education, all get the exact same job. Essentially they are all the same except for the hours they work and the amount of money their spouses make. One works twenty hours, one works sixty, and one is in between, all for the same wage. They all buy houses in the same run down suburb. They decide to refurbish the neighborhood, and even set up a gate to protect against crime (prudish much?). The wealthiest brother expects them to all pay equally, but he instead found himself paying the highest amount, because he had more money. This little yarn is meant to convey why unequal taxation is morally corrupt. This is a completely fair and justified point of view, and we can begin implementing a flat tax just as soon as one fifth of people in the U.S. don't own four fifths of the country's wealth, and individual people don't have more money than entire sovereign nations. We can also begin condemning lazy people right about the same time everybody has access to the same education, and earns the same amount of money for each hour of work, as the story indicates. I actually found that aspect rather curious for a proudly conservative media source, because the idea of equal wages is a socialist concept that doesn't occur in a competitive capitalist society, but I digress. More importantly, it's actually somewhat horrifying that there are still people blissfully ignorant of the different opportunities people are given at birth.

Another major criticism in the early twenty-first century that the government is overstepping its grounds is affirmative action. A lot of well-to-do folks complain that affirmative action is flawed because it gives an unfair advantage to minorities, or because hand-outs make people dependent on help. There is some merit to that, and it's certainly true that people who are qualified for certain tasks should always be chosen because of their abilities, not just because they happen to look a certain way. People need to be driven to get what they want; and they need to work hard for greater affluence. However, it's not always that simple. The history of the United States goes back before even the declaration of independence, all the way the first settlement in Jamestown in 1607, and the for the next four-hundred years people from all around the world would be discriminated against as immigrants, and that certainly made it more difficult for them to find work. The people of African decent are so often highlighted above other groups of color simply because for the first two and a half centuries, they were owned as property. They were specifically bred for harder working genes, they were beaten and raped, they had their children and family sold like common handbags, they weren't even allowed to read, and uprisings for justice were met by fire and steel. It took an entire century even after they were freed for them to get access to the same schooling as those of European decent. On the other hand, the Natives of North America barely had any chance to defend their homeland. Some of them even kept the first settlers alive, but without immunity to smallpox and other diseases they dropped like flies, just for shaking hands with the colonists. There are so few of them left that people often forget they were in America centuries before the fight for independence or that some natives and slaves even fought for the British, precisely because of colonist oppression. And throughout that time, women were all told equally to, "be quiet, and stay home". So while these days it may seem unfair that some kids get free college tuition or job support because of how they look, and there are certainly people of European decent living in poverty, its important to remember that male white protestant men have had a three century head start to accumulate wealth and pass it on, while everyone else has had just a few decades. Thankfully, with every passing generation, people are rapidly realizing that the freedom their country was founded on applies to everyone, and people are taking steps to rectify inequalities through diversification.

The government of the people, by the people, and for the people has come a long way since its birth in 1776. Nowhere is this more apparent than the social relations between African Americans and European Americans. In school, students often learn this history through actions made by the government, specifically in terms of policies. However, the laws and court cases only superficially show how the government treated African Americans. More than anything, they chronicle the development of race relations in the United States as a whole, from slavery to civil rights. The potency of the struggle faced by African Americans through history is only understandable when one realizes how much progress they have made, from being treated as property to becoming citizens with equal

rights to whites, something that some still have to fight for. The state of slavery was exemplified in the Supreme Court case of Dredd Scott v. Sandford in 1857, where it was ruled that Africans couldn't sue because they aren't citizens, regardless of whether or not they were born or lived in a free state. Even after they were emancipated, officials continued to marginalize black freedom, but progress was still apparent.
Andrew Johnson actually vetoed the civil rights bill of 1866, which provided nationwide equality of people regardless of race, but was overturned by a two-thirds majority in congress, and this led to the fourteenth amendment being passed in 1868, which extended citizenship and due process to all people born in the United States. The fight continued as southerners attempted to segregate black communities and force them to remain poor second-class citizens. This is another critique of weak central government; delegating law enforcement meant that state governments could perpetuate racism and enact Jim Crow Laws, making the Supreme Court come into play instead of just dealing with the issue immediately at the federal level. Although Supreme Court cases are pretty monumental, they too make mistakes. During the Plessy v. Fergusen case in 1896, they enforced segregation with the justification that blacks and white were "separate but equal", the highlight of southern hypocrisy. It took a century after emancipation for black citizens to be allowed to go to the same schools as whites in the south, thanks to the Brown v. Board of Education case in 1954.

Incidentally, when addressing how liberal folks associated with the left view United States history pessimistically, conservative right wing folks claim to be more optimistic. In this spirit, our friend Prager also cites that slavery was practiced everywhere, and the United States was one of the first nations to abolish it, thanks to their Judeo-Christian ideals. Barring the fact that Christians used the Bible to defend slavery, Canada, Argentina, and Haiti all abolished slavery around fifty years before the United States. This argument also radically oversimplifies the various cultures and lifestyles different slaves lived under. For instance, even though Islam only bans Muslims from becoming slaves, it at least recognizes the humanity of non-Muslim slaves, something Judeo-Christian plantation owners sure didn't acknowledge. Even the slaves of ancient Greece were welcomed to homes with fruits and were even allowed to serve in the military. American slaves couldn't do that, because the conditions they lived under were so inhumane that they would have revolted. Nearly all foreign forms of slavery were a far cry from the chattel slavery of the cotton plantations, which were still less gruesome than slavery in the Caribbean, where managers kept hatchets at the ready to amputate hands caught in sugar mills. I hate to beat a dead horse with this issue, but it's better than putting the horse in a corner and ignoring it. After all, one of the biggest stains in the Republican Party since J.F.K. and L.B.J. has been racism. Granted, virtually no genuine republicans are racists, it just so happens that all of the racist southerners switched teams when the Democrats granted black people civil rights in the sixties. The nation accomplished this despite race mixing being labeled as communism. And that leads us to the second greatest catalyst of stupidity in American history.

I'm not going to claim that the tensions between the United States and Russia were unjustified, even though with a cosmic perspective I think that all conflict is ultimately a waste of time. However, it is important to understand that people living at the time genuinely believed the other side was out to get them, and that made it true on both sides. But after Stalin died, a great deal of the rhetoric used by the U.S. and U.S.S.R. was nothing more than a result of grievous miscommunication on both sides, and the entire human race faced mutually assured destruction as a result. So while it's certainly true that Stalin was a malevolent asshat who needed to be opposed by both The United States and Russia, as a whole the cold war was a wash. Since the twentieth century, totalitarian communist nations like China and Russia have given far more freedoms to their people to buy luxury goods, while capitalist nations have begun adopting socialist policies for the good of the populous, looking to countries like Sweden and Norway as examples. That's another common tenant in history, extremes are nearly always wrong, and it takes both perspectives to reach solutions.

Despite being at "war" with communism, the only instance of direct conflict between the United States and a major communist power was at the beginning of the era, during the Korean War form 1950 to 1953. We didn't even face Russian troops, we fought against Mao's China. It's even debatable whether anything productive came out of the conflict. While the United States was able to push North Korean troops back to the Chinese border, China ensured the border between North and South would remain at the thirty-eighth parallel. However, like Japan, South Korea does at least demonstrate how the combination of American capitalism and Asian exceptionalism can result in some of the fastest growing and most prosperous nations in history. This was a different story than the second armed conflict in Asia, and the only war the United States definitively lost. Vietnam was a French colony before the Second World War, and like many other former colonies, was faced against European dominance that they wanted to free themselves from. The problem was that they democratically voted for a communist government. The United States couldn't have that, and between 1961 and 1975 the U.S. continued to engage in a losing battle with the Vietnamese people who were fighting for the same ideals that the United States itself was founded on. And much like the American colonists fighting the British, the Vietnamese overcame the American military machine through guerilla tactics and sheer willpower. That just goes to show that even powerful governments sometimes have no idea what they're doing, why an educated populace is so important, and why politicians should remember what human beings are.

Worse than the armed conflicts, the greatest tragedy of the Cold War on the part of the United States lies in the long-term effects of promoting corrupt governments in the spirit of fighting communism. The fight in Vietnam is just one example of this. American efforts led to a destabilization in Cambodia that allowed the Khmer Rouge to come to power. This party led by Pol Pot outright murdered over a third of the Cambodian population, exhibiting the most gruesome and detestable crimes of not only the Cold War, but of human history, stacking up mountains of human skulls. They exemplified the fears of Americans toward communism, but their crimes weren't a product of communism itself. They were a product of a totalitarian regime. The United States not only led to the rise of many corrupt regimes, but also blatantly supported the election of corrupt governments into power worldwide so long as they also hated communism. The list is wide, and includes many horrible politicians, from Mobutu Sese Seko to Augusto Pinochet. But the worst governments birthed in the developing world were all run by inexperienced politicians who were put into power without any knowledge of how to serve their people. Many parts of Africa and South America in particular are still recovering from the cold war. Several developing countries also ended up with militaristic and repressive regimes, and are also more prone to civil wars than interstate disputes, further hindering their development. This further sets them apart from European states, which historically always fought with each other. Interestingly, many developing nations that emerged in the twentieth century grew increasingly unstable due to the poor governance of a military. Interestingly, many United States presidents were also military commanders, including the first one. In fact, the diplomatic face of the country is also the chief commander of the military. The difference of course was that the new developing governments were often totalitarian regimes, with one or few people making all of the decisions and hoarding power. In the United States on the other hand, other politicians and the population check the president's power. Yet despite this, the United States as a whole continued to support those same authoritarian regimes in developing nations because they were scared of the color red (and communism too I guess). Despite their sense of justice, the country was still heavily segregated and patriarchal during the cold war; showing how irrationally focused the white male government was on communist ideology. No country is perfect.

But the American government on the whole isn't bad. For one thing, it does take very good care of its employees, which is likely the case for most nations. The United States government is also especially good at leading fair employment and equal opportunity by example, since they're the ones who make those laws. During my time in high school working as a clerk for the Bureau of Reclamation, I saw first hand the strengths and weaknesses of government work. There were groups for Native Reservations, a decent balance of male and female employees, as well as minority groups, and while I can only speak from limited exposure, from the way the employment policies were written, I am fairly confident that equality is more strongly enforced in the government than in many private companies. Security is also very important to the government, and they take privacy VERY seriously. But they don't tend to keep information secret unless it regards things that you wouldn't want other people knowing. Since the scandals of the late twentieth century, they seem to have become much better at trying to earn the trust of the people, making most things as transparent as posible. The best example of this is seen in scientific research. There are few organizations as transparent as NASA, at least in terms of research that benefits society. There is a crucial point to make here. While a small handful of the government—mostly the talking heads—are wealthy and have little to gain from the prosperity of the people they govern, the vast majority of government workers owe their livelihoods to their jobs. They depend on their jobs the same way everyone does, to feed their families and pay their own taxes. If they want to keep their jobs, they need to do them to the best of their abilities, or someone else will just walk in and take them, and their job is to make sure citizens of the United States are as prosperous and secure as possible. There are problems to fix, just as there always have been, and just as there always will be. But there are people actively trying to fix them, and they do. You might dislike this politician or another, but they do not represent the majority of the government. Citizens do, and they also happen to have certain opinions about their politicians of choice, just like everyone else.

A major reason for my country's greatness, I think, rests in the master craftsmanship of its constitution. Taking influence from The French Revolution, and the idea that laws come from the people, The Constitution of the United States is the shining example of why a weak central government is useless. The Constitution was ratified on June 21, 1788, with a Bill of Rights added in 1791, replacing The Articles of Confederation, which were in place for a decade but tossed out because the government was unable to collect revenue. Again, without taxes, there are no government services, so there may as well not be a government. The ideals of the Constitution were made so well, with such pompous perfection and boastfully arrogant equality that the history of our country has essentially been trying to prove that our constitution is accurate. Everyday, we have to wake up and remind ourselves to fight any violation of equality and liberty. That drive to achieve perfection and maintain prosperity is built into the American Dream. It's actually a somewhat socialist concept, but with the caveat that treating people as equals doesn't necessarily translate to how much money people are willing to give you in return for doing something productive. It's a good system, at least in that it promotes creativity and innovation. If you want to be successful and make lots of money, you need to work hard. In the United States, you have to be unique to make money, and do something that nobody else can offer. That creates more jobs, because nobody can build a multibillion-dollar empire without people to work for them. And if they don't keep their employees happy, then they might quit, sue, or start a competing company instead with their own innovations. We all rely on one another in that way. That's also why the idea of American exceptionalism is slightly hypocritical. We profess that all people are equal, but are so arrogant that we constantly refer to ourselves as "Americans". When every country in South America has the stability and opportunity the United States and Canada do, then we can all be proud Americans, instead of proud Part-of-Americans.

Finally, that leads us to the president, the respected head of state and de facto mascot for our country, the number one citizen as it were. Being the president of any democracy is hard work, not because of policies, but rather because they have millions of people constantly squawking at them about what they want, and it's their job to filter through the wants of their people to decide what's in everyone's best interest, a task they may or may not be good at, depending on who they support. Dictators have it easy. Nobody ever criticizes them, because if they do they get shot. So if nothing else, I am very thankful to live in a part of the world and in a time where the president's duties include being able to take a joke. But that's also why people need to work with their political officials, who are representing them to the rest of the planet, so that playful banter doesn't rot into hateful slander. It's a good thing the presidential term is so short. The president, no matter in what generation, has essentially been the most stressed out celebrity in the nation, even before celebrities were such an ingrained part of our culture. The most famous presidents weren't the ones who just represented their party, but the entire American populous. Look at the before and after photos of any president, especially the wartime or digital age ones. I can't imagine that wearing the face for tens or hundreds of millions of people is ever an easy task. That's also a reason why the people are so passionate about making sure the right face gets into the oval office, even more so than the Congress, which actually makes the laws and were intended to be closer to the people. But the popularity contest aspect of the whole thing is also a good power check for the person that commands the armed forces. Not only that, but presidential elections are some of the only times when American citizens actually care about politics. Every four years or so, tensions heat up, holidays become awkward, and rhetoric becomes extreme on all sides. There are even talks in politics about building walls to keep foreigners out, but this nationalist ideology that, "we can take care of ourselves and other people will just make our people worse off" is the same childish isolationist policy adopted by Nazi Germany, North Korea, and the United States in the 1920's. Extreme views are fine, only so long as they are able to compromise with opposing groups. Otherwise, one extreme will run rampant. Just as with life and chemistry, opposites are best when they reach equilibrium. Otherwise they are useless.

Presidential races are not only good at galvanizing people for certain causes, but also at bringing issues to light that divide American citizens. What people so often forget, however, is the great privilege they have in being able to contribute to the political process. For the majority of modern history, the right to vote was reserved for landholding men, and in the case of the United States, that meant rich white men. It has taken minorities of all walks of life nearly two centuries to get voting rights, which is something many people still don't have in other countries. It's one of the greatest achievements of human rights in history, and yet it is so commonplace that people seldom have the motivation to participate. There might be ways to fix this however. For instance, political debates are important, but it would be really nice to also have informal debates amongst politicians that double as drinking games. If nothing else, it would humanize political figures and make them more popular with the public, unless of course they turn into bigoted monsters when they're drunk. This wouldn't necessarily be a bad thing though, because it would weed out unfavorable representatives, and any unfavorable citizens who may support them. The government should also give people chocolate and suckers when they vote. The United States can even have tiny chocolate American flags and sugar pops. The world may be coming to the end (for 4-8 years), but at least we have pure biochemical energy in the form of candy. On a more serious note, citizens who committed felonies should be given a path to re-earn their voting privileges. The United States only perpetuates its embarrassingly high criminal population by ostracizing them for what society has raised many of them into.

Growing up I'd say I spent about as much time working around my elders as with peers, and I've already mentioned how they have all been nothing but nice. I'm sure that if those people were to get together in a political debate, they would find themselves divided. However, the fact that their politics mattered so little in their daily lives gave me the impression that those views actually didn't matter much at all. Politics only seems to arise when there is uncertainty about the future, something that nobody can predict fully. But on a day-by-day basis, it seems like people are far more likely to just want to get back home safe. Regimes may rise and fall, but people forge onward. That is why it is up to the people to fight despotism and for the government to protect those people. One of the most tenable lessons in history: if you don't keep your people happy, you don't get to keep your power.

Agriculture

Agriculture was the first big thing that human beings did to set themselves apart from every other animal on the planet. It allowed us to develop society, thanks to nothing more than the fact that we didn't have to spend so much time finding food. Agriculture allowed us to cook fine cuisine, giving us the likes of the beloved Julia Child, whose spirit could save the world with a global dinner table. Indeed, food is what unites families and diplomats alike, just imagine every politician on the planet getting together with dishes made in their homeland as a means of fostering peace among nations. Not only that, but as we've discussed, agriculture has intimate ties with something else that fosters bonds between people: humanities favorite drug, beer! With fine food and drink, I think even bitter enemies could become great friends. Now in the new millennium, advances in genetic biology and chemistry have allowed us to develop agriculture and sustain an even greater population with even less work. Bon Appetit!

Of course, agriculture is equally to blame for every human endeavor to come after it, both good and bad. For one, economy is rooted in agriculture. The first money came when people had a surplus of food, and began using that excess to create wages. But this comes at a cost. When crops and livestock get diseased (with diseases being heightened by the formation of cities) or otherwise fail due to poor weather, everybody goes hungry, and economies can fail from the bottom up. This wouldn't have been an issue, pre-agriculture. According to anthropologists David Graeber and Caroline Humphrey, before there was money there was bartering. However, only strangers and enemies bartered with one another. In close-knit communities, people are often very familiar, and that made it difficult for people to swindle each other. Instead of bartering, they were able to mostly pay one another based on debt, or in other words, trust. Originally, money was used to keep track of resources and grains, but it wasn't circulated amongst common people. Interestingly, this is opposite to the trend we have seen in global economics. We need money to exchange goods with foreign powers, but with the advent of credit cards and digital money, people are becoming more and more reliant on trust-based loans for everything from national debt to cars, homes, and college.

The link between money and power with food is one people seldom consider. Many of us don't know how fortunate we are to not die of starvation, one of the slowest and most dreadful ways to pass on, especially when it is out of your control. This fear for survival has driven many individuals to raise serious concerns about human overpopulation. One of the most outspoken and famous voices on this topic was Thomas Robert Malthus, a cleric and scholar, and the author of *An Essay on the Principle of Population* in 1798. Malthus is famous for raising fears that humanity's need to grow and reproduce would eventually result in global famine that would end the human race (or at least cap the population). On the one hand this is a chilling proposition. The problem is that it isn't grounded in reality, and in fact has been used as justification for the preponderance of the factually inaccurate idea of Social Darwinism. It is very rare for famines to arise purely because of unseen circumstances. In fact, the most harrowing famines in recent history were the result of dehumanization and preventable actions. Famine was seen not just as a result of the collectivization schemes of Mao and Stalin, but also in places like colonial India where the British continued to tax its citizens despite droughts, all for nothing more than wealth and because of world wars also grounded in stupidity. What's more, Malthus could never have predicted how advances in science would contribute to society far beyond expectations, resulting in the twenty-first century having the lowest number of hungry people in history. Yet despite advances, there are still so many who grow fat while others grow thin. But we can prevent so many tragedies if every person is simply well fed. If you don't want to share your resources with as many people, then you should probably be in favor of helping people live more sustainably, so they don't have to worry about their children dying. Then the only reason people will have for popping out a dozen children if they're following a (extra) crazy religion that abhors condoms, like those people from Kansas that picket funerals and gay pride parades. Of course, if more people live the same lifestyle we enjoy in the first world, then humanity will have to make serious advancements in the way we produce food and energy. Otherwise society really will collapse. But there is plenty of support from history to be hopeful.

Perhaps one of the greatest contributions to agriculture, that has saved the lives of billions of people, has been nitrogen fixation. You'll recall the need for nitrogen not only in DNA, but also in all the amino acids used to make proteins. As it turns out the vast majority of the air we breathe is comprised of nitrogen (78%), but none of it is available because nitrogen atoms pair in strong triple bonds, making them almost impossible to wrench apart for crops to use. The only organisms able to break those bonds and "fix" airborne nitrogen are prokaryotes, and this is why fertilizers and decomposing organisms are needed for good soil. That is until Fritz Haber came along. His story is perhaps one of the most fascinating in the history of science. The genius of the Haber-Bosch process, which he helped develop, was in taking nitrogen in the air and forcibly bonding it with hydrogen gas in order to create ammonia, which can then be cooled and condensed into a liquid. The process is remarkably simple as well; one only needs heat, pressure, and a catalyst to lower the energy barrier of the reaction. The Haber process has allowed biochemists to develop nitrogen fertilizers that have allowed the human population to explode by quadrupling agriculture yields. Thanks to these innovations, Haber received the Nobel Prize in 1919 in chemistry. These advancements haven't been without consequence however. Fertilizer runoff gets into oceans, makes lots of algae grow, and when they die decomposers sap oxygen from the water making vast dead zones. Interestingly, Haber didn't develop this technology to save lives. No, that was just a byproduct. He figured out how to fix ammonia to murder millions for the German war effort in the Great War. He was a devoted German nationalist who relished the use of chemical weapons to drown soldiers in the trenches with asphyxiating chemicals. Ironically, despite his dedication to the fatherland, he was a born Jew, and was excommunicated along with millions of others as Hitler came to power. More than most people, Haber's life and work demonstrate that science is neither good nor evil. It is simply true. His tale is a reminder that the will to use our knowledge for good is in our hands alone.

As it turns out, fertilizers aren't nearly all of the ethical issues surrounding our use of the plants and animals we need to survive. One that we have already mentioned is of course the use of genetic engineering on our crops and livestock. Many people have raised concerns about not knowing what goes into their food, and rightly so. However, there is no reason to demonize genetically modified organisms, because they too have saved many lives. By altering the genes of our food, humans have been able to grow crops that stay fresh longer because they don't give off as many rotting chemicals, and even make such things as featherless chickens with more meat that hasten poultry production and reduce the numbers of livestock that need to be slaughtered for us to make fried chicken. Not to mention seedless fruits, which I consider to be bioengineering magic! Not only that, but we can actually produce plants that are more resistant to disease and hungry insects, minimizing the need for chemical pesticides, which unlike GMO's are actually harmful. We can even add vitamins to crops that prevent people in developing nations from suffering through nutrient deficiency.

Genetic engineering has promising research that spans beyond our food as well. My favorite example of this is in bioluminescence research. The same genes in bacteria and animals that allow many deep-sea life and fireflies to glow can be implanted into other plants and animals. Not only is this just plain awesome for perspective pet owners, but there are also economic benefits to this research as well. We could one day reduce our use of electricity by replacing light bulbs in homes and streets with glowing garden plants and even glowing trees! This pulls double duty by reducing carbon emissions as well as increasing carbon sinks, and did I mention how cool glowing plants are? It's strange how much people hate GMO's, despite them being in virtually all the things we eat. They are no more or less organic than any other living thing. In fact, even the Prager YouTube channel has a decent and scientifically accurate video on the matter, something I feel like I should mention for criticizing them earlier. Now, I understand the concern with not knowing what's happening to our food, but that's exactly why citizens must be educated in the nature of these products, in order to ensure that these advanced in technology aren't misused and are fairly regulated.

Another controversial topic—also amongst folks who are left leaning—is the ethos of factory farming. Many people take issue with the callous, frankly repulsive way we treat the livestock we eat. Many put our treatment of farm animals on par with the Holocaust, and some even say that it's worse. There is abundant justification for this as well. The problem isn't with killing animals, the problem is the blatant disrespect of animals that are strung up by their feet, have their throats slit while they bleed out crying, have their bodies stomped on and beaten, and get their tales lopped off as babies without anesthesia. This isn't just disgusting because of how cruel it is, it's repulsive because of how unnecessary it is. I'm not sure who got the idea that people need to eat meat at every meal, because nearly zero percent of humans throughout history have had access to meat more a few times per week or even per month. Also, how strange is it that people drink cow's milk, or the milk of any other animal for that matter? It's even stranger to consider how complacent people are about stores lined with animal flesh, something that requires huge amounts of crops and water to maintain, crops and water that humans could be consuming instead. The problem with this behavior is that it's wasteful and just not sustainable. In fact many aspects of developed society don't serve a particularly practical purpose in the long term. One example is in the use of paper towels and other products that could be replaced with washable fabric. Granted, they are sanitary, but they just don't need to be used casually when alternatives are possible.

However, the environmental movement has been in full swing for many decades now, and there are many developments in new sustainable technologies. As just one example, the Saltwater Brewery in Delray Beach, Florida recently released edible and biodegradable six-pack rings. This will help solve one of the most painful side effects of wastefully disposed plastics, minimizing the number of animals who are forced to slowly suffocate while being caught by human products and unable to eliminate them from their environment. In addition to new technologies, many people have simply given up eating animal products altogether. I was vegan for a month, and it was remarkable easy. Now I'm no longer vegan, but I find meat always tastes better when saved for a dozen or so special occasions throughout the year. Unless it's turkey, which doesn't satisfy my palate. Maybe if you fry it. In bacon grease. As a general rule of thumb, the sooner something was in the ground, the healthier it is, and the longer it was refined inside of other organisms, the tastier it is. This is why you could scrape by on a mostly potato diet, but putting bacon bits on anything is like magic. So much magic in fact, that it makes your entire circulatory system implode. That's the key, an all plant diet isn't tasteless, but it is lackluster, just as an all meat diet is savory, but it also saps your life away. And if you're rich enough to eat foie gras and veal—which are the epitome of excess—you should be rich to buy the tastiest vegan food, or maybe 3D printed meat.

One of humanities greatest challenges now is not only being less wasteful, but also spreading wealth and food to those not fortunate enough to have been born affluent. For instance, I think most people in the United States can agree that we don't need as many fast food restaurants as we do. But they are heavily useful when there isn't much food for a long distance, like if you're driving across our vast country from state to state. They could serve a similar purpose in developing nations, and goodness knows we've given fast food chains enough money to expand their empire over the land. They're certainly low on the nutritional hierarchy, but it's still better than eating dough for eight weeks straight with a few local fruits and vegetables. People in developing nations could at least have the same access to tropical fruits, but first they need refrigerators. The excess sugar in our diets might even give them an extra boost to their daily work lives. But if the nutritional value really becomes a crucial problem, then they can just lobby for researchers to innovate cheap nutrients. Who knows, maybe the same mastermind that was able to put so many ungodly calories into twenty minutes worth of an average person's wage in the developed world could develop a pill that immediately gives you all the nutrients you would need in a day. Right now it's only a full day's work in a developing nation to get a fast food combo meal, but I'd gamble that people would be willing to pay if they got to eat a hamburger every day. It would also help people in the developing world lose weight if resources were redistributed. For the same reason, if you're on a diet just send a portion of your calories to a local shelter, or save time and just send them a portion of the grocery bill. Or just make the big corporations do it, or both, whichever you prefer.

There's an argument to be made that turning crops and animals into food and rock into fuel and forest into architecture is bad for the environment. But, technically, it's all the same thing, nothing more than bundles of protons neutrons and electrons. Is it just as bad that stars must die in order for planets and people to exist? Is it morally different when humans rearrange those chemicals to feed themselves and learn about the universe? A retort to this might come from the fact that other living organisms literally die in a way that stars only figuratively do, and they are often mutilated for barbeques and sometimes skinned alive for clothes. But when animals die they become no more sentient than the stones around them, so perhaps people could compromise by humanely killing animals and use less energy. Even better than that, we could just get better as using what we have more efficiently, with artificial meat and fabrics. Here's a summary for why agriculture is important. Without agriculture we don't have food. When there isn't enough food, the wealthy begin to hoard resources, and people begin to starve. When people don't eat, they get cranky, and when people get cranky, they won't hesitate to rob or stab each other if that means getting a loaf of bread. If we don't find a way to maintain our current food surplus and find a replacement for fossil fuels before we run out, people will go to war, and then it's back to square one. Speaking of fossil fuels..

Industry

Bright red fire, it is humanity's crimson flower. It is the source of our watt power, our dominance and willpower. For no other creature that ever swam or crawled, was ever able to harness fire. Using oxidation, we continue life's eternal tradition, taking entropy, and turning it into productivity.

First of all, let's not demonize carbon dioxide. You put more into the atmosphere every time you breathe. That's just what animals do. We turn sugar into carbon dioxide by breathing in oxygen, that's how we've done the whole living thing for the last few hundred million years. It's by no means the only way. The plants have done just fine for twice as long using the most regular source of food in the universe, which basically makes them storehouses for cosmic radiation. It's how they make delicious tropical fruits and nutrient rich roots. Fungi break down both plants and animals when they stop cycling energy and get put in the dirt, and then recycle the carbon through the life machine all over again. That's all the carbon cycle is: prokaryotes, animals, plants, and fungi all living and dying over and over and over again since abiogenesis. They all create a variety of toxins, and build different complex structures with different proteins using different DNA. But at the end of the day, every organism from human to fungus to ferns is just playing hot potato with carbon in different forms. We put out greenhouse gasses that plants eat, the plants put out sweet stuff that we eat, and the fungus eat both of us when we die, and then get eaten by us and give nutrients to the plants. It's the same circle of life, all the way back to our greatest common ancestors. So let's make this abundantly clear, carbon dioxide is not the problem, the problem is humans who keep lighting dead plant matter on fire to turn on their lights and drive really fast and fly through the sky inside giant metal birds, and we have become so much smarter thanks to harnessing that energy that we've overturn about two thousand years of ignorance in just about eight generations.

Humans born today are living in a magical fantasy world relative to pre-industrial peoples from the crown, to Rome, to Athens, to Timbuktu, through Egypt, long before Mesopotamia, and so much farther back into the past, that every single human being existed within a ten-mile bubble inside of Africa. The world of every human being was infinitesimally diminutive compared to the sheer immensity of the vacuum surrounding them. They only got glimpses for half of everyday, when most animals like them would be sleeping to prepare for tomorrow's hunt. For two hundred thousand years, twenty thousand generations of human beings have lived and died under the pretense that all of creation lied within their fingertips and as far as they could see. As it so happened, after they figured out what things do when they get hot, how plants worked, and what happens when old dead rocks get really hot, all they needed was an industrial revolution in order to fully grasp reality. Humans learned that their grip was too large, and their vision too short.

Since we started lighting huge lumps of carbon on fire, we've put out more carbon dioxide into the air than a trillion animals could in a single day. Humans have become so advanced so quickly, all thanks to a rock made of old dead plants and their carbon. We are progressing so quickly, that in less than two centuries, we have been able to create machines to peer into the birth of the universe by making atomic fireworks in order to understand what they are truly made of and where they come from, making us the first animal to ever be able to do so. Everything else alive on the planet is having trouble keeping up. Climate change is occurring so quickly, from quintillions of exhaled breaths, because humans are so smart. Life had been wallowing in the mud for billions of years, and in less than a quarter million was able to leap onto the moon. Nobody can possibly deny the power of the human animal's great and adaptive mind, and nobody can deny that our god-like evolutionary pace—from pointy rocks to nuclear bombs—has allowed us to become the best organism at harnessing the very fabric of existence itself.

However, some green young children dare to deny that our great adaptive minds are not capable of changing the climate of the entire thin lithosphere we live in, and that lighting carbon rocks on fire does not increase the Earth's temperature, when the same laws of electromagnetism allow for the existence of life. Those foolish children dare deny the power of the human race to cause global warming and change the climate enough that no living thing can adapt as quickly as we. All humans needed was a little extra food and a little extra fire, and here we are today. All of the wild plants and animals are simply unable to migrate fast enough, and breed quickly enough, because we turn their habitats into shopping malls and split continents apart to make canals for our ships. We add so much exhaled air that we melt polar ice, and acidify the oceans as they soak up carbon and force oxygen in the water to send free hydrogen loose on aquatic life. Mountain ranges of sugar grass and hamburger flesh have been born and slaughtered to sustain the global human radiation. That's what all those sheepish liberals keep whining about, and what those pig headed conservatives are scared to admit. The human race is awesome, so awesome that we need to adapt a new means of being fed without hunting and gathering or mass slaughter, just to prevent the food that our food eats from going extinct. On both political spectrums there are children who keep fighting over something neither of them is educated in, and they rarely try to teach themselves or listen to the other side. My parents always said people were stupid, but I think that's only true if they choose to be. People are actually the smartest current living organism, and the proof is literally in the air (and ocean), making it very warm everywhere. So if you hate being human, then fine, human made climate change isn't real. Now go play with your little electrical toys powered by coal. If you're an adult human who actually cares about what happens to your children after you die, you may continue reading.

Humans are causing global warming, which is making the climate of our planet shift, and animals who are living in different environments are dying, and places that had too much water are running dry, while others are flooding. Crops and the animals that maintain balance in the ecosystem are dying, and if they die, we lose the game of life. We will either have to go back to caves and savannah, or work together and prevent it from ever being a problem, coming out of these scary times with more security, prosperity and happiness. We can do this, together; just the same way we got into trouble in the first place. Now, let's talk about how to save the world. At least for humans. Extremophiles in thermal vents are just waiting for their time to usurp our reign, after another four billion years of evolution. As I've said frivolously, the essence of why pollution from humans is killing off biodiversity and weakening global ecosystems is because the rest of the living world hasn't been able to keep up with human progress. Human development is outpacing natural selection. Not only that, but the process of acquiring fossil fuels also harms the environment. As just one subtle example, sonic blasts that test for oil reserves negatively impact aquatic food sources in places such as the arctic. It confuses native animals and causes them to migrate away or be unable to find their natural food source. The native Inuit people rely on those food sources, because the markets there have a monopoly on groceries, allowing them to mark up prices to ludicrous levels. If they are unable to hunt, they will be forced to pay for overpriced food supplied by greedy corporations. But that's just the beginning of the problem.

Electricity also means that people actually don't know what the night sky looks like. Have you seen the Milky Way and all six thousand stars in the sky? That's a privilege that most people in the developed world have sacrificed, but in return we have safer suburbs, highways, and protection from wildlife. Even though the cosmos are so well understood, the night sky has become an enigma to industrialized people more than it was to nearly every other human to ever live. For many city folk, seeing the true night sky is probably what it feels like to see the Earth from space. From up there the city lights burn like a thousand miles of fire, a mirror of what they sky once was from the surface. It's the glimmer of life, after four billion years bright enough to shine through the dark. But the price we pay for this night light is the natural glimmer of the very stars were the original harbingers of every circuit in your home. Ironically, that's the reason North Korea is so striking when orbiting the Earth at night. Sadly, it is one of the darkest patches of land on the modern planet. On the other hand, that means they must have the most beautiful view of the sky. It would be a perfect place to do astronomy.

The genius of the industrial revolution is in how it calls from several laws of nature. The first step is finding out how to move electricity. Metal works well, thanks to all of its loose electrons. Next you will need a way to move current through that material. This is where Faraday's law comes into play; all you need is a magnet and a loop of whatever metal you're working with. Here's the tricky part. Now you need a way to rotate the loop of wire through the magnetic field to induce current. Most of the techniques used industrially involve water. If it's a hydroelectric plant, one could use water rushing from a waterfall or river. One could also heat the water and use the thermal energy in the steam to rotate a turbine, like in coal and nuclear plants. One could also use a different fluid, namely air, to spin a turbine. However, if humanity wants the most potent form of energy, then we must learn to capture the power that has been raining down on us for the last four and a half billion years, the source of all other energy on the planet capable of running modern society for an entire year with nothing more than a few hours of its output. Unfortunately, solar power has been extremely challenging to harness.

Current means of capturing photon energy are quite remarkable to say the least. They utilize not only the laws of electromagnetism, but also the quantum effects that inherently unite electrons and photons. Like many batteries, solar cells are made of two separate layers of various chemicals. There are actually several combinations of molecules that can be used to make a solar cell, one example of which is based on silicon. You might recall that silicon is similar to carbon in that it has four outer electrons, giving it a very stable, flexible structure that can easily fit with other atoms. Like any battery, a solar cell is split into two separate halves. The silicon is then doped on both sides by adding boron to one diode and phosphorus on the other. If you have your periodic table handy, you'll notice that phosphorus has one valence electron while boron has three. The neat thing is, since silicon normally stacks in squares, adding in these elements causes the chemical mixture to gain charge. Since boron has nearly the same structure as silicon with one less electron, that mixture has "holes" that can be filled with electrons, and since phosphorus has a full shell if it has one less electron, it stacks neatly with silicon by doing just that, giving the mixture a free electron. I'm sure you can see where this is going. When a photon hits the diode with a loose electron, it gives it enough energy to move through a circuit to the boron-doped side. The best part is that this process doesn't wear out the solar panel, and so can be used for decades. Coupled with the fact that the Sun hits the Earth with 1365 watts on EVERY square meter, one can quickly see why so many researchers are excited about the solar industry. Unfortunately, this won't be an easy task.

Despite how much energy is available from the Sun, there are still several obstacles that need to be overcome before we can fully transition from burning rocks. Cost and efficiency are the two main political issues, but the government could easily subsidize solar energy and the research needed to make it more efficient, so that's a lame excuse. The money could just come out of the military budget, like a dozen other things. There are however genuine logistical issues based on science that we will have to consider. The most important task is to devise a way to store energy. After all, the Earth is dark half of the day (sometimes longer thanks to our tilt) and if you have clouds, that means no power. However, means of storing energy will be critically useful for every part of human life that requires electricity, be it in space or at home, so there is a lot of potential money to be made on that front. It seems that there must be a way to store energy in chemical bonds that could then be broken through the use of some sort of catalyst. Another issue is finding the space to put solar panels, which is a far less scientific dilemma. Seemingly, it would be a simple matter to just line every roof with solar panels for basic needs. This would be especially useful for those in developing nations who still don't have reliable access to electricity, and we certainly don't them want to waste their time on carbon-based fuels. For large-scale projects, one could imagine using desserts in the tropics where sunlight is most abundant. There have also been ideas proposed to place floating solar farms in the ocean as well. I could easily envision solar cells along the Savannah and Middle East, providing not only energy, but also well needed shade for both people and wildlife, especially as the planet inevitably warms. In the far future—when state borders become archaic—one might even imagine a society with globally connected network using transformers where the half of the planet that's illuminated sends electricity to the dark half, which would further mitigate the hindrance of cloud cover. But before we can do any of this, we need to put our differences aside to halt the current climate debacle.

Global warming is a very sensitive topic to a lot of people. It doesn't help that we hear the phrase, "climate change" so frequently that it becomes a useless buzz in the background of people's minds. As a result, those who try and warn about global warming turn into broken records. People also find that believing in the climate observations of professionals is associated with their politics, for some reason. People still make false claims about some sort of disagreement among people who study climate for a living, having the gall to claim that they still disagree, even though the only people who still bicker over climate change are politicians, bureaucrats, and scientifically illiterate (but well-meaning) citizens. While it's certainly the case that some professionals think that the alarmism of climate change is ill-founded, other climatologists have also said that the entire human species will already be extinct by 2030. But regardless of how they react to it, any educated person knows that it's happening with the same certainty as the Sun rising. It's almost the most ideal problem for humans. It impacts everyone, but it's so easy to ignore. Historically people have always had a face to put on evil, which were almost always people that looked different. The major problem is that climate change doesn't have one face; because it's caused by every single person who's life depends on modern technology. Climate change also doesn't have any one direct effect, because every aspect of the planet is tied to the climate. Its rapid change causes power shortages for people dependent on hydroelectric power when dams dry out, coastal flooding from melted ice sheets, water shortages to cities, monsoons that threaten farmers and fishers, and decreased tourism (ergo money) to nations who's whole income is based on their climate. It even contributes to increased natural disasters like wildfires and hurricanes, which are fueled by heat on land and sea respectively. It's our first global issue as a species. So will we face the challenge as a planet, or just go the way of the dodo?

The fact that people still don't understand how their actions affect the world they live in has become indefensible. I've seen so many excuses for climate change that I almost want to believe it's a conspiracy myself, and I probably would if I didn't have the good fortune of learning physics. So I'm going to try and tackle some of the excuses that one might pose against human accelerated global warming. To reiterate, weather is NOT climate. First off, some claim that the Sun's temperature could be the cause global heating. However, solar data from The Max Planck Institute for Solar System Research and the National Solar Observatory strongly suggest that the Sun has been cooling slightly since 1880, if not just oscillating between the same energy output as it does every eleven years. Volcanoes also don't justify it. Geologists can measure volcanic activity, and burning fossil fuels still puts out far more gas than the Earth alone. This is one of the few cases where data is actually dangerous; there are so many graphs and figures that they may actually seem fabricated. The general public, including myself, have just become numb to all of the squiggly lines and bar graphs that are somehow supposed to convince us to stop driving our fancy cars, unplug our televisions, and stop eating bacon. Sweet, succulent bacon. Finally, the main defense is that the science isn't settled. But that's the beauty of science; it's based on unchanging laws of nature. If you're scientifically literate, you can decide for yourself whether other scientists' claims are reasonable. All you need is a little bit of history, chemistry, and electromagnetism. Then, human accelerated global warming goes from being a political hot topic bathed in ambiguity to being a blatantly obvious fact.

But maybe scientists are all conspiring, and fear mongering to make people do something evil like not leaving their cars running, because it's too hot and we need air conditioning. Except gas costs money, so why would people want to waste it anyway? Or maybe climate scientists are pushing a malevolent agenda to take away our right to eat fried animal carcass basted in honey mustard or barbeque sauce, slathered in bacon. Obviously these climate scientists are so ethically against people driving up to a window to eat the succulent flesh of a cow that they've never seen, that they want to strip us of our sovereignty over animals, who only exist because of us. How dare they think that they can force people to eat disgusting meat substitutes made of plants, even though they taste identical? Who wants to eat plants? Just feed them to animals and we'll eat those instead right? It's not like you can get all of your vital nutrients from a balanced diet of mostly fruits, nuts, leaves, and grains, with access to meat only once every few weeks—like the majority of humans throughout history—that's just barbaric. Climate scientists want us to devolve back to a time when people actually had to kill what they ate, which were usually plants because they don't fight back. All of this because climate scientists have the audacity to think that our right to drive cars, use electronics, and eat cows that were strung up by their legs, having their throats slit while they writhe in pain, after spending their entire lives eating corn that could have fed starving children, and belch several tons of carbon dioxide, is less important than ending the human race. How dare they?

Oh right, it's because climate scientists are human beings too, and they don't want to live in a world full of pig headed idiots who get offended by being told that all of our wasteful actions are detrimental to humanity. It's because climate scientists are actually good people who want your children and mine to live a better life than their parents. They aren't even asking for much. Just turn off the lights you don't need, wear more clothes in the winter and less in the summer, ride your bike every now and then, recycle your junk, and don't clog your arteries by eating meat every day. I don't even understand why people argue about climate change, the solutions literally just save people money. Unless you're a coal miner or oil refinery, then using more plentiful sources for fuel would definitely cost you money and jobs. Luckily there would be a huge market boom for engineers to built innovative wind turbines and solar cells, so I'm sure you'll get back on your feet. Ultimately, the only sensible excuse for debating climate charge is money and laziness.

People often state that pollution and climate change wont fix themselves, but that's somewhat misleading. The planet can return to equilibrium on its own with the help of microorganisms. Earth and life in general don't care what happens to the climate, or people. Living organisms dug their heels into the Earth four billion years before we ever got here, and they're in it for the long haul. When scientists are calling for action, it's because climate change from wasteful consumption could possibly end the human race. Every light bulb you run, every device you charge, every piece of meat you eat, contributes to several tons of carbon being put into the atmosphere, which heat the planet and threaten to eliminate the marine life and agriculture our societies are grounded on. If all humanity facing calamity is too hyperbolic for your tastes, or if you just don't care, then remember that fossil fuels are called such because they took millions of years to make. All of human history would have to occur a thousand times over before more of it becomes available, so it will almost certainly run out. And getting more of it requires burning more carbon, deforestation, shrinking biodiversity, and crops being fed to livestock that trumpet even more carbon into the air. So even if you are scared to lose money from oil, remember that if you enjoy airplanes, Las Vegas, refrigerators, or light bulbs, then the search for sustainable energy directly benefits you. Of course, the fact that climate change is gradual rather than immediate understandably causes people to dismiss it, because they aren't able to envision how their actions will affect the future. People also quite enjoy procrastinating. If you don't care about climate change, that's fine. But if you aren't going to do anything productive yourself and keep eating animal carcass three times a day, at least give microloans to people who also aren't doing anything because they have to walk several miles every day just to get clean drinking water from a well that's probably drying up because you're burning fossil fuels. Even if you personally don't have spare change, you can help by making your government or local businesses do it. They're getting your money anyway, just tell them to use it to educate developing nations. Give them fluid dynamics and atmospheric data from Venus, so they can figure out how to stop climate change, make lots of money, and get us one step closer to eating 3D printed steaks in space. Bonus, we would have a planet to come home to if we survive long enough to leave.

Although as much as I preach for it, I suppose it's important to note the dangers of globalization as well. People lose their jobs to cheap labor in other countries, labor which may have dreadful conditions akin to early industrialization in Europe, where people depend on sweatshop labor just to keep their own economies afloat. This can be seen as a hidden blessing though. People who were doing jobs that can be done more efficiently elsewhere will be forced to learn new skills in areas of development not possible before. They can become experts in fields like technology, data security, or even become music stars. What's more, being aware of when companies are using unethical labor will drive people to force governments and corporations to create safer working conditions. They can be aided by the experience of developed nations that have already gone through the pangs of development, which can help them circumvent unwanted side effects of industrialization, such as pollution, or people getting their hands chopped off. I think we can all agree that hands are pretty nice to have. We can do all this with optimism and a little help from our friends in foreign countries. After all, borders are nothing more than artificial, and people everywhere want to have better lives for themselves and their children.

It's also important that globalization occur gradually, so that people have time to adapt to the new economic environment. Which it more or less has since Christopher Columbus in 1492. These ideas translate to immigration also. Many people fear foreigners because they take peoples jobs away. The thing is, more often than not they take jobs people don't want to do, which frees up natives to pursue work that they are passionate about and benefits society in new ways. If you were really indispensible at your old job, you probably wouldn't have been replaced. Creative and educated people don't have to be afraid of losing their jobs, because they can always find something else to do. People are smart, and they will make ends meat anyway they can. If you really don't want to lose business to foreign competition, then you had better innovate your product to make people want it. People will always go with the cheapest option that satisfies their needs, so the innovation will be different case by case. You might make your product cheaper, more durable, give more personal customer service, or just sell customers 3D printed bacon with your product. Regardless, you will either need to have a viable method to compete, or pack up and find a new niche to fill, neither of which is impossible for an innovative human. We just have to make sure people have the opportunity to express their creativity, lest they become desperate and begin stealing from others. Or perhaps you're afraid of foreigners stealing your culture. Well, as we discussed in Chapter 8, culture is usually indicative of the environment and whatever keeps people alive. All the rest is usually just idiotic filler.

On the matter of population, it seems that there may still be too many people on the Earth. My grandmother says people are like ants in that regard. Population could be controlled if everyone had food, birth control, access to education, trustworthy governments, and only two kids. My parents always said that I was enough trouble, and many poor mothers have to have multiple kids to keep up a farm and because many of their children may not survive past five. The population would easily stabilize if no generation had more than two offspring. Factor in all of the couples that don't want kids or could adopt, add in people only eating meat a few times a week, and then we would have enough resources for the planet to be well fed.

One last important note. The single thing that periodically bothers me enough to care about anything other than space and other people is

Yellowstone. The United States would probably like to have friends in other countries where people could live IF it erupts anytime in the next century and covers a large portion of the land with ash. The volcanic winter might help with the whole global warming thing, but it's very bad for crops. The only way I could think of that would mitigate it would be to dig giant holes into the ground to let some steam out. I can't conceive of what something that grandiose would cost, nor if it would just incite an eruption. Perhaps a geologist has done a calculation on that already. I have a hypothesis that Old Faithful and other sources of steam could also be used to fuel a natural electric power plant. The same technology could be used to harness Jupiter's gravitational energy through the volcanoes on Io, or from the geysers of Enceladus and Triton. They could serve as solar system fuel stations. I'm sure if a supervolcano does erupt, since millions of homes would be at risk people would readily band together to either get people to safety. Although, many people would probably be too stubborn to move, and would just tough it out for a few decades. Humans are developing means of controlling the weather as well, mostly to increase rain. Perhaps something clever could be engineered to disperse the ash and volcanic gas. Either way, it would really be nice if people in developing countries could live in real houses and have concrete roads, so that they can either be prepared for immigration, or so their citizens could engineer a means of preventing the volcano form erupting at all. Or we could just go to another planet.

Exploration

Humans, without question, are the most successful invasive species on the planet. We can live basically anywhere, all we need is a little dry land, some clean water, and maybe some plants and fish. There have thus far been only two places we have yet to fully explore, the depths of the oceans—the first home of life—and space, the first and current home of all things. First and foremost, we need to ask ourselves, why should we even bother leaving the Earth at all? If it really is so inhospitable out there, then why should we spend trillions of dollars on space exploration, especially when we have so many problems already here on Earth? Even Bill Nye has his doubts about putting long-term colonies on Mars, for good reason I might add. Would you want to colonize Antarctica without breathable air?

The answer to all of those questions is twofold, one being historic, the other being blatant (but not apparent) common sense. First, let's look at the past. Did you know that there was a time when humans could only walk place to place? As a matter of fact, the very first humans migrated to every corner of the planet on nothing more than their own two feet. Some lucky Eurasians had access to horses, which they could train and ride, but that was the fastest any human being could travel for tens of thousands of years. People on different continents had no idea how big the Earth was, they assumed the entire universe was only as large as the distance they could see and travel across. But after the dawn of agriculture, it wasn't long until the world gradually became larger, and people moved to fill in that space. Merchants traveled for months and years just to get a sampling of a foreign food, spice, cloth, or technology. Eventually they built boats, at first only large enough to fit a small party, at least until there were enough shipwrights that people were able to craft whole armadas of ships, each capable of holding several tons of minerals, food, and even livestock. They sailed for years an decades just to get access to more land and resources. Empires with a strong enough navy then had the power to extort and pillage foreign continents with an iron fist, exacerbating wealth inequality. Then, we learned how to fly.

People cite how humanity went from flying to space travel in such a short time, from the Wright Brothers first flight on 17 December 1903, to Apollo 11's moon landing just sixty-six years later on 20 July 1969. This is a severe understatement. HUMANS CAN FLY THROUGH THE AIR! The only other animals that can accomplish that have to do so using their bodies, like bats, birds, and pterosaurs. Those chumps have to devote the whole of their anatomy just to get off the ground. We humans, on the other hand, evolved to run on the savannah. But with a little teamwork, diesel, hydrogen, oxygen, and math, we figured out how to soar through and above the heavens. That's inconceivably remarkable! That's like a dolphin popping out of the ocean to take a stroll through Manhattan. What's more, remember that modern humans have been around for over 200,000 years. That means that for over a hundred ninety-nine thousand revolutions around the Sun, people were stuck on the Earth walking around like common beasts. For nearly all of recorded history, the fastest way to travel was by using another animal, either horses or in some awesome instances elephants. After a few millennia of doing that, the British dug some black rocks out of the ground, and developed steam technology that birthed automobiles and locomotives. Then in half a century and with less computation power than a cellular phone, we went, once more with feeling, TO THE MOON! This is the same species that exiles people for not believing in some voice in the sky that supposedly does magic tricks, kills each other in mass genocides, and makes celebrities famous for having sex, something that literally all human beings and most multicellular organisms do just to survive. That same species built machines that take people into, and way farther than, heaven. Why? Because we can. Also because of a half-century playground rivalry between the U.S.A. and U.S.S.R., which could have potentially ended all of humanity. That's right, one of the most remarkable accomplishments of any living organism on the planet was accomplished because some hairless apes on opposite sides of the planet were afraid (correctly) that hairless apes on the other side were out to get them. Apparently we can only do anything interesting when we are fighting with each other, or when death seems immanent, even though in truth it always is.

This is obviously ridiculous. Friendly competition can be far more lucrative than a fear induced arms race, if nothing else because both sides get to party together following big accomplishments. What's more, unlike European colonization, nobody lives on asteroids or most of the gas and dust falling around the stars, so morality isn't an issue for extracting resources, not that it mattered the last time. That being said, we are much wiser as a species now, because we've made several mistakes by abusing our power during colonializing, really just to get rich and smoke some tobacco or eat some raw sugar. Now, if we want to leave the planet and continue our constant history of expansion, the only way to do so is by working together as Earthlings, to do away with petty differences and treat each other with respect. Or we could just keep our fake national borders and continue hating each other. The problem with that, however, comes from the second reason we probably want to go to space: The Sun is going to die.

Now there's no reason to get dramatic. We all know that stars eventually run out of their nuclear fuel, and when they do they expand, potentially engulfing planets like the Earth, evaporating all water off of the surface, making life on the planet impossible. That's common knowledge. That sort of thing also won't happen to the Sun for another four and a half billion years or so. That's enough time for life and humans to evolve all over again! So what's the big deal, it's not like we don't have time to get around to it right? Well, not quite. First and foremost, if you had the opportunity to orbit another planet, or maybe even another star, while staying healthy, and you just decided to ignore that opportunity, then you are remarkably bad at being human, and your grandparents would be ashamed of you. Perhaps that was a bit of an exaggeration, people are actually much better at procrastinating than being interesting, and your grandparents probably wouldn't have left the house as much if they had Netflix and Sports Center either. But there's a second (better) reason why the end of our solar system requires immediate action, besides the fact that being stuck on a tiny pebble for the next billion years sounds absurdly boring. Namely, space is enormous! Sure, having a luxury resort on the Moon, Mars, or Venus sounds nice, but just getting out of the solar system will literally take decades. That is of course, unless we figure out a realistic method within the laws of physics that will make it easy to create wormholes or Alcubierre drives. The fact of the matter is, the cosmic speed limit—the rate at which all massless particles move and the rate at which all information and causality travel through spacetime, able to circle the Earth seven times in a second—is painfully slow. Everything solid in the universe is so radically far apart that traveling through space cannot be done within a modern lifetime. Even if everyone on Earth built an ark and left the planet in 2100, we wouldn't be able to make it to the next star system for decades. The fastest spacecraft sent by humanity thus far has been New Horizons, clocking in at 58,536 kilometers every hour, and it still took a decade to get to Pluto. That's a distance that light takes just over five hours to cover, the same length of time for a casual road trip. Space is so painfully vast that traveling becomes a game of endurance, or a game of ingenuity. But physicists—just like the rest of us—are impatient, and so there have been numerous ideas proposed that might make distance a non-factor.

The key to becoming a spacefaring species is harnessing the abundance of energy within the universe. We've already been doing this since the first humans discovered fire. Since then, we've only continued to advance society by learning how to use larger amounts of energy with greater efficiency. Mastering electricity has been bar none one of the most impressive human accomplishments, and all we had to do was twirl some metal around, connect it to other metal stuff, and voila! Free shiny things. Well, not free, we need to either burn rocks or capture energy puked out of the Sun to power all of our iceboxes, hot fans, and indoor Suns. Otherwise it's back to the farm, and we're running out of black rocks and sludge. We're going to need more power if we want to go back to the stars from whence we came. Luckily the sea of spacetime has plenty of star vomit riding on its waves, and if we're clever enough we might even be able to make our own tiny stars to run sparks through our shiny things while we're taking a siesta on an island planet, or visiting the home world. But when you're adrift in the sea of space, you have to watch out for neutron lighthouses, stellar tsunamis, and black whirlpools. If you get too close to those last ones, they'll fling you so far into the future that your newborn will have graduated from school. So there are plenty of dangers to space exploration, but far greater rewards as well.

For as long as humans have understood the structure of the universe, creative thinkers have been devising ways to energize a galactic voyage. One of the most popular is a common science fiction trope, teleportation. Most often, teleportation is explained as a means of encoding a person's being, converting that code to light, and beaming that information to a detector on whatever planet or moon you want to go to. Then, that information gets decoded and you get rematerialized, ready for your vacation to the geysers of Enceladus! Presumably, if information is being beamed as light, then your physical body is being dematerialized every time you teleport, which would mean either the forces binding your cells, molecules, atoms, and nucleons, are all dissolved and shot through space, or your information gets read by a computer, you get murdered, and then new information is shot through space. How either could be done is anyone's guess. Even if it were possible to do this, it would still take anywhere from minutes to years to go someplace interesting. Although, you wouldn't age on the way, so it still beats a cross-Atlantic pilgrimage in that regard. There is one last problem with teleportation, which has to do with the very nature of light. That is, it loses energy the farther it travels. For the same reason a candle is a quarter as bright twice as far away, light becomes diffuse as it travels. Even lasers shot to the moon grow to be over a mile long. A beam traveling to a distant star or planet might grow as large as a city or continent. It would have to be relayed between closer satellites, like the beacons of Gondor in *Return of the King*. Not only that, but it would also experience redshift if the destination is moving relative to the starting point, and it could be scattered by any gas or dust lying in between them as well. Either way it sounds horribly troublesome, and difficult for a large group of people. Some might inquire about using quantum entanglement for teleportation. From what I gathered through my first quantum professor and McIntyre's textbook, I'm fairly sure information doesn't travel between two entangled particles, and I'm not confident that one could build a machine that could instantly entangle every one of a person's octillions of atoms at once. Although, I could have just as easily taken very poor notes during that lecture, that is also possible. So we won't bother entertaining that idea, at least for now.

My second favorite means of space travel involves making space do the moving for you. Recall that nothing in the universe can break the cosmic speed limit, but space itself is not obligated to abide by this rule. Presumably, through clever manipulation of Einstein's field equations, one may be able to bend the fabric of existence to actually travel through a real warp drive. The first option is a wormhole, a sort of passageway through the universe. The basic idea is the same as connecting two points on a piece of paper by folding the paper and bringing the points together. Of course, since our universe is four dimensional, and not a flat sheet, that would mean a wormhole would be a five dimensional rip. I could speculate about the properties of such a thing, especially the obscene amounts of energy needed to make one, but as of now the mathematics are so complex that some of the keenest theoretical physicists are still grappling with it. The same is true for the famous warp drive devised by Mexican physicist Miguel Alcubierre. In essence, it works by negatively warping the spacetime in front of a starship, and positively curving the spacetime behind it. Now, negative curvature is the same effect that gravity has, which means the warp drive requires some sort of anti-gravity in order to surf on the fabric of spacetime. Since energy (like you, me, and planets) dents spacetime, it is often said that a warp drive runs on negative energy. Simply creating such a thing does not seem to be possible based on our current understanding of the universe. However, it doesn't seem inconceivable to make space positively curved. Take for example a bowling ball on a taught surface, like a trampoline perhaps. If you immediately remove the ball, the surface will recoil, momentarily curving opposite direction. Hypothetically, it may be conceivable to do the same thing with spacetime; assuming that matter puts strain on it the same way a bowling ball strains a trampoline. Of course, this would require a large amount of matter being collected and then spontaneously removed from that space. Perhaps opening a wormhole near a supermassive black hole could accomplish this. It could be a method for us to travel between galaxies. Perhaps we could travel through space if we could somehow convert the energy within the vacuum to power a craft while simultaneously reducing the distance we would need to travel. Under any circumstance, such things are still far in the future.

Alright, so let's take a step back from interstellar travel, we'll leave that for the next few unborn generations to deal with. Let's focus instead on what we can accomplish this century. You have to walk before you run, and you have to colonize a few worlds before you can surf along cosmic waves. In addition, traveling throughout our solar system will give us the invaluable experience we will need when it comes time to move to a new system. There are countless obstacles that we are going to have to face when we leave our home behind, beyond merely covering vast distances. The same diversity we find amongst humans seems to also hold true for planets, there doesn't seem to be a single place exactly quite like Earth. Let us also not forget about the satellites orbiting around planets as well. There are several moons just within our own solar system that are larger than

Mercury, several of which with thin atmospheres and Titan with a thick one. As a matter of fact, there are more moons in the solar system with liquid water than planets. Saturn's moon Enceladus has cryovolcanoes of liquid water, and Jupiter's second moon Europa likely has more liquid water than every ocean, lake, and river on Earth, and both of these worlds have organic molecules within them. We must consider all of our options when it comes to shopping for future worlds. This of course isn't to say that there aren't any planets similar to Earth. Obviously within the immensity of the cosmos there is very likely going to be a nearly identical world, but on that same token they may be so far away that it would take millennia for us to just get to them. There are so many critical factors at play, nearly all of which we take for granted.

First we must consider what kind of star the planet is orbiting around. Obviously, planets around high mass stars have shelf lives billions of years shorter than those around smaller stars. However, we have observed planets around even the dead remnants of supernova, including pulsars, so it may not be inconceivable for a planet to find its way from one system to another blue star, then be bombarded with comets, and after all that be able to sustain life. So we will assume that habitable worlds can exist around any type of star. The question then becomes a matter of how different stars would affect the habitability conditions of different worlds. Not only does the size of the star dictate its brightness and its lifetime, but it also determines the subtler trait of the kinds of light it emits. Different stars produce different light, and different light means a different planetary environments. As with all blackbodies, stars emit photons of all wavelengths, but they do not distribute light equally along the electromagnetic spectrum. So not only do red stars emit fewer photons than blue stars, but out of the light that they do emit, a higher percentage are lower energy, longer wavelength light. In other words, when visiting planets orbiting larger stars, one will face greater danger from high-energy radiation like x-rays and gamma rays. This effect would mean that planets could potentially have different atmospheric compositions and still be able to sustain life, and future explorers would need to adapt to those conditions. Furthermore, not only are more massive stars brighter than less massive ones, but the increase in brightness grows exponentially. A star with two times more mass may have a brightness that is eight or sixteen times greater. What this means for the habitable zones around stars is that planets must be exponentially closer to a low mass star in order for water to remain liquid on their surfaces. But being very close to any star puts planets at greater risk of being blasted by stellar wind and ionizing radiation. So while larger stars may put out more high energy radiation, it may also be that planets around smaller stars still get more radiation overall, simply because they are so much closer to their stars.

The size of the star also plays a role in the specific conditions of habitable worlds that may orbit them. One of the basic questions that need to be considered is whether the planet has an atmosphere or not. Not only that, but what is it made of? If it doesn't have sufficient levels of oxygen, could we change it? We know now how oxygen was only released into the Earth's atmosphere by living organisms, so potentially uninhabited worlds could be planted with cyanobacteria or archaea that could alter the chemical composition of a planet's atmosphere. However, this would only be possible if there were already sufficient levels of water to feed them, not to mention that using microbes presupposes some sort of atmosphere on the planet. If a planet could already support some sort of microbes, it would seem as though life would already be abundant on the planet anyway. But what if the planet just doesn't have an atmosphere? Or what if it has too much atmosphere? This is the interesting case we find ourselves in with our own system's planets Mars and Venus. Both of these worlds provide ideal case studies for us to experiment on how to bring life to barren worlds.

In many ways our brother and sister planets are true polar opposites. Mars has nearly no air, while Venus has too much. Mars is frigid, and Venus is scorched. Interestingly, many of their differences also derive from similar traits, the most obvious of these being that they both have weakened magnetic fields relative to Earth. The case of Venus is particularly fascinating. It makes sense for a small object like Mars to cool off more quickly, but something future explorers will also have to consider is whether the planet is even rotating fast enough to generate a strong field. Mars may have once been habitable, losing that ability over time, but Venus would likely still be habitable if it was just spinning more. The issue of magnetism brings to light many other complexities. The European Space Agency released a report in 2012 stating that Venus still has a magnetic field due to how solar radiation interacts with its atmosphere, and a separate study showed that Mars still retains magnetism on half of its surface. Not only that, but even Mercury still has a slight magnetic field. We all know that magnets retain their fields, so it is also important to remember that objects that were previously non-polarized can gain magnetic fields and retain them even after the original source is gone. The fact that there are so many factors at play with all of these planetary traits that we observe on Earth opens the imagination to an immense variety of different planets. One may see this as evidence that the Earth is indeed an anomaly, with so many things having to be just right for life to arise here, but this complexity can also be seen as evidence that if we are clever enough, we can alter the entire landscape of a planet by taking advantage of the sheer diversity of celestial bodies. Who knows what techniques future innovators may devise to revolutionize terraforming? And let's not forget about the moons.

As we have learned from our solar system, moons can have just as much variety as planets. It also seems quite feasible that there may be more moons than planets, considering that it is easier for gravity to pull together smaller objects, and because this is what we observe in our own system. If that is indeed true, then the laws of probability would suggest that there are more varieties of moons than planets. We also know that moons too can have atmospheres, they just need to be orbiting a large body if the moons themselves are going to be as large as a small planet, as is the case with Saturn's moon Titan, which I'll remind you is larger than Mercury. To support the diversity hypothesis, we can also observe the differences between Titan and Jupiter's Ganymede, the largest of all solar moons, which does not have an atmosphere, but *does* have water, albeit frozen water. It may very well be that the future of humanity lies on the moons of hot Jupiters, and not just Earth-like planets. Titan on the other hand not only has an atmosphere, but rivers and weather of the organic molecule methane, which as you might recall is based on carbon.

Regardless of the type of object, or what it's orbiting around, there is one glaring problem that I feel should be discussed more often when we consider choosing our next home. It has nothing to do with what the object is made of, but nearly everything to do with how life evolved on Earth. It's gravity. The one thing I never hear about in any science fiction movie, or any alien documentary, is the fact that all life on our planet has developed feeling the exact same amount of force on our bodies. When we travel to the stars, even if the world's we visit have different atmospheres, we can deal with that. If we never figure out how to terraform the entire surface, we could at least build shelters with controlled atmospheres and protection from radiation. But there is no way to change the mass of a planet-sized object, short of forcing two of them to collide somehow. Yet that is by far one of the most universal conditions that we share with all life on the planet. We have the same chemical composition, and we feel the same force of gravity that signals those genes to grow, divide, and move. There are no objects in the whole solar system that have the same mass as the Earth either, which makes sense because it would seem completely random how much matter is available to an object when it's forming.

Venus is the closest, at 90% of Earth's gravity, and the next closest is Mars, at just 38%. That's not something I remember Andy Weir mentioning in his beautifully written book *The Martian*, but that should have been far more detrimental to Mark Watney's health than just scarcity of resources. The most reasonable means I could imagine to counteract this long-term would be to genetically modify humans to be born adapted to lower gravity, altering their natural size, weight distribution, muscle mass, etc. This is the exact same reason astronauts today can't stay in Earth orbit for more than several months. This leads us to the difficulties we will face just getting from point A to point B.

The first step is getting off the planet, already quite an obstacle to surmount. Not only does one need to travel over eleven kilometers *every single second* to get off the planet, but the fact that so much fuel is needed provides immense risk for any living thing that may be aboard a spacecraft. Many brave, honorable humans and animals have given their lives for the sake of humanity's exploration. Even the very first Apollo mission was a failure, exploding on dock and tragically claiming the lives of all three crewmembers. Luckily NASA learns a great deal from every miscalculation, and was able to accomplish several successful missions, but danger was always present in each one. This was plainly evident in Apollo 13, dubbed "NASA's most successful failure", where crewmembers managed to make it safely home despite one of their oxygen tanks rupturing, although they couldn't make it to the moon. Of course, successful missions in the late sixties didn't nullify the dangers of leaving the planet, as illustrated by the disasters of the 1986 Challenger launch and 2003 Columbia re-entry.

Once in space, astronauts still face multiple dangers even within their spacecraft. Their air must be highly regulated to prevent CO_2 poisoning. Solar and cosmic radiation must be mitigated during spacewalks, as they have the potential to damage astronauts on the cellular and genetic level. New spacecraft must also watch out for debris from past missions, as there are over 500,000 pieces of space junk traveling thousands of kilometers per hour around the planet. If launched on the wrong path, spacecraft might risk catastrophe if a tiny piece of debris puts cracks in a module. There are also curious "tin whiskers" that can grow between electronics and could potentially short them if they grow large enough. There are also major medical concerns associated with space travel. Astronauts face bone and eye deterioration, losing one percent of their bone density every month. There may even possibly be detrimental effects on the brain, as some research mice exhibited fewer dendrites in their brains in space, although that may just be a result of the mice not getting enough stimulation. There are also logistical issues, like checking our maths. If you thought I was kidding about how absurd it is that Americans don't use the metric system, then just look at the 1999 Mars Climate Orbiter that burned up in the Martian atmosphere because the acceleration software on the Earth was using imperial units while the craft was using metric, wasting millions of tax dollars. There are also psychological issues involving cramped spaces and constant danger. Then, there is the great physiological mystery of how humans will be able to reproduce in space. The first long-term astronauts will need to be trained like special forces operatives and be able to cope with fear and stress like Navy Seals.

There is also an ethical issue at play, which involves our hypothetical encounter with extraterrestrial life. As I have already said, I wouldn't be afraid of sentient life, given that it only seems likely that a civilization could become advanced enough to travel through space by working together and understanding the rarity of life. However, should space farers encounter microbial life, they may risk *SPACE DISEASE* like the fictional *Andromeda Strain* written about by Michael Crichton. Microbes have no noticeable sentience, and disease has killed far more organisms than any war has, especially when people travel to areas where there are microbes they don't have immunity for. This could potentially be the quickest way to end a voyage. There is another dilemma as well. So far we have only considered finding life that is far less complex than us, or far more advanced, but what if we were to encounter sentient life on par with primitive humans? Many fans of *Star Trek* would recognize one response to this dilemma as the prime directive, which forbids advanced civilizations form getting involved in the affairs of alien species. This has been the guiding principle thus far for governmental agencies that consider looking for life in the oceans of Europa or in the sands of Mars. Many people in fact are very afraid of planting life inadvertently on alien worlds. I however, find this sentiment to be paranoid and irrational. I say this not only because the crew on the Starship Enterprise disobeys the prime directive constantly, but also on a more pragmatic basis. It's one thing to sterilize robotic reconnaissance devices to check for signs of life. I have no qualms with that. However, it is simply not possible for humans to travel to alien worlds and not carry their microbes with them, we sort of need some of them in order to survive. When we set foot on the Moon, Mars, and every other object in space, we become an invasive species on those worlds. Life is too rare in the universe for us to be sensitive about killing off or being killed by alien microbes, it doesn't seem to matter as long as something is alive. The issue of sentient life is more complex, but we again have an example of this at home. Consider the many isolated people on the planet still living as hunter-gathers, like the people of North Sentinel Island in the Bay of Bengal. Of course, they are still human, and so one may not consider them to be an apt comparison to aliens. But the sheer fact that they have intelligence should be enough to constitute humanity, regardless of what planet you come from. I would argue that when the gap in technology is so small, that it would only benefit us to offer them a lifestyle with modern medicine and bacon, so long as they choose. It may actually be immoral that we have left so many of them behind. I see no reason why they could not add their values to the already eclectic global culture, especially if that means adding more innovators who could contribute to society.

With all these obstacles ahead of us, it is no wonder so many people are pessimistic about space travel. Why bother? I think the poor science guy has lost hope from dealing with people who are too lazy and callous about slowing climate change to be optimistic about space travel. But it's thanks to him that so many people in my generation have hope for a brighter future. How would we feel if our ancestors were too scared to travel out of Africa, to sail across a seemingly infinite ocean, or trek across vast desserts, mountains, and tundra? What would our lives be like if nobody had the curiosity and drive to unravel the mysteries of cells, atoms, and the stars? We would still be suffering through plague, bad weather, and tasteless meals. It's only thanks to exploration that staple foods traveled the world in the seventeenth century; like tomatoes to Italy, curry from India, potatoes to Ireland, corn to European farm animals, or Asian tea to Britain. Those explorers accomplished the seemingly impossible, facing odds they saw just as daunting with courage and fortitude. The drive to expand goes beyond human curiosity; it is the nature of life itself. Our genes have already made it from the bottom of the ocean to the moon, both of which will be gone eventually, so why stop now? Plus, just think about how cool going to space would be!

Now, at this point I would be remiss to not mention the equal need for us to explore the ocean, the cradle of life itself. Clearly as a student of space, and not a marine biologist, I have a bias toward space exploration. But there are actually several parallels to deep sea and deep space. The most obvious one, of course, is that we can't, well, breathe there. We need to build submarines or spacecraft to make micro biomes where we can survive. However, opposite to the empty vacuum of space, the ocean is far denser than the air we move through on the surface. Going to both of these domains is utterly terrifying, as death is at most a few meters away from you at any given time, and at least a few centimeters. The main difference, though, is that there are actually living things in the water. There are no squid the size of small buildings roaming around between Earth and Mars. Granted, we probably would have noticed anything more intelligent than a whale by now, which would be far more terrifying. Despite what you might see on click bait websites or fake documentaries, a mermaid with any curiosity would have ended up on Twitter by now. Still, it's rather unnerving that something might actually try and eat you down there, at least astronauts only have to deal with things like lethal radiation and bone deterioration.

Humans are also ill-equipped to for both environments. But again, space is inhospitable to anything alive. The last time we had a common ancestor with a fully aquatic organism was nearly half a billion years ago. We've been on a path toward lower density for a while, so it isn't a trivial matter to just go back and forth between land and sea. Granted, a few mammals ended up back in the ocean, but they still need to come up to the surface to breathe. Although, with so little of the ocean being explored, it may very well be that there are actually whales that adapted some sort o pseudo gills. For us apes, it seems like going outward is more in tune with our own branch of the evolutionary tree. That being said, it's still rather pathetic that we just don't know what's going on under seventy percent of the Earth's surface. There is crucial data in the deep blue that may be critical for inspiring future technologies that can help us live, both on Earth and beyond. That leads us to probably the most terrifying thing about being underwater: you can't see ANYTHING! In space, with a good enough telescope, you can literally see the universe as it was before the Earth even formed. If you go far enough underwater, there is no light, at least from the Sun. There are plenty of animals that look like floating Vegas casinos. That's not a complete problem though; all we have to do is bring some flashlights. I know there are plenty of passionate individuals who feel most at home in the aquatic cradle of life, and I look forward to the day when we have underwater luxury resorts and research bases in the murky depths, and we light up the sea with bioluminescent algae. But one can only cast their view in so many directions, and my gaze is firmly pointed toward the stars.

To lower costs, we must develop innovative fuels and make our travel more efficient. I'm no rocket scientist, but I have heard about a nifty idea to use a series of nuclear bombs to incrementally propel rockets, known as the Orion Drive. This method has been demonstrated to have the potential to be an efficient means of escaping the Earth, and calculated to even be able to reach 5% light speed. However, there are far more efficient ways to travel once in space, including solar sails and potentially the use of nuclear fusion, an advancement that would not only benefit space travel, but would revolutionize the entire human experience beyond the extent of harnessing coal power. The cost of building rockets shouldn't be a deterrent either. There were points in history when both automobiles and airplanes were rare and expensive, so why couldn't rockets be mass-produced one day? The same is true of extraterrestrial bases on the Moon, Mars, and beyond. As 3D printing continues to advance, we may one day be able to build infrastructure remotely before ever sending humans to colonize objects or conduct research. Training individuals will also only become easier and easier, thanks to video game technology no less. You might be surprised to know that the programs used in many planetariums derives from gaming software, and I don't see why the government couldn't use planetariums or something similar to simulate space travel (in fact they probably already do). Lastly, genetic engineering may allow humans to live for centuries, but people will still likely want to have sex in space. It is what keeps bonobo society so peaceful after all. However, scientists tend to be very awkward, which may make research difficult. It may be easier to just have professionals in the adult film industry conduct research in that specific field. I bet they could make a lot of money off of that as well. Regardless of the enterprise, space travel has something for everyone, even if all you want is to be able to accurately map the galaxy. Who could blame you? Maps are awesome!

The only way to reduce costs is to conduct research, and the more research is done, the faster technologies will advance, and there has been almost no research as fruitful as that of the International Space Station. The I.S.S. may be the most important endeavor of all human history, uniting nations under the banner of science, and more stations would only continue to help bond nations and increase prosperity. Of course, the price tag for such an endeavor is steep, being $150 billion for the first station. However, relative to the some $200 trillion world wealth, this is chump change, worth less than a tenth of a penny from the world's dollar. Furthermore, they don't have to be orbiting the Earth. We could spread the research by placing stations along the Lagrange points of the Earth-Moon system, or the Earth-Sun system. Every two orbiting bodies have 5 of these points, where the gravity is flat enough to allow stable orbits; one in front of and one slightly farther behind the satellite (e.g. between the Moon around the Earth), one opposite the orbit of the satellite, and two others along the path of its orbit, a sixth of a circle ahead and behind it. We already use these spots for several satellites. Since one in the Sun-Earth system always faces the Sun, we can put solar satellites there so that they will always be facing that direction, ready to report dangerous flares. On the flip side, the Lagrange point always facing away from the Sun was the perfect spot to map out the cold glow of the Cosmic Microwave Background. A space station could conceivably be placed on these spots as stops for farther travel, or just simply research.

The Moon could serve a similar purpose. I'd even say that the best use for the Moon is as a jumping off point to go somewhere else, like a cosmic truck stop. It even has weaker gravity, which means it will be an easier location to launch subsequent missions. Plus if you miss on the way there you could just crash land at home. It will also be extremely lucrative, because everything you need for an Earth civilization you will need on a lunar one. That's essentially a free opportunity to make money and create jobs. Plus people will readily pay for a chance to spend their holiday on the Moon. We can mine water on the poles, as well as other vital minerals like gold and iron. Imagine getting an engagement ring made from Moon gold, or making a decorative sword or rifle made from lunar iron. Just as with cruise lines and airplanes, only people of affluence will be able to go at first, but eventually trips to the moon will become so commonplace that shuttles will become as normal as planes, trains, or busses, which have become as common as walking for some. Sure people may not be able to live in that microgravity, but it could be used as a resort. It could also work as a punishment. Lunar prisons could have people serve time by mining resources on the moon. We would just have to make sure to integrate them back into society, so we avoid that whole slavery thing again, or a lunar revolt from Earth. Execution would also be easy; all you would need to do is send people outside. I doubt that would need to happen though, since nobody would be able to escape anywhere. If nothing else, seeing the whole of the Earth would be a fantastic way to humanize criminals. Regardless of who goes, people need to be on the Moon. Of the 536 people who have been in space, three people completed only a sub-orbital flight, 533 people reached Earth orbit, 24 traveled beyond low Earth orbit, and only 12 walked on the Moon, and none of those were women! We can do better than that.

Space exploration has already demonstrated its ability to contribute to society. NASA has 6,300 patents, including cordless tools thanks to Apollo, the aural infrared thermometers used by doctors in our ears that replaced mouth thermometers, scratch resistant coding that allowed glasses to not be made with real glass, enhanced water purification filters, memory foam, and the list just goes on. Furthermore, imagine the social implications of being able to communalize space travel. Ritzy people are always looking for ways to flaunt their cash, what better way than going to outer space? Rich people fund an enterprise by making it look cool by doing it first, then it becomes cheap for everybody, and before you know it, we'll be partying in space, drinking a moontini on the moon or a red sand sarsaparilla on Mars. Imagine doing a waltz in the clouds of Venus, and while we're there maybe we could plant some extremophiles on the surface. Imagine vegetation on other worlds. The fact that plants are green on Earth might suggest that vegetation naturally reflects the highest intensity photon coming from their host star(s), which for the Sun are green. If that's the case, alien plants might be blue or violet on planets orbiting blue giants, and red or brown on planets orbiting red dwarfs and old giants. Plants may actually be able to evolve this way naturally as star systems evolve. You can even go to space if you still believe in God(s). Why wouldn't a person of religious faith not want to go to space, and actually travel through the vast expanse of their god's creation? It would seem to me that the ultimate expression of faith and distinction from animals would be to survive without the Earth and study the universe.

For all these reasons and more, space travel takes unbridled nerve beyond any voyage at land or at sea. In the vacuum, one is enveloped entirely by death itself. But people have still done it. Not just the select few pilots who left Earth, but also the countless scientists, engineers, and taxpayers that made it possible. So don't feel too bad for astronauts; despite the stress they must endure, they are the only living organisms to experience what it's like to ride on nothing but cold calculation and the smooth tides of space and time. Exploring the world unites all kinds of people. Scientists from The United States worked around the clock the get a man on the moon before The Soviet Union, and it only took ex-Nazis, women supercomputers, and a Kennedy to promise we were going to do it. Now people from Russia, Canada, Japan, the United Kingdom, and other European nations are all doing science together in a little capsule perpetually falling over our heads. I'd like to see whales do that! And who knows, one day we may even find monotony in that majesty.

13 | Entertainment

"Buried deep within you, beneath all the years of pain and anger, there is something that has never been nurtured: the potential to make yourself a better man. And that is what it is to be human. To make yourself more than you are. Oh, yes—I know you. There was a time you looked at the stars and dreamed of what might be."

— Captain Jean-Luc Picard, Star Trek: Nemesis (2002)

Applause! Applause! Applause for those who are admired, applause for those who are ridiculed! Applause for the few human beings who have found themselves, and who proudly stand in the spotlight for all to see, and allow others to judge their every thought and action. Free cheers for all who are brave enough to put themselves into the spotlight. Truly, they are the pinnacles of confidence and showmanship. We've already spent some time talking about culture, and even how it affects art. But we all know there's a very special place in our hearts for those of us who perform. The glamorous. The courageous. Those who are able to get up in front of millions of people and say, "Here I am!" Of all the crucial jobs and careers that we depend on to build society and innovate the future, few have the pressure associated with living in the public eye. Entertainment is what allows us to escape the harsh realities of life, to dream about the future, to take us outside of ourselves and live for just a moment in the life of another. That's why we revere our celebrities, our actors, our musicians, our athletes, and even some of our virtual icons. We put these individuals on pedestals, and bestow them with our hopes and dreams. That's why they get paid so much. So let's take a moment to honor these most hallowed of human endeavors. If you're sentient and you know it, hoot, holler, and make muffled sounds with your thumbed extremities!

Music

There's something vexing about rhythmic sounds. Music excites every corner of the brain, even more so when you are playing it or dancing to it, which is why hearing your favorite songs bring about the same emotions as your real life experiences. Music gives voice to people who feel beaten down and tossed around by the universe. They help to remind us that other people feel the same way we do, and as a result music can help people express themselves as individuals or in like-minded groups. It's thanks to the Rolling Stones that we learned that every cop is a criminal and every sinner is a saint, and The Beatles taught us that all you need is love. Performance in music, dance, and acting hold a special place amongst the pantheon of the arts. Rightly so, having people stare at you can easily make most people cringe. This of course extends to athletics, but we'll get to that later. Music is how Public Enemy, Run DMC, and James Brown inspired countless African Americans to be black and proud, and also fight the powers that be. Music from the era of Frank Sinatra is romantic, nostalgic, and has the magic of flight associated with it. If you prefer classical music, you can still find it as the perfect backdrop to many modern media. *In the Hall of the Mountain King* is constantly used for - holiday or dramatic commercials. Many are also used on television, like the *William Tell Overture* in races, and one of the most famous Bugs Bunny scenes featured *The Barber of Seville* overture. Some songs are indistinguishable from auditory poetry, which is why *Bohemian Rhapsody* is most excellent and upmost righteous. Others have lyrics that don't' mean anything at all, like Charlie Chaplin's solo in *Modern Times*. But no matter what they say, if they say anything at all, music always speaks to our souls.

Just like cuisine, music reflects the culture from whence it came. For instance, Japanese music is more intertwined with television than in the United States. In fact, the soundtracks to many animated Japanese media are more popular than the shows themselves. Unlike in the States, they use music from modern artists for the intros and outros of shows, which are much longer, as well as a way to add another layer of complexity to their programs. This may come as a surprise to those who see the high number of animated programs from Japan as juvenile, but anyone who has seen an anime convention knows that there is no shortage of mature content in Japanese television. In many ways, the fact that they are animated allows many shows to explore major issues in ways that shows with live actors simply can't. One of my favorites is *Hellsing,* which explores *t*he meaning of humanity and the morality of war using The

Vatican, the British government, and Nazi vampires. Then there are the protagonists, headed by a woman referred to as "sir" and the main character who, unlike Edward, is a real vampire with class. Another series close to many hearts and my own is *Cowboy Bebop,* with motifs about existentialism ennui tied with spiritual and technological ethics, but in space! It has Jeet Kune Do, and episode titles like Waltz for Venus and Asteroid Blues. It's the 90's show made for the modern day space enthusiast, with espionage reminiscent of classic spy movies. And the one thing both of these series have in common is an exquisite soundtrack.

But even in the United States we still venerate our musicians. After all, music artists are some of the only people who can still be considered as royalty in the West. Michael Jackson was the Prince, and so was Prince. Rock 'n' Roll fans all know to say god save the band Queen. Speaking of which, we all know Beyoncé is the Queen B., and Elvis is the King, thank you very much. Even if they aren't given the title, we still let them think highly of themselves. Take Kanye West, who is actually a very deep thinker, especially after his near fatal car crash. Yeezy realized the only certainty in life is death (stop me if that sounds familiar), so now he lives with utmost gusto. Although he does seem to agree far more with Kierkegaard than I do. He does a lot of talking, but at least he can sort of back it up, like a Teddy Roosevelt who doesn't speak softly. At least he speaks for social justice, when he isn't focused on death and flashing lights. Even if they don't call themselves royalty, we can see examples everywhere of the blurred lines between musicians and celebrities. That's how we learned that Eddie Murphy's girl wanted to party all the time. It's also why Jared Leto of Thirty Seconds to Mars was also in *Fight Club*. By the way, do you think Marky Mark Walberg would rejoin the Funky Bunch on Mars if we get around to going there in his lifetime? Or was it Matt that went to Mars? Regardless, we need all the star power we can get. We can even ask Queen's Brian May for help. He is an astrophysicist after all.

So we clearly love music, but what's the science behind that love? When I began my planetarium career, my very first unique task wasn't actually in the theatre speaking to public, although I did start out as an usher like a lot of people. I actually didn't end up giving shows until the end of my second year. For the first year and a half, I facilitated birthday parties. Planetarium parties are always a guaranteed way for kids and their folks to have a good time, because planetariums offer fun learning activities and better-than-IMAX quality tours of the universe. Planetariums are space, on Earth! That's one of many reasons why astronomy and physics are the coolest sciences; no other field offers an existential experience like them. We did countless activities for the kids, from building better telescopes than Galileo had, to private laser shows and occasionally launching rockets, the same activities we offered to school groups. However, there was always one activity that was by far the most popular: the sound lab.

One can do so many activities with sound. Usually I would begin by using a slinky to demonstrate the properties of waves, and the difference between a sinusoidal wave and a longitudinal one. Then I would go over the difference between amplitude, which for sound waves is volume, and frequency/wavelength, which for sound is tone, or if you're musically inclined, the note. We had many different demonstrations we could use, but I had two in particular that I always liked to do. One of them was taught to me by my first physics teacher, and it involved using a tuning fork and a table. I would hit the tuning fork, and then place it on the surface of the table, making the note louder. This was meant to show how musical instruments work through resonance, like how a guitar string vibrates the body of the guitar. Since the table (or guitar) has a larger surface area, it produces a louder sound, but it's more or less the same sound that the tuning fork (or string) was making. The second demonstration was also a lesson in resonance, and involved using a drum with a Ping-Pong ball attached. If you have another drum with the same diameter, it will make the same sound, and so when you play on the second drum, you make the first one vibrate, and you can see the ball bounce! That's my favorite part about science: you can actually make it fun and real through demonstrations that others can participate in.

From an evolutionary perspective, it makes sense why certain sounds are pleasant while others are not, and how a mixture of different sounds can illicit different emotional responses. In nature, humans would have needed finely tuned senses, like all animals. Think about the sounds dangerous animals make. Some of them hiss, others growl, and still others make low rumbling sounds. These noises often involve rough and disjointed sounds that clash with one another, and can cause fear or panic. Contrarily, constant and rhythmic sounds, like exact notes, can be quite beautiful. The sound of water flowing, birds chirping, and cats purring can all be very relaxing. On the topic of purring, studies show that it can be not only calming, but can also contribute to bone growth for both you and the cat, which just shows how influential vibrations can be. Animals would agree as well. In that sense, dogs are good for emotional health, and cats are good for physical health. But music is found throughout the animal kingdom. Various birds, amphibians, and insects sing to find mates and different species each have songs unique to themselves, hence the muffled chirping so many of us are used to hearing in nature. Those beautiful sounds are animals looking for sex, but who would argue that continuing life shouldn't sound beautiful, if not a little annoying?

Of course, if we aren't singing to music, we are dancing to it. Many birds, bugs, and lizards dance to attract mates. Bees even do it as a primary means of communication. Dancing can help humans find mates as well. It demonstrates strength and rhythm, attributes that may have helped ancient humans find prey and avoid predators. Music moves us, and that's why we dance. Not only that, but just like dance, music has no borders. Rhythm speaks all dialects. You can't stop the beat, cause the world keeps spinning round and round, and as your heart keeps time to the speed of sound, you'll get lost in the drums as you find your way. Although, we could hypothetically stop the motion of the ocean, or use those ocean currents to drive turbines for energy that we could use to make electronic dance music. Regardless, you can't stop the beat. And unlike paintings or sculptures, which are like snapshots captured in a moment, performing arts are dynamic and organic. Moreover, each physical performance is slightly unique, at least for non-professional dancers. I presume that people who dance for a living are far more consistent in their performances. Even so, I'm sure there are subtle differences in each piece that only the performers notice. During my days as a competitive martial artist, my most active area of competition was performing *poomsae*, or forms. In many ways these are similar to dance pieces, except for the fact that they are meant to simulate combat. That style also influenced me during my very brief time as a dance choreographer. Both dance and martial arts are rehearsed and strenuous sets of movements that require precision, dexterity, and memorization. And no matter how much you rehearse and practice, you will almost always screw something up. I can only assume this is just as true for performing dancers as it is for performing martial artists. This leads us to our next topic.

Athletics and Recreation

In many ways, sports are simulated war, and athletes modern warriors, masters that meld mind and body. They harness anxiety, sometimes even pure dread, to create victory. Often, this involves turning off the brain, and letting experience and instinct take over. Being, "in the zone" and meditative is actually the best way to focus, to harness to most primal areas of our brains, and reach a sense of unity through the body to achieve a singular task. Plus, exercising makes you very attractive. Sports are key tools for building camaraderie, amongst athletes, the people they represent, and between rivals as well. It is not, however, something that should be tainted by celebridization and money. Everyone should be an athlete in some fashion. After all, physical activity is good for your brain. Traveling for sports is also a great opportunity, and should be encouraged. But that money needs to come from somewhere.

Nobody can deny that the work of professional athletes is very demanding, both mentally and physically. They must be in the best possible shape at all times. Most modern athletes work out all year, both during the season and in the off-season. They must be able to perform their jobs at the highest level to stay relevant. Professional athletes also face the constant threat of injuries that could end their careers. For these and other reasons, this kind of job can be quite stressful. It's admirable, without doubt. But I'm still not sure how all of that hard work and dedication is useful in any way. If they focus too much on athletics in their youth, and neglect their schoolwork, then what do they do when they stub their toe? Watching sports is even worse in terms of the health of fans, especially if they just sit around on a couch with enough calories to feed a village. If nothing else, athletes are at least keeping their bodies healthy. But from my experience, most college students (and a lot of adults) just use sports games as an excuse to get drunk and yell for a couple hours. Which is fine, everyone needs some sort of catharsis. I just never understood why people wanted to do so by sitting on cold bleachers while some people run around hitting each other in a way that can only be done if one dedicates their whole life to, well, throwing a ball and hitting each other. Their talents would be better used for making them heroes of space exploration.

Don't get me wrong; I have a deep and sincere respect for human athleticism. Believe it or not, my entire life has been deeply enthralled by sports. My father was a world-class power lifter, and while I didn't inherit all of his muscle mass, he did teach me to have a spiritual respect for my body and the effort that athletes must put themselves through. I also have a deep fascination for the psychology of various sports, and how many diverse kinds of games people can create. There are many highly athletic sports that don't get nearly the publicity that ball-based sports get, such as ultimate frisbee, rowing, and lacrosse. Rowing specifically is one of the oldest sports, being part of the Olympics since 1900 and even practiced in ancient Egypt. One of my roommates in college was actually a rower, and I have to say, rowing might be one of the most strenuous sports that people can participate it, and rowers some of the fittest people out there.

I have also been deeply inspired by the wisdom of many athletes. I greatly admire the words of Muhammad Ali's poem *I am the Greatest* and Bruce Lee's philosophy and film career. I actually find them quite similar as well. Both were famous for their ability not only to fight, but also to entertain. They also lived in a country that distrusted them for the way they looked. Ali lived during times of great racism, which also drove him away from "white" Christianity to Islam. Lee too became famous during the cold war, when his Chinese ethnicity could have branded him a communist, and only white actors were playing oriental film characters. Muhammad Ali knew to float like a butterfly and sting like a bee, and Bruce Lee understood that he needed to be like water. Both of these great warriors understood that in life or a brawl, one needs to be flexible when moving, but rigid when striking. So I do love sports, I just don't understand why the salary of an athlete can be over a hundred times greater than that of a custodian or grade school teacher. People may have plenty of trophies, but a lot of times they just end up collecting dust. The real pleasure comes from putting forth the best within you and working to achieve a goal. All else is secondary.

I too was once an athlete of sorts. As it happens, it was just an accident that I ended up becoming a long time martial artist. My parents wanted me to do some sort of sport, and I said martial arts without even knowing what it was. Then I did it for over a decade, became an instructor to many, and a student to all. I'm glad I did, because I personally believe that martial arts transcend simple sports. They aren't about performing or working with a team, they are about self-defense and self-growth. That's where the name comes from; they are militaristic and artistic. A martial art means the same to me as it did for Bruce Lee, who said that in its truest form, martial arts are the honest expression of the self. One can act cool or phony with ease, but true expression is arduous. Many other sports teach people to be strong, but martial arts teach one to speak softly, and carry great restraint. Also like Bruce Lee, I do not believe in styles. That's not at all to say that athletes don't need self-discipline, of course they do. But the fact that martial arts stem from the practical purpose of self-defense, and require you to harness every atom of your mind and body for maximum efficiency, makes practitioners of martial arts unique amongst athletes.

I should also say, that while most people think that the hierarchy in martial arts is based on the belts, this is actually a myth. Nor is it based on age, ethnicity, or creed. But perhaps where you come from and how you grew up may affect your affinity for the high discipline required by the art. Primarily, ranks are based only on dedication and perseverance. In truth, the most respected students and instructors were the ones who could hit you the hardest. And in my experience, it was never the thicker or taller folks, or even always the men. Sure they had inherent muscle or natural mass, as well as length. As a matter of fact, most of my competition was comprised of tall thin gentlemen who were as thick as carbon, and sparring them was always a good challenge. You see, I'm quite low in stature. However, out of all my senior instructors, the vast majority of them were short, or women, or people of color. The Brazilian and South American students especially were some of the most ferocious warriors I have ever seen. I actually have a sneaking suspicion that they may actually have to use it, although I doubt it. They are fiery and passionate in many ways, but no more violent than any other group of people. There were of course masters of all shapes and sizes, and they were all wise, powerful, and every one of them commanded a sort of respect that I feel is only slightly less formal than a true military commander. The only Olympian substitute I know was probably the tallest person I have ever met, very thin, and my demo team instructor, but I've never been hit harder than by our chief and senior master instructors, and every one was shorter than I was. I may just be slightly biased however. You see, I'm also quite meek in size. Above all else, the most respected people amongst the colored and back belts, were always the wisest individuals, although those types almost always become black belts eventually. I respect those people, and others like them, most of all.

The original martial artists all started with pure white belts, so fresh and so clean. They trained, forged their bodies for self-preservation and combat, but also for mental stability and inner peace. Through perseverance, their blank white belts gained color, in many ways matching the cycles of the day and year. White was morning twilight, or from another perspective the dead of winter. For my style the belts also represented the life of a tree sapling, which as you know begins in the Earth. Next comes orange, the crack of dawn. Spring breaks as the sapling begins to emerge. When the yellow Sun has risen, the sapling has grown as leaves and spring branches begin to bud. Next come a camouflage of mixed leaves and dirt. The sapling is hidden amongst the taller pines, and while it is eclipsed and covered by elder pines, it begins to plant its roots into the ground and fight its way upward. Next is green, and it is midday. The tree is midway through adolescence by mid summer, and now begins to peak among other smaller pines. Next is purple, early afternoon. The tree begins to get pushed toward the mountains, where the blue sky and the pillars of stone ahead blend into a violet blur. Next comes blue, early evening. The sapling has reached toward the sky, now elevated amongst the other, younger pines. Following is brown, around suppertime. At this point, the tree has firmly rooted itself into the Earth, and as the cool autumn air blows, some of the trees begin to uproot and collapse. As the crimson sunset approaches, the tree has survived its first day. The last few colors of fall disappear, and the trees rest for the winter. The next day, when dawn has broken, the sapling awakens as a full-fledged pine. Fewer trees survive their first winter, and only the hardiest trees survive their first revolution around the Sun. With the beginning of a new day, the now firm trees shepherd smaller pines that are beginning their own journeys, and on it goes, until the last syllable of recorded time. By the time a martial artist has gained enough wisdom, and worked strenuously through many hardships, only then do their belts gain color. When they become true masters, their blank belts merge all of the colors, evolving into a new permanent black. This marks their dedicated hard work, and it commands the respect of all who meet that person, they whose wisdom guides others, in return for their own knowledge. That is the basic core of what it means to be a black belt.

For my organization, becoming a better black belt with higher rank correlated directly with increased leadership, at first with young students, then young adults, then colored belt adults, and eventually over time black belts of all ages, so long as you outranked them. That's how I ended up having students and peers who were also my wiser-in-life elders, including a chiropractor and massage therapist, people who are very useful in any active sport. The first degree black belt followed the same story as the colored belts, just with more wisdom and maturity. Their job—like adult trees—is to begin to plant seeds for the future. The second degree was about developing a new permanence in the way you act, and striving to always act with nobility. My favorite was the third degree, "peace of mind and tranquility" which I spent nearly half of my career as, since you weren't allowed to test for your fourth degree until you were old enough to be drafted, gamble, and drink if you're in Europe. By the way, being a peaceful teenager isn't easy, as anyone who has ever survived high school knows.

Then after that comes my rank. Like the purple belt, the tree has come higher toward a mountain, except this time the summit is made of granite. Firm roots or something like that. I'm not sure about the fifth degree, but the sixth degree—where one has the closed option of taking a yearlong route to become a master—means "long life". Rightly so, if you become a master they never let you leave, and as a young person I was scared of commitment. Had I kept training, I'd have my fifth degree in about a year and a half or so. After fourth degree, you can only test during national or world events, with around several dozen people. They put on a big ceremony, but only after rank testing. They're still setting up when that happens. Promotion may or may not happen after you have fun at the event, and maybe compete. We also had a fitness exam consisting of push-ups, sit-ups, and three minute-long bursts on training bags, one for kicks, kicks and punches, and a cool down with punches only. I had a close friend who went to a non-international, but more authentic school. Rather traditionally, they had more quantized requirements, while ours were timed and based more on merit and personal success. So it goes with added diversity from multiple ages and abilities. All I know is I've never done anything that has left me more exhausted. I hated it because the merit based system always forces you to push yourself farther, even as you get stronger. In my best performance, I hit eighty pushups by sixty seconds (remember, I'm quite small and so have less to lift, and Coloradans also train over a mile above sea level, which is why real Olympians visit there to train), and I was never not out of breath by the end of the exam. I hate them more than any human being I have ever met, but they are a good de facto workout. Another contrast that international and traditional organizations have is that the old school studios still use real wood and concrete, which I have to admit is pretty hardcore metal. Since we have a wider student base, we use standardized plastic boards to save costs (and trees), only using thin wood for demonstration. There were different strengths for different age groups and body types, with the most common adult board being roughly equivalent to some sort of hard wood, and naturally after some time they all grew soft. This is why they always unwrap fresh boards for every high rank testing. It's a scare tactic as well. That's the other tricky thing, quid pro quo. Any hulking goliath could push a thick board open, but only one who has mastered proper technique can break the thickest of boards, and masters come in all shapes and sizes.

No matter who you are, you can be a black belt in mindset always. Perhaps in your job, perhaps at your church, mosque, synagogue, or at home with your family, these are all perfect occasions to act just like a black belt martial artist. It's no easy task, even people who already earned their rank forget to act like black belts when they aren't wearing them. If you always do your best, and act with the utmost respect to your juniors and seniors, you too can be a black belt human being.

But out of all the astounding feats of physical prowess and casual pastimes, there's only one that doubles as the cause of billions of homicides throughout history: guns. Ha! I bet you weren't expecting that topic to come up here of all places. I'm going to be honest; I was planning to talk about it in Chapter 12, but I figured that section was long enough already. Moreover, wars have been going on long before the invention of gunpowder, and even today a large number of people use firearms for non-lethal competition, self-defense, and hunting. I would even go so far as to say that the vast majority of gun owners don't use them to kill anything, even though they are clearly made for the sole purpose of murdering things. As such, I see guns less as a tool for murder (which they still are), and more like toys that people get a sense of power by using and don't want the government to steal. Guns are kind of like booze, incredibly cathartic, but not really useful for anything. If you do like using them, fine. Be safe and try not to kill anyone. Guns themselves aren't evil, and can actually be used for good. Sometimes they don't even need to kill anything, like when army veteran Jason Galvin saved a juvenile bald eagle stuck upside down with twine around its leg with a .22 rifle. Of course, guns do usually kill things, but even that isn't necessarily bad. Take for example the controversial ethics of animal hunting. The main problem with trophy hunting is that many poachers hunt animals for phony medicines, useless trophies, and basically contribute to the extinction of many species. However, even though poaching is deplorable, taking a hard line against hunting overall doesn't take into account the potential benefits it has.

The fact of the matter is not only that people enjoy hunting for sport, but that the permits sold by governments can raise tens of thousands of dollars that go toward conservation. For example, according to a 2005 paper by Nigel Leader-Williams and colleagues in the Journal of International Wildlife Law and Policy: Leader-Williams describes how the legalization of white rhinoceros hunting in South Africa motivated private landowners to reintroduce the species onto their lands. As a result, the country saw an increase in white rhinos from fewer than one hundred individuals to more than 11,000, even though a limited number were still killed as trophies. Hunters also seem to be open minded and cognizant about how they only hunt for fun, as well as staying out of tourist locations and only hunting problematic game that are detrimental to a population, are too old to breed, or have already bred. Hunters also do something most people in developed nations don't; they actually handle the carcass of animals they eat, instead of just picking up the refined edible parts from a grocery store. People who take a hard line against hunting are probably no less likely than the average person to purchase meat from chickens and beef who have been mutilated far more than hunted meat, although I haven't come across any surveys that address how many conservationists are also vegetarian. The last point to be made is that many people in rural areas—like say small farmlands in Nepal—use guns to defend themselves from dangerous animals, and they also use them to hunt. By and large I truly believe that people prefer peaceful alternatives over life risking violence. That's the heart of the primary defense for gun ownership: the desire to maintain peace and the right to self-defense.

So let me make this clear, nobody wants to take away your guns, but they do need to be HIGHLY regulated if people want to play with them. That is especially true in the United States. It doesn't necessarily mean the government needs to regulate them, because the citizens are the ones who ultimately have the power to do so. It's heartbreaking to know that people from around the globe are afraid to live in my home country because from their point of view, everyone wants to own a hand-held murder device. But it's in the constitution, and it's a part of the culture. The most pragmatic way to do this is obviously through serious background and psychological checks, because we all know owning a weapon is serious business. Actions never change when ideas are left alone, so if Americans want to reduce gun violence, they need to change the way they view firearms. As far as killing devices go, guns are sort of cowardly and lazy, but mostly just dissatisfying. As a person who's spent years training with people to use their body to take down and potentially end the life of an assailant, I find people who use firearms maliciously to be quite pathetic. There's no satisfaction in something that happens instantly. If you want to end a life, you need to earn it. It's also quite unnerving to think that, while incredibly unlikely, an entire decade of training could be rendered useless in a nanosecond because someone wasn't intelligent enough to imagine others complexly. Everyone has a plan until they get a cap busted in their spine.

According to the FBI's *A study of Active Shooter Incidents in the United States between 200 and 2013*, published September 16, 2013, out of 160 total shooting incidents, 21 ended when unarmed civilians restrained the shooter, and only 5 were stopped by an armed civilian. Some may see this as evidence that many people don't exercise their right to bare arms, and that if more people did, then they would outnumber criminals, and hence deter crime. However, the majority of the heroes that took action to defend their fellow citizens (nearly half of whom were education teachers and students) didn't need a gun at all to quell the situation, which should be a testament to how capable people can be without a firearm as a handicap. Furthermore, in the heat of the moment, it's possible that a person with the intent of defending people may miss and hit a civilian, or just not shoot anything useful at all. People likely wouldn't be able to even grab their holster before an innocuous person had already killed several random people. What's more, it doesn't seem sensible that a room full of people firing guns is any more productive than one person doing so. This is especially true considering that proper firearm use and emergency training take decades of dedicated training, something no average person has time for. Again, this study found 160 large-scale incidents of gun violence, which means an average of one every month for thirteen years. With news outlets covering nationwide events, one may become convinced that mass shootings happen all the time, which isn't false. However, this data spans nearly four million square miles and tens of thousands of communities. Over such large scales it's naturally going to become more likely for anything to happen, that's how probability works. 160 mass shootings in a decade is also infinitesimally small relative to the rest of human history. After all, China's warring states period lasted 254 years from 475 to 221 BCE. That may not be totally relevant, but it does demonstrate how much more pervasive conflict used to be in the past. Furthermore, England and France were at war for over a century from 1337 to 1453 CE, and their countries are tiny. As much as news media want your attention by overly selecting negative events, don't forget how much better the world is now than in the past.

On an individual citizen level, in 2013 The Center for Disease Control cites 11,208 homicides due to firearms. On one hand, this is clearly unacceptable. That's eleven thousand people, just like you and me, with friends, family, and their own hobbies, gone, poof. But it's important to remember that the total population is over 318,900,000. That means relative to the total population, essentially nobody dies by getting shot, only .0035% in 2013. It's very easy for the news to convince us that gun violence is common, but people really don't need to be so paranoid about it. They should be more afraid of falling to death or getting into a car wreck. What's more, the same report cites 21,175 suicide deaths from firearms. Based solely on the figures, you are more likely to kill yourself than someone else if you own a gun, which isn't exactly supportive of the idea that people should have a right to defend themselves with firearms. It should be noted also that after mass shootings, a majority of the perpetrators committed suicide before police could arrive, suggesting that they aren't maliciously targeting a specific group so much as just giving up on humanity, like cowards. Many of them are also quite young, and almost exclusively male, which shouldn't come as a surprise. Now, we must also acknowledge that since basically nobody dies due to firearms in the first place, many gun owners will never fire off a single round. So the issue really seems to be less about the government taking away people's peace of mind (and their toys), but more about education. Education about actual statistics and human empathy are both necessary for peace of mind. It's perfectly acceptable for people to blow off some steam at a gun range or to shoot competitively, so long as we live in a society that doesn't motivate people to kill each other. In order to do that, all we have to do is be nice to each other, and not assume that other people care enough to do us any harm. Thankfully, people seem to be getting better and better at this as time goes on.

Here's the thing about the second amendment, and the Bill of Rights; they were added to the constitution by the Anti-Federalists to protect citizens from their government and its military. The founders wanted create a balanced system where neither the government nor the governed had too much power. They wanted to protect the citizens from tyranny, and they also wanted to protect the nation from the uneducated riff raff within the population. You have the undeniable right to own the equivalent of a musket or a bladed weapon, just in case the government tries to seize your property. We have a right to form a militia to protect ourselves from the federal government. Those militias go by another name, the police. They may get paid through taxes, but they aren't federal employees. They aren't like the secret police of the kremlin, nor warriors born into nobility and class. They are average American citizens. Any one of us can learn to become an officer in the law, it's just another job like any other, and some people are just more predisposed to help others.

In all honesty, this is even truer for our military. Those individuals are the best and brightest citizens in the country. Our military is exonerated, and are amongst the least prejudiced and most noble of all our population, which is why we must respect our veterans. They have seen horror beyond imagining and the worst of humanity. They are by and far the people least likely to commit mass shootings, and some of the few people who should be trusted with modern weaponry. Of course they carry their prejudices along with their badges and medals (especially if they're from the south). But as society grows, and as people mature, and as armies modernize their training, prejudices fade away. Trained law enforcement are the only people who should be allowed to own firearms, because using such a powerful tool of death requires a lifetime of dedication to justice and months of focused training. Real people in real hostile situations are notoriously clumsy shooters, and even police are slow to draw when faced with a knife running toward them. This is true for all weapons; if you don't know how to use it properly, you have no right to put other people in danger by owning it. By the way, Alexander Hamilton, a federalist who supported the second amendment, died from a gunshot wound.

People might be able to take classes, or wait until they became a homeowner to own a gun. Every expensive house can come with a ridiculously massive cannon if that makes you feel better. It will do the job, after all having big guns at home is the same reason other developed nations are afraid to visit our country. Guns, they deter theft AND economic growth! If only people shared instead of stole, begged instead of hoarded, or drank instead of fired. Then our problems would become mute. Young people especially should get the highest restrictions on lethal weapon use without sufficient safety training. People need to have a stake in humanity as well. If they don't see a viable alternative, or if they suddenly become deranged, they will easily find a way to murder with or without a firearm. At least if trivial disputes can't be ended in a heartbeat, then crazy or impulsive people will have to work harder to commit crimes. After all, everyone knows how lazy people are. If they have to go all the way to the store to buy rope, duct tape, and a sharp knife, they'll more than likely prefer to just sit on the couch and get over it. Guns never need to be involved, period.

This might seem like a violation of rights, but all people have the opportunity to train and earn the privilege of owning a firearm. It's no different than owning a vehicle. People need to go to multiple classes, take permit tests, and train with a vehicle for several days and months. You do this because obviously you can kill people with a car. In fact, it would be easier for you to kill more people with a car than an AK-47, you already have a means of escape, and bullets run dry rather quickly. Nobody ever seems to think about just driving onto the sidewalk or into a crowded building to mow down innocent people. We actually care about our vehicle rights, because we've earned them. And if you act dangerously, you don't get to drive anymore. Yet somehow, there was a time in history when any deranged psychopath could purchase a gun at a convenience store or when some brash teenager could snatch their parent's bedroom pistol to finally get back at all those kids with the pumped up kicks. It's beyond moronic. If you own a gun to protect yourself from other people with guns, that isn't your right, that's called cowardice, and it's more dangerous for everyone. It sows the seeds of distrust that leads to gun crimes in the first place. Unless you're an individual who still hunts their food in a developed nation, for some reason. But you don't need a conceal carry license to own a hunting rifle, everyone can see it, like a more lethal fishing rod. Large weak guns are less of an issue, because it displays your passion for blowing things up to all the people around you. There's nothing wrong with shooting cans in a field, it can actually be quite entertaining, albeit primitive. But in general, guns only belong in the hands of people who make a living by protecting others, and even then they are rendered useless when people learn to treat each other like human beings. The digital age has given us toys that work much better than them anyway, like Tasers. If regulation doesn't work, the answer to solving the gun issue is simple: if you don't have a college degree or military level training, you can't own a killing machine. But if even that doesn't work, my favorite alternative would be Chris Rock's idea to make bullets cost $5,000. That way, you'll have to make investments and maybe take out a second mortgage if you want to kill somebody.

Comedy

Comedy is right next to acting as being amongst the highest exalted of human endeavors. It does absolutely nothing to feed people, clothe people, or put roofs on their heads. No. However, having humor—which is crafted by façades built upon a bedrock of lies—is perhaps one of the most human things you can do, because you need to be smart. Comedy is intelligent, and its job is to quell the mind melting existentialism that happens when animals have food warmth, and safety. Many animals laugh, form apes to rats. But in humans, it shines a light on the cruel jokes that come with being alive. It calls to light social injustice, and makes us feel better about our short, meek existence. Comedy is what happens when you look dead into the eyes of the farthest abyss of oblivion and say, "screw it!" in front of a crowd of people. It's kind of like politics, except it actually makes you feel good, and it makes people less phony instead of more.

Having the power of laughter sort of allows people to say anything they want. It acts as an anxiety buffer of sorts. That's why I think comedians are bar none the most effective public speakers, thanks to their wit and tact. In fact, among my list of things I wish every politician had training in, comedy is first, and astronomy is second. A comedian could insult you and a thousand other people in an auditorium, and as long as people are laughing, they will actually get paid to do it. Comedy is liberating, it lets people look at issues from the outside in while simultaneously protesting social injustice and cultural stagnation. When people laugh at a joke they are, at least partially, validating what the comedian said or did. Perhaps a joke was simply just ludicrous, or maybe people can relate to it, but many times comedians make fun of normally grave, sensitive topics. Comedians are also hugely intelligent and cultured. As a matter of fact, I think the most accurate religious film I have ever seen was the British comedy *Monty Python's Life of Brian*, and I should know, my father had me watch a cartoon nativity film several years in a row on Christmas when I was a child.

Comedians are actually how I learned about many social issues growing up. Black comedians like Chris Rock can crack wise about the abuses of slavery, and some white comedians like Louis C.K. can openly acknowledge white privilege and joke about crimes that most people didn't even know occurred because white people did them. One of Jeff Foxworthy's main shticks revolves around helping people identify if they're a redneck or not. As an obvious redneck hick and extremely kind and funny person, he can do that. Then there's Eddie Izzard, the executive action transvestite, who makes a hilarious tale out of religious history. Gabriel Iglesias amazed me with his story about a trip to do a comedy show in Saudi Arabia. He talked about how people told him that the reason everyone looks angry is because it's so hot there, how happy air conditioning makes people, and how not everyone in the Middle East is the way Fox News portrays them. It makes perfect sense why he's the number two comedian in the region, right after Jeff Dunham, who actively makes fun of terrorists using a puppet of a dead suicide bomber, and people in the Middle East find it hysterical!

I'd love to keep talking about comedy, but my dog just turned blue and I have to paint it green again.

Vicarious Living

Actors are some of the most celebrated of celebrities, despite the fact that they get paid to be fake. Essentially, they are professional liars. But we know they're lying, so it's okay. It can work both ways, some actors fill niche rolls that suit their personalities, or they can become famous for playing a great array of characters, and both forms show their humanity. The important thing is that they are convincing. They don't have to be glamorous movie stars either. Part of the power of Steve Erwin was how genuinely excited he was teaching people to love and respect animals, and teaching animals to tolerate goofy apes messing with them. The great thing about film and television is that it allows us to escape from reality into the lives of someone else, someone cool, relatable, or just plain unique. They allow us to be somebody who can do what we want to do and get away with it. That's probably why we call them stars, because they represent our ideal selves.

In many ways, film and television are the natural progressions of a long standing human tradition of story telling. Before there was *The Godfather*, there was *The Bible*, and before that there was *The Iliad*, which was passed on through oral tradition. Now, there are also video games, which are one of the great crowning achievements in the quest for peace. It's why Buckner, Garcia, and arcades everywhere had *Pac-Man Fever*, and why people strategically play League around the globe. They have grown and diversified to fill a wide array of niches that few entertainment mediums have. Unlike books or movies, games put you in control of the show. Different games vary on user input from guided story to open sandbox, and everything in between. Some let you play as your favorite icons from movies and television, while others let you completely customize avatars to suit your ideal self.

Video games have also evolved to tackle more and more serious issues, beyond collecting pellets and fleeing from ghosts. One of the best examples of this is the *Fallout* series. It shows exactly what would happen if the worst fears of the cold war came to fruition, and tells a complex and personal story about what it would be like to live in a post-nuclear holocaust world. It shows the power of propaganda, the corruption of government, and comments on our increasing dependence on technology and drugs. It even has a religion based on Elvis, which actually wasn't that unusual in the real 50's, the time period the games are based on. Another famous series that speaks volumes about humanity is the *Pokémon* franchise. It's a world famous brand, extremely popular, and fun for all ages. But when you think about it, it's a very strange in concept, because when looked at literally, it is at worst condoned slavery, and at best regulated animal fighting. But the pocket monsters never die, at least since the first generation of games, so it's okay. But that leads to perhaps the most unusual aspect of gaming; the fact that it turns dying into a minor inconvenience. The hallmark of gaming is that you have the opportunity to learn through trial and error, which is really the whole reason why video games are good for your brain, at least in moderation. Games that penalize you for failure are especially good for critical thinking because they force you to puzzle out solutions and learn from past mistakes, and that's something no book or movie can offer.

Regardless of what you do to pass your free time, we all need something to entertain us. This has been the case since the earliest days of humanity. Nay, it has been the case long before humans, since the first organism was born with enough neurons and time to play a game. It's good for our brains, and that's why the best entertainment simulates life. The key factor that makes some things entertaining and other things blasé is urgency. We don't care about unlimited firepower; we care about someone with only one bullet left. That's why television loves conflict, why shows end on cliffhangers, why the best movies have twists (Vader means Father in German, and the Empire are Nazis), why the best sports films have an underdog that makes a comeback, and why the best video games are the ones where you've died a million times. It fills a need for adversity and challenge in a world that is increasingly safe and mundane. That's how we want to live, by getting knocked down and forcing ourselves to get back up. That's what makes us human, and that's what makes us great.

14 | Dissemination of Knowledge

"The only person who is educated is the one who has learned how to learn and change."
— *Carl Rogers*

"Education is not preparation for life; education is life itself."
— *John Dewey*

My entire life, I have been an educator. As such, I spend all of my time thinking about ways to better gain and disperse knowledge to others in a way that makes learning exciting. My life revolves around disseminating knowledge. Of course, a good teacher must also be a good student. I guess my perspective is unique because I've simultaneously grown up both being a student and an educator essentially since middle school. For this reason, I would like to dedicate an entire chapter on education, the thread that ties us not only to past generations, but to society, and ultimately everything in the universe.

Education is the number one most crucial issue during any point in human history, period. It is the source of prosperity, and the lack of education is the single greatest plight facing humanity. I would even go so far as to argue that a lack of education is the root cause of all other misfortunes, including terrorism, poverty, and accelerated climate change. It isn't a matter of kids learning algebra or how to do a literary critique of *Madame Bovary*. Those things alone are boring. It isn't about young adults getting degrees in mechanics, computer science, or any other piece of paper in the hopes of lucking into a high paying job. Half of the time nobody makes money working in the field of their college major. And it most certainly isn't about a bunch of old people living in the past catching up with modern technology. Those people are on their way out, and if they've managed to live that long, it's highly likely that they've already found a respectable way to contribute to society, and they probably hate all the changes that have happened since they were young. All of that is important, but it isn't why education is important. The reason education is the number one top priority of every human being on the planet is simply this:

In order for each individual of our species to survive, we must contribute to society as a whole. The only way to actualize this singular objective is for each human being to be caught up on the last ten thousand years of human development, the four and a half billion year history of their planet, and the near-fourteen billion years of existence, so that they have the fullest ability to find a calling, to find a problem that needs to be solved. We are born with zero knowledge, we die the same, and everything in between must eventually be left to the next generation. The only way to prosperity is to understand the world as best we can from beginning to end.

I don't know what sort of qualities you attribute with a person of intelligence, what sort of background they have, what they wear, or what they look like, but you're wrong. Dead wrong. Anybody can be smart. Everyone, even with cognitive disabilities, has the potential to learn more about how the world works and achieve things nobody else thought would be possible. Whether you consider yourself intelligent or not, you are. You can do things that have taken eons for humans to develop, things that nobody else can do. There are also people with skills you don't have, just as you have some that they don't have. Everyone has things they're good at, and bad at, but with practice you can become a master where you were once an apprentice. Whether you live in Cambridge, Mumbai, or North Korea, you have the same capacity for knowledge and compassion that took matter four billion years to achieve. You can do something that could help everyone, and don't let anyone tell you otherwise.

Knowledge is also subjective. I could teach you how the planets go around the Sun, but a Polynesian sailor could teach you their millennia-old methods for traveling across the Pacific ocean using nothing but the stars in the sky. You too could teach me things about your profession that I would have never imagined. To me the greatest tragedy is when people become so carried away studying a single thing that they lose sight of the larger picture and stop learning things that have nothing to do with what they do on a day-by-day basis. I detest when people dismiss certain topics as useless, effectively shutting the door to new ideas. One never knows when a single sentence in a historic piece of literature could influence the next technological breakthrough, or when a mathematical equation can inspire innovative art or medicine. Most importantly, you never know when you might learn something that completely contradicts everything you've ever understood, which is something that will happen less and less the more you know. Your brain will filter what is useful to you, but nobody should ever lose their hunger for new information. There's simply too much in the universe to learn about.

While it is certainly tragic when people let go of their curiosity, what's more deplorable is when they never get the opportunity to learn. Throughout the world and as long as history has been recorded, people have lived their lives being told what they are allowed to say, what they are permitted to do, and what they are entitled to think. Sometimes there are good reasons for censoring speech, and for the most part people are justified in regulating actions in the name of safety. But when people are forbidden form thinking, and most importantly questioning, that is the most grievous crime against anyone's humanity, especially without justification. Indeed, the most toxic form of knowledge is that based solely on authority and without justification, and we should all strive to purify what we are told through empirical evidence and the experiences of others. To see the dangers of knowledge based solely on authority, one need look no further than the totalitarian regimes of kings, queens, and dictators. Of course, being the decedents of god, European royalty and Asian emperors alike had no trouble maintaining power. This cultural façade of born royalty ran so deeply that even peasants believed that they were predestined to work the land and serve the crown. So many people have lived and died having never thought of bettering their lives through hard work, instead relinquishing their fate to where they were born. In societies where the poor can never become noble, neither faction has any motivation to better society as a whole. The rich squander their power, and the poor bend over in submission, all because god has willed it. That's the problem with divine will: nobody will dispute it.

As societies developed, and the world laid witness to two meaningless global wars, religious faith waned, and governments needed to develop new tactics if they wanted complete control over their subjects. This is seen in the violent totalitarian regimes of the twentieth century, where individual thought wasn't punished by church based governments, but by fully militarized ones. Whether it was Soviet secret police, Hitler's anti-Semitic hate speeches, or Mao's little red book, one can find countless examples of censorship and severe punishments people have faced for criticizing their government. But while threatening your citizens with death for thinking freely is an effective tactic, it becomes increasingly difficult as time moves forward and civilization becomes more connected and liberated. In the age of the Internet, the only way totalitarian regimes can survive is either through strict censorship or strict punishment through violence, as seen in North Korea and parts of the Middle East and Africa. But as information becomes more widely available, and people see what they could have, even those petty governments will be unable to quell the freedom cries of their citizens. The Berlin Wall can testify to that.

Being silenced by your government is one thing, but being ostracized by your peers for thinking freely is another beast entirely. This is the key weakness in religious institutions throughout history. We have already discussed the inability for the Catholic Church to accept the heliocentric model of the solar system. However, a more harrowing example of the uselessness of faith is seen in the Black Death. In just a few years, nearly half of Europe's population was wiped out with no warning or salvation. Without any explanation, people were convinced that the world was coming to an end. As with many other calamities, they blamed their unwarned blight on the amoral actions of the population. Ironically, when faced with immanent death, they lost all morality and confidence in the church, and lived as if the second coming prophesized in revelation was nigh. As usual, humanity recovered, and we eventually realized that the plague was nothing more than a mere bacterial infection spread by flea-ridden rats. But instead of having any sense of what to do, all Europe had to rely on were folk remedies and empty hopes that piety would appease god's wrath. I often lament how much time has been wasted by people simply trying to pray away their misfortunes, instead of actually going out and trying to solve their problems. Religion may not have caused the Black Death, but it sure wasn't useful in preventing or solving it. People have so little confidence in their abilities that they just acquiesce to fate and allow suffering to go unabated. I frequently wonder how many lives could have been saved had people spent less time reading their holy books as a source of knowledge and more time studying nature itself. Faith is all well and good until calamity strikes. After that, for those that live based on superstitions and rigid morality, there is no one to blame but each other. Or the Jews, who are all too often the scapegoats for ignorant prosecution. That's how witch-hunts begin, which only leads to more death and suffering. Those tragedies only occur when people become so complacent with the way the world is now, that they don't work to improve it, and as a result they are ill equipped to adapt to disaster.

It's curious how freethinking tends to rewrite what religion means to civilization over time. Take for example Martin Luther's new German Bible (with the new testament published in 1522 and the old in 1534) that allowed common people who couldn't read Latin to read the Bible. This eventually led to new Christian religions and infighting. Roman Catholicism was the practiced form of Christianity in western Europe (but not Eastern Europe, Africa, or Asia), in which only the church was allowed to read and interpret the bible, while everyone else just had to take their word for it. Most people would go their entire lives only meeting one person who could read the Bible, because the Church was so deeply interwoven in society and had owned over a third of the land (ergo food) in Europe during the medieval period. This lasted the same amount of time the Roman Catholic Church was in power. It's thanks to Martin Luther that modern Christianity is fueled so strongly in the phrase *sola fide* (only faith), the idea that one doesn't need to actively practice or simply do good deeds to earn salvation, but that all one really needs is faith. Before that, the church was so corrupt that it was selling reduced time in purgatory in return of tax revenue. They didn't care about god, they just wanted money, something that holds true for many people of power. Luther spat in the face of millennia of corruption by claiming that people didn't need preachers, but instead that every individual should contribute to a priesthood of all believers. This is just one step away from the first amendment of the United States, which gives protection to practice all religions, not just one of the several dozen Christian faiths that developed after Luther's reformation. This demonstrates how fluid religion really is, and how it varies based on people's needs and circumstances. This also illustrates the ignorance of the masses, and how easy it is to swindle people, especially after you've earned their trust, and how faith without scrutiny is so often misguided and gullible. People will even go so far as to call the Pope—noted as being the holiest person on Earth—the antichrist, because the church didn't give them free speech.

Finally, it should be noted that while criticisms of the church have arisen multiple times, the key reason reformation was so successful in 1534 was the invention of new technology: the printing press, which allowed huge numbers of German speakers to read, interpret, and further translate the Bible. And ever since then, nobody has been able to agree on what the darn thing is saying. The Protestants, Anglicans, Methodists, Baptists, Anabaptists, Puritans, Presbyterians, Lutherans, Catholics, the Mormons, the Quakers, the Bakers, and the candlestick makers all became sure that the exact same God was going to send all of the other sects straight to Hell.

This doesn't even include the Islamic faith developed by Muhammad between 622 and 623 CE, which also has the same but different rendition of God. Islam even acknowledges Moses, Abraham, and Jesus as prophets; it just denies the resurrection of Jeezy Creezy (like Judaism). But even they can't decide amongst themselves which version of the faith to adhere to. It was almost as if giving a book of faith to a billion people creates a billion distinct versions of God. It seems as though religion is based more on how people view the world, not how the world actually works. However, as a result that means it can develop and evolve right alongside civilization.

Now it should be made clear that faith is not the sole cause of the ignorance that leads to disparity. Religion only causes ignorance when people are intolerant of other people's faiths. If one wants a secular example of this, take the famous broadcast of the *War of the Worlds* directed by Orson Welles and Written by H.G. Wells. Even though the Halloween broadcast in 1938 was relatively benign, the same radio drama caused riots and seven deaths in Ecuador a decade later. Clearly, people will act ridiculously when given minimal knowledge under any circumstance. What's more, there are several examples throughout history of religion and death being completely independent of one another. Despite having their fair share of spilled blood, both the Persian and Mongol Empires had the benefit of religious tolerance. The Mongol Empire especially was renowned for keeping their empire safe and encouraging trade, revitalizing the Silk Road. As a result economies flourished, cultures and ideas spread across Eurasia, along with the aforementioned plague. But people still only surrendered to the Mongols because of how ruthless they were. Religion neither causes nor inhibits human tragedy, and this is plainly seen by how people of all faiths and creeds take part in ignorant violence. The problem is that it promotes complacency and ignorance. Ignorance in turn breeds violence, and complacency gives people the false impression that there is nothing they can do to improve their situation. All too often are people so scared of their god or their government that they are unable to speak freely and learn about how the world actually works.

On the surface, it would seem obvious that democratic regimes are more conducive to education. They are often more stable, and since most usually agree on the value of knowledge; education seems to naturally come out of an institution that is built on the views of the public. However, this philosophy should hold regardless of the regime, if one assumes that the need for education is inherent to human survival. Then on the same token every political regime should have institutions that support higher education. But unfortunately that isn't the case, and so one must reconcile what exactly makes a secular democracy more successful at supporting school enrollment and performance than other forms of government like oligarchies, autarchies, and monarchies. By analyzing the varieties of social byproducts associated with these regimes, we might be able to see how something as inherent as schooling can still be sub-par or simply absent under certain governments. It can be seen that many non-democratic governments reorganize their priorities in such a way to ensure they get and keep power, which takes resources away from education. For one, many non-democratic developing governments in the twenty-first century are often militaristic in nature, where the government forces its citizens to either obey laws at gunpoint, or to serve long term in the military, thus keeping them out of school. Many developing countries have historically been subject to corrupt governments in this way.

In order to make school accessible to citizens, it is often pragmatic for the state to fund them. This funding, just like all things the government must spend money on, ultimately comes from its citizens in the form of taxes. For corrupt regimes, where only a small group or even an individual holds all the political power, tax money can be redistributed to the regimes small winning coalition, either for personal affluence, or to maintain influence and power. On the flip side, if the state is simply unable to collect taxes in the first place, then again there is no funding, not just for schools and teachers, but also for infrastructure such as roads and healthcare. Without fundamental things like buildings and medicine, it is obvious that a country could want higher education and not be able to attain it, just as much as a regime that simply doesn't care enough about its people to give them education. Hence, governments that are more highly influenced by the public (i.e. democracies) are less likely to allow leaders to squander resources on themselves, and more likely to have it used on the people who are funding them. It may very well be that a poor country has some of the most educated people on the planet. But if that education is delegated solely to the ruling class, it means nothing for the development of the country as a whole.

In order to further understand the relationship between democratization and education, hence development, we can look to the nation of Chile. By many accounts, Chile is seen as one of the most successful developing nation not just in South America, but the whole developing world. Chile today is home to one of the most successful Universities in South America, and is also home to—thanks primarily to its atmosphere and terrain—some of the worlds most used astronomical observatories. This makes it a hub for intellectuals all around the world.

Chile is also an interesting case study because even prior to its official transition to democracy in 1988, it was in many ways democratic for the majority of its history after its independence in 1818, but had small periods of political uncertainty and even militarism. This allows one to compare how regime changes over small periods can affect a state.

In order to understand what makes Chile so successful as a nation, we must first understand its history of politics and education. Since independence, Chile has been under several constitutions, but has always seen an importance in education. The earliest of these constitutions, ratified in 1833, made public education a responsibility of the state, much like in the United States. In 1842, Chile became the first Republic in Latin America to establish a system of public education, and to have a standardized normal school for teachers. Later on, The Organic Law of 1860 made primary education free and provided groundwork for establishing additional secondary and normal schools. This would echo on to modern day, as Chile currently provides vouchers for secondary education. Chile was also a leader in decreasing segregation in schools as well. In 1877, Chile became the first Latin American country to pass a law admitting women to higher education. With more educated citizens, Chile has long been able to sustain higher development, which would not be possible if only wealthy males were getting schooling. However, despite these seemingly good policies, like much of South America Chile devolved into a system that gave priority to the ruling oligarchy through much of the 20th century. While elected presidents occasionally served the middle class, there was much tension between democrat and socialist parties, and this tension along with military rule kept reform to a minimum. This lasted until 1964, where an absolute majority elected Christian Democrat Eduardo Frei Montalva. Frei had begun major reform in housing, agriculture, and education, but was halted early on by opposition from all sides of the political spectrum. There was much good done to enhance education though. In 1965 primary education extended to eight grades, ideally designed for ages six to 13. Policies during this time raised the length of primary schooling from 4 years in 1920 and 6 years in 1929.

This long period was to be followed in infamy by the militaristic rule of General Augusto Pinochet, lasting from 1973 to the end of 1989, whose policies still have ramifications that echo to the modern democracy in Chile. This is yet another example of how cold war policies in The United States had negative consequences to the rest of the world, as the American government played a role in inciting and supporting the coup that led to Pinochet's rise to power. His rule is often closely associated with the death of hundreds of thousands of his political opposition, but there was also a key educational policy enacted just before the end of his rule. This was The Organic Constitutional Act of Teaching, enacted on March 7, 1990 enforced on March 10, the last day of Pinochet's dictatorship. Despite being widely criticized by students and teachers, it has remained largely unmodified since the restoration of democracy. This policy, in essence, makes it so that the state has limited influence on the education system, making schools financed through private means, or through smaller local governments. This has resulted in massive student unrest, most notably in the pingüino protests in 2006 and 2011, where students attempted to gain increased access to higher education and get the state to fund education directly. Chile today is still in a transitionary phase of acquiescing to public sentiment, and overturning policies left behind by past militaristic government. But progress is being made, as seen in how students are getting more access to school vouchers.

We have seen throughout human history that having knowledge of our world and ourselves allows us to prosper. By extension, it stands to reason that higher levels of education in a state would correspond to higher development. Of course, it is not a simple matter of just sending kids to school. Proper education requires funding and structure, which are only possible in stable institutions. As seen in Chile, which has a long history of good education policy, stable governments where the people have power often have good education. When those regimes are plagued with civil unrest due to poor governance, however, education and development can be severely hindered. Therefore, it would seem that, at least in the case of Chile, democracies provide support for good education, while other regimes often hamper it. This seems to also be the sentiment for the Chilean people as well. In the words of Cristián Larroulet, former Minister General Secretariat of the Presidency,

> "One of the most fundamental factors in decreasing inequality levels has been the reduction of educational gaps, mainly due to the increase in higher-education enrollment, which has risen from 200,000 students in 1985 to more than one million nowadays."

Part of the enigma of education is that nobody seems quite sure how to "fix" it. There are different strategies proposed around changing funding, school hours, etc. But none of these seem to get to the heart of the matter. How do we improve the global educational system anyway? There doesn't seem to be a shortage of teaching jobs as much as a deficit of motivation on the part of both students and teachers. Many people often say that teachers should be paid more, but this is an oversimplification of the issue. In the U.S. the average teacher's salary (grade school and university) is around $60,000 a year. This isn't phenomenal, but remember that education is also funded by the government through high school, which means that grade school teachers are technically government employees. As I already mentioned, the government takes good care of its employees, and teachers do get higher benefits than the average private worker. Teachers also don't work all throughout the year, but do work on weekends, so it's hard to compare exact hours. Raising pay might also draw in more applicants and keep old teachers from retiring, and teachers should want to do so because they have a passion for it, not just for money. But just because teachers are getting paid doesn't mean schools have the funding or infrastructure for quality education. These are the main difficulties in building new schools in developing nations. Either new schools are built without motivated teachers, or good teachers are put into rundown buildings without access to good learning materials or a safe learning environment. Perhaps educator income could increase slightly, but what I think is more crucial is to create a paradigm shift that glamorizes the benefits of education. We need a cultural revolution, making the market for teachers become more competitive, leading to more rigorous schooling. We all have to work hard eventually, so why not get it out of the way early with a good education?

Another one of the greatest difficulties for students around the world is funding education. In many developed nations, the government funds basic education, usually through high school. Unfortunately, universities in developed countries and some primary schools in developing nations still require parents and students to pay for their education, which cripples humanity's ability to progress. Beyond the ghastly fees, it is just generally difficult for many universities to fund all of their facilities through government subsidies. Interestingly, the National Center for Educational Statistics cites that in the 2013-2014 school year, public universities got about 42% funding from federal, state, and local grants, 20% from tuition, and 25% from "all other revenue", which includes gifts, private grants and contracts, auxiliary enterprises, hospital revenue, services for educational activities, etc. It may seem that the government is already subsidizing a portion of public colleges, and when one also looks at what universities spend their money on, this issue becomes far more complex. Namely, universities are far shadier about the way they spend their money than the government, and on many websites are intentionally vague about the diction they use to describe revenue allocations. This is also reflected in student bills. I got through all of college without ever knowing exactly where the "tuition and fees" portion of my payments were going. This added bureaucracy is one major reason why paying for college is such a pain. The biggest step universities could probably take to reduce student debt would be to simply increase the transparency of where their money is coming and going. Besides tuition, another one of the biggest costs for a student after tuition comes from their books. Paper textbooks should be recycled for new editions. In fact, with the advent of online documents, fewer textbooks may have to be printed at all.

Across the board, and especially for lower grades, school needs to be more fun, and an enthusiastic educator makes all the difference. The best teacher is able to retain some sense of child-like wonder. They must expect greatness from the students, but give them kind support when they struggle. At a good sermon, a preacher will tell you what to think. At a good lecture, a professor will force you to think for yourself. Don't just give students facts, force them to think about them, to defend their ideas, and encourage them to be wrong, not right. People are so scared of being wrong that when they actually make a mistake, they delude themselves into still thinking they are correct. Students should also have the opportunity to challenge themselves at any school, especially through the International Baccalaureate Program. Advanced classes are nice, but what's more important is to teach students that having a global mindset is crucial! That was the key benefit to being in the IB program, we were forced to think about how people around the world would look at different problems, and use those perspectives and empirical evidence to find the best solutions. But even if they weren't the best solutions, we still had to find a way to support any opposing argument, because knowing how to think like people who disagree with you is the best way to find compromise. Taking into account different perspectives is a big reason why in this book I juxtapose my opinions and make factual statements (or statements I think are factual) with conviction. My education taught me that we live in a global community. As such, knowledge has no borders.

It's important to remember that some students legitimately have a harder time learning than others. However, by and large I have found that even the rowdiest and most challenging students can be taught a passion for learning, they just need personalized guidance. In college, I joined a club that went on outreach trips to high schools that didn't have the resources to fund strong science programs. On our first trip, my teammate and I gave a presentation on cosmology. Her and I did demonstrations on how the universe was expanding, and on later trips we got to supervise other students, and during each trip one thing was always abundantly clear: kids are smart, they just need a good reason to care about what they're learning. Even today, grade schoolers already know what an atom is, making it all the more easy to teach them how electrons emit light when they jump between orbitals. Of course adults are smart too, but young people seem to innately have more curiosity, and they certainly have less false information that they need to override.

I think this is a highlight of one of the key ways to help the school system. We need to remember that all people are born with great capacity for knowledge, but that many students of color are handicapped at birth by being born in a society that has privileged people of European decent for the last few centuries. Segregated schools and voting in the United States were only eliminated a half century ago in 1964, some two generations, after numerous marches, protests, and the assassinations of both a young president and a young preacher in the 60's. Our educational system still seems to be recovering from a scar slashed into our history a century prior in 1860; a scar called slavery, which itself goes back to early colonists in 1619. We also can't forget about all the other minority groups that too are looked down upon. Those kids still don't see people like them represented in highly educated positions, and even if they have the drive to succeed, low quality institutions that are continuing to fight to catch up handicap them. Then there is the extreme case of "uncivilized" groups like the Sentinelese. Again, integrating them into the twenty-first century is an open matter of debate. Clearly they are getting along just fine without the global community, but on the other hand they might appreciate modern medicine, fast food, and video games. Regardless, we must remember to never think of anyone as savages who are less than human. Just like everyone else, they are potential engineers and doctors that could contribute to the human endeavor, should they choose to do so.

Education is the number one deterrent of crime. I remember going to work with my mother, and seeing kids in juvenile corrections. They were usually somewhere between fourteen and eighteen, but many only have fourth or fifth grade reading levels. For so many of us, our parents are our heroes. But there are some people in the world that grow up without that guidance, and they may have trouble fitting into society. Instead of love, they might grow up with mentally ill parents addicted to drugs that molest them or chase them with butcher knives. In every moment we should be sympathetic to allow people to rebuild themselves, while at the same time never being complacent with unacceptable behavior that endangers others. Do whatever you want to yourself, but nobody's going to be pleased if you cost them money or just make them unhappy, and the more unhappy you make other people, the less likely they are to let you do what you enjoy. If that joy comes from hurting others, fine. Just don't expect society to let you live the rest of your life as a human being, assuming they aren't kind enough to put you to sleep like a sick puppy.

On the same note, I also think that education is the most important tool for fighting terrorism. It isn't as simple as building quality schools with high quality teachers, although that is probably the most important method. That only works for children, the adults might still try to blow up the schools. Adults must be persuaded to change their views, a much more difficult task than teaching young people. For adults to change their minds, they need to be convinced from within, by their peers and constituents. This too is no easy task, and implementing it in less time than it would take for the adult terrorists to simply grow old and die is the current dilemma facing western politicians today. Hypothetically, abduction could provide an interesting way to fight terrorism as well. We could send them into space to see the Earth and show them just how insignificant and misguided their struggle is. Or we could just leave them on the moon with no way to get back until they promise to behave. Again, that is a matter for the politicians of the future.

Giving people safe havens that promote peaceful atmospheres can also fight violence. Out of all the horrible things you've seen or heard of, how many of them took place in a library? I can't think of one, except when they get burned down by Nazis. But even the Nazis didn't kill people inside of libraries. They did that at camp. Crimes don't happen in libraries because the only people who go to libraries are curious intellectuals. Smart, peaceful people. A library is the only place where the cultural and sociopolitical law is to not be noisy. Being loud is simply not acceptable. So people just stop talking and act like respectful adults. Of course, these are also the reasons why libraries are so mind numbingly droll and soul crushingly boring. Maybe if they added some thoughtful music, it could be a more favorable model for a global society. Although these days people usually have headphones so that they can bring their own sound without bothering anybody. Many school libraries also have glass rooms there people can talk freely. But the point still stands; if you want to be treated by a human being, find a building where the walls are paved with knowledge and wisdom, from magazines to epic novels.

In addition to physical safety, students also need data security and Internet safety, at their parent's discretion and with minor regulation. Cyber security and personal records need to be protected both by the citizens and those who need personal documents. Developments in quantum computers with constantly changing passwords will assist with this as well. Regardless, everybody should have basic cyber security training. This extends to the population as a whole, but governments also need to make sure they provide security to their people. You never know who might be able to hack into your devices from around the world, and so everyone needs to do their part to keep their personal information secure, although in an ideal world nobody would try to rob people of their personal information.

We must strive to do this for people all around the world, because as much as we may want to focus on citizens within our fake borders, we should take steps to see all humans as being citizens of the Earth. People in developed countries have access to more knowledge about how the world works thanks to centuries of dehumanization and extortion from other countries. Now people have an opportunity to use that education to develop new innovative ways for people to prosper and contribute to humanity, and give back to the people who's countries were left in the dark. There are so many things in this world to know, and so many people that can help us learn. Don't be stingy with your knowledge; say what you think you know so that you can be proven wrong. More importantly, listen to what other people think they know, so you can prove them wrong. When neither of you is able to prove the other wrong, you can help each other reaffirm and modify how you understand and live your lives. Most importantly, always converse with your mind unclouded with hate, fear, or malice. It's as simple as that.

It's important to also remember that progress is often a gradual process. In developed countries, we can take steps to accomplish this by continuing to increase diversity, not just in students, but also in educators. I believe that my peers and I had a great privilege to receive some of the best education available to humanity at the time. But when I think about it, I have taken no more than two courses taught by a person of color. One was my fifth grade teacher, and I distinctly remember her as one of the kindest and wisest teachers I've had, bless her soul. In addition, I have had increasingly less female instructors as I entered more into the scientific field. To be fair, people who hit balls or walk on a runway for a living get paid more than most instructors, but the overall lack of diversity is quite uninspired. Luckily, my peers where far more diverse, at least in who went to school with me, less so my actual classes. But I think they were all pretty good people. Even more, everyone should be given access to the Internet and all of its educational resources, where they can hear non-European, non-rich, non-male perspectives. In 1775 Patrick Henry asked for liberty or death. Well I think that education is liberating, so give my people quantum chromo-dynamics!

We get our power from our fellow humans. Great men don't make history, history is made by the people that gave them money, power, and the position to take blame for everything they did. People gain wealth when they produce a good or service people want. The problem with this is that people can leave behind their wealth to their children, who will subsequently be born into affluence without having done anything to earn it, thus giving them less motivation to work hard and make a name for themselves like their parents did. If the children of wealthy people do any work it's often just to carry on the family name, although others do try to distinguish themselves out of their parent's looming shadow. Perhaps one day people will relinquish their wealth back to the masses once they die, only giving enough to their children for them to eat, receive a quality education, and get out of the nest prepared to better the world. Many successful people have already pledged to do this after their passing, notably including Bill Gates and Warren Buffett. This could be considered a sort of compromise between capitalism, which encourages people to compete and rewards people who work hard but may also lead to greedy hoarding, and between socialism, which encourages people to work to help the community and a cause larger than themselves but also marginalizes people who really do deserve to make more money than others. After all, any wise person knows that money doesn't lead to happiness, but you do need money if you want to live. Like all things it's a delicate balance, between bettering ourselves and helping each other. Both pieces are crucial, and should not be undermined.

For all of its importance, it's strange how expensive and time-consuming college, and school in general, is. Clearly, it benefits every single living organism on the planet, but its effect takes a long time to return. Humans just take so much time to develop from their tiny larval state. This may perhaps be why it often doesn't get the attention it deserves. But make no doubt that people are getting smarter all the time. Community colleges allow for people to get degrees quickly and start making money and running society, and long-term graduate schools can allow people to tackle big long-term problems. Even doctorates are on the rise, with the degree being a bare minimum requirement for some positions. And with the Internet it's easier than ever for people to communicate information, so that we can all get smarter. Ultimately, education is just as selfish as it is altruistic. Don't want cancer? Help big universities to develop in Malawi, so that they can build research facilities for doctors. Love your air conditioning too much to deal with global warming? Get clean drinking water to Chad, Ethiopia, Afghanistan, India, and Haiti, so they can develop cheaper, cleaner energy. Give poor youths quantum mechanics and supercomputers so they can figure out how to harness the weather and the ocean tides. Put a planetarium in impoverished neighborhoods; inspire kids to make a meaningful impact on this enormous universe. If we work together as a planet, we could harness the Sun's energy, colonize The Moon, make 3D printing cheap, or even make fast food nutritious! Anything is possible when more people are tackling more problems to help everyone live better more prosperous lives. If people are sufficiently curious, they won't want to fight each other, and jeopardize their health. If you want liberty, then you want at least somebody who can unify general relativity and quantum mechanics and explain it simply to other people. They can make stable economies, and keep them afloat. Educated people who can communicate well could also share their culture, giving people broader perspective and even more knowledge.

So here's the deal. The only times I have ever been in a conflict or have seen people get into conflict, without a doubt each and every time it was because of a misunderstanding of the truth. When people don't understand why something is happening, they get scared, and they get aggressive. People act with stupidity only when they become frightened of not knowing something, when there are high degrees of uncertainty. And by far the safest places in the world are the places where there is open communication or high intelligence. Schools and universities are the only places where I could walk with confidence that everyone present was by far more adult and human than those who no longer have curiosity and wonder for the world. That is why people need to be educated. If conflict is born through dehumanization and stupidity, then the only way to rehumanize us is by understanding the natural history of all things through critical thinking. People, places, and the laws of nature that make them all go, every one of us must strive to know and understand as much as they can, for the sake of themselves, their ancestors, and their descendants. Everyday is a new opportunity to flex your brain and grow your mind, the most powerful and complex thing in the entirety of cosmic space and time. Put a good school everywhere on the planet where anybody can attend, and peace will come to all.

In short, the only people who disserve discrimination are those that purposely refuse to educate themselves or who deny others the opportunity for education. You gain greater humanity by living longer, experiencing more, meeting new people, and constantly redefining your existence. You must constantly acquire knowledge to help you sculpt your unique place in the cosmos. I often reflect on why school was so boring as a child. Everyone tells you that a good education is the key to your future, and that better education leads to greater prosperity. But no one ever tells you why. The reason is simple, being educated is why humans are the most successful animal to walk the face of the moon. Without the context given to us by our ancestors and the world we live in, we can aspire to be nothing more than children, no more in tune with the universe than rocks.

Part IV: What In the Spacetime is Going On Here?

Epilogue | Welcome Home

"Learn from yesterday, live for today, hope for tomorrow, rest this afternoon"
— Charles M. Schulz, creator of Peanuts

"One's life has value so long as one attributes value to the life of others, by means of love, friendship, indignation and compassion."
— Simone de Beauvoir

Carpe Diem! Seize the day; gather your rosebuds while you may. You've done it, congratulations! Buy yourself some chocolate truffles. It's quite an accomplishment to have read this far. As a matter of fact, I don't think I've even fully read more than a handful of the books I was examined on throughout school. I commend you, who has more dedication than I. But even though I don't like to read myself, I still think I turned out okay. So what have we learned?

We have learned that all of us are family. It is unequivocally written into the cosmic tablet, which is made of non-empty space, mysterious lumps of spacetime, and ionized hydrogen. However, it is equally impossible to object to the simultaneous fact that we are also unique, each one of us snow flakes folded by years of life history and quantum fluctuations, from birth to death ever more wise. Basically, we're all just the same person living under different circumstances. There are far more human beings on the Earth than any other point in our history. Every one of them wants to have a full stomach and a safe home with a bed to sleep in. Everyone has different needs and wants, and as more and more people gain voice through the Internet, it's going to be ever more difficult for people to sift through all the noise. That's why it's in everyone's best interest to take as much time to educate themselves and listen to each other, and spend less time trying to simply force others to think a certain way. Never think in absolutes, because you will almost certainly find a way to contradict yourself. Nothing is black and white, everything is circumstantial, and in general, generalizations are a bad idea. Lastly, one should never be afraid of learning something new, for in the words of Bruce Lee, "All types of knowledge ultimately mean self knowledge". That self-knowledge is the key to success. I can't say that I can quantify success itself, either for myself or for others. But what I do believe is that people are far more likely to succeed when other people have faith in them. When people count on you, you have an innate need to not disappoint them. That's why we need to put our faith in each other, and realize when we are already dependent on one another. Youth need to be trusted by their elders to learn about the world and solve future problems just as youth are obligated to learn from the wisdom of past generations. We need to recognize all the people and organisms that make our lives possible, from custodians to executives, spanning to every creature in our global ecosystem. Our well-being as life forms is directly tied to how well we serve and rely on each other. There's no need to be full of oneself. Other people will tell you when you are number one at something, and the more people agree, the more likely that it's true. Never assume you're the best, just do your best. Only then you will be satisfied.

What humanity needs is a cultural revolution. Kind of like in Mao's Communist China, except with less hate, less death, and less vandalism, and more kindness, more understanding, and more empathy. Humans need to understand the full scope of the cosmic stage they are acting in, and not be preoccupied with borders, cultures, or genetics, all of which are subject to arbitrary change. When it comes to life on this planet, there are no borders, period. If we master solar power, we can cover every household on the planet, which will be enough juice to power the air conditioning we're all going to want in the mid twenty-first century. If we master nuclear fusion, we will be able to transform the planet and forever eliminate the fear of losing energy. If we master our genes, we will eliminate the need for needless suffering and death, so that we can instead focus all our energy on wonder and discovery. Finally, if we are able to master our consciousness, we will alleviate the soul crushing weight of an infinitely apathetic universe.

We've learned that an enemy is one whose story you have not yet heard. Imagine your worst enemy. Imagine the person in your life who has wronged you the most, perhaps to the point that you would consider actually harming them. Now consider where you live, on an insignificant pebble with a paper-thin atmosphere that harbors all known life in the universe, made up of less than one thousandth of one percent of everything in the universe. For hundreds of parsecs in every direction, the universe is ready to assimilate you back into the uniform spread of raw energy from whence we came. Every moment of every day, natural forces are cycling through the cosmos that would turn our entire planet into nothing but atoms, and everyday we all have to wake up and push back against entropy to make tomorrow possible. People, and all organisms, have been doing this for billions of years. Anything catastrophic that has ever happened in your life has undoubtedly happened to someone else, quite possibly in a worse way. But after all you've endured, here you are, reading this sentence and pondering the great mysteries of life. Nobody exists on purpose, no one belongs anywhere, everyone's going to die. Now image your worst enemy again, and justify to me why it's even worth your time to ever hate or dislike anyone ever again.

It really shouldn't be too difficult for people to get along. As a general rule, people are very good at maintaining peace, so long as they can speak to one another. World peace just has two requirements: realize that everybody is as complex and nuanced as yourself, and be excellent to one another. That isn't to say that we should lazily assume everyone has our best intentions in mind, sometimes it's simply more beneficial to focus on oneself. But by that same token, we also shouldn't assume that other people want to harm us either. Truth be told, we don't really seem to care about each other all that much in the first place. You care most about yourself just as you are your greatest critic. Just keep making yourself better, and always help other people when they want to improve themselves, which they usually do, because they're just like you. Remember to have *Sola fide*, faith in life and people. Things like guns and bombs, they are archaic, obsolete. Nothing but childish remnants of a human race that was too young and full of hubris to see the universality and rarity of life. Only children use these things, because only children seek immediate solutions to long-term problems. After all it's much easier to just get out of a situation using entropy than putting in the energy to repair negative situations. People can still defend themselves by forging their bodies through athletics and self-defense training. Furthermore, with advancing technology, large-scale defense can easily replace bullets with non-lethal methods that are still completely debilitating. Until death is no longer a certainty, there is no sensible reason for people to intentionally allow each other's already short lives to be cut shorter by fear and mistrust. But people will never trust each other so long as they believe that they are greater or lesser than one another. Peace cannot exist when people feel subjugated, and the people who oppress them remain unaware or uninterested in their crimes. We must remember that everyone feels like they are under attack from something, be it a faceless institution, or a person seen in passing who's story we've never heard. However we shouldn't need to know each other's individual history, because all of us share the same story. Every visible thing in the universe, from stars to stones to sapiens, they are all nothing but unique combinations of the same ingredients. We need to trust each other, and just as with any healthy relationship, we should spend every moment of every day earning each other's trust, and work to counter social entropy. After all, the one thing all creatures that hold the spark of life have in common is that we don't want to die. Inherently we all just want to eat, procreate, feed our families, and maybe find a purpose worth living for. There is not a single living thing or person in the last four billion years that doesn't have this basic need to survive encoded into them at the cellular level. This is why people are so selfish. Nobody cares about anything you say or do unless it directly harms or benefits them in some way. That being said, people often worry about trivial things and people that have no effect on the way they live. For that reason it is crucial that people are able to look past their petty walls, and be able to communicate with one another, so as to avoid miscommunication that can lead to deceit and distrust.

In order to speak, you must first listen. In order to find your voice, you must first hear what the world and its inhabitants are telling you. You and I inhabit a universe vast beyond imagination comprised of cogs unfathomably minute, and we share it with life forms broad and diverse, and we are all in the same boat. Black, brown or white, man or woman, both or neither, infant, child or adult, human, plant, or single cell, we're all just trying to get by. So let's cut each other some slack every once in a while. The universe is chaotic enough without trivial conflict. It is a world both terrifying and beautiful, and the only way to get the most out of it is to move forward fearlessly.

The easiest way to treat people well is to realize that we all likely experience the world as being worse than it is. We can grow closer by speaking the pain in our hearts, and sharing the burdens we carry. Or, if you don't like talking to people, simply remember that all of us have the capacity for sorrow. In knowing that pain is inherent to experiencing life, it becomes much more difficult to dislike other people. It's time for our species to grow up, stop bickering, and quit needlessly wasting resources on unproductive boredom. We need to work with each other and all forms of life, get our rears in gear, and start utilizing the universe to protect life for as long as possible. Things could be worse, things could be better, so don't make such a big deal about everything. Go out into the world and enjoy it, and help other people enjoy it with you. You only have a few decades to catch up on your billion year-old story. Now, armed with a richer understanding for our strange, beautiful world, why don't we take a crack at answering those questions from the beginning of our conversation?

What the hell is going on? Who knows! Certainty not me, just do your best, try not to think about it too much, and be happy you're thinking at all.
Why is the sky blue? It's known as Rayleigh Scattering; the molecules in the air absorb bluer light and disperse it in every direction. The more atmosphere you look through, the more low energy light is scattered, which is why Sunsets, Sun rises, and lunar eclipses are red.
Why are you hungry? In short, you survive off of the death of other living things, which also ate each other, eventually leading back to plants, which have the wizard-like power to turn photon energy into sugar.
Why did a lion eat one of your siblings and why did one of my parents fall asleep forever after getting cut by a tiny rock? Lions have kids to feed too, and it sounds like your parents didn't have access to modern medicine. You have my sincere condolences.
Why is there evil? Poorly worded question. Things just are, and they are only good or bad relative to all of the other octillions upon octillions of things that happen to be occurring at the time. Also some people just want to watch the world burn (i.e. give in to entropy).
Why do stars shine and thunderstorms roar? Nuclear fusion for stars, and acoustic vibrations in the air sparked by static electricity inside of clouds for thunder.
Why do we love? Well, as of yet, organisms haven't evolved a better way to survive and pass on their genes.
Where did we come from? We came from our parents, from their food, from the Earth, from stars, from energy, which may or may not have always existed.
Who put bop in the bop, shoo bop, shoo bop? Ask Barry Mann.
Is the cake a lie? Yes, so you need to bake your own. Teleport to the grocery store and get some flour immediately.
Do we even exist, and if so how can we be sure? If thinking was good enough for Descartes, then it's good enough for you. Also, any funeral service will tell you that somebody still needs to deal with all of your junk when you die, so presumably everything else would exist even if you weren't around to think about it. The party goes on without you, as Christopher Hitchens would say.

So what's the meaning of life? Well, ultimately that's something that each individual has to decide based on his or her own experiences. My answer: be full, be warm, be safe, and remember that everyone you've ever met or heard of is just trying to do the same. Learn and experience as much as possible before your time comes and you assimilate back into the cosmos once again. You are the universe looking back at itself. So why in the Milky Way is it taking us so long to work together? I'll leave you to ponder that next time you meet somebody you dislike. I have greatly appreciated the time you've given our conversation, but I think it's time we both got back to work. There is always something that needs to be done.

So long, adieu.

References

The Internet! Also every person I've ever met, either directly or vicariously. I understand that this may not be a very satisfying answer for some. My generation is among the first in all of human history to have access to the sum of human knowledge at their fingertips. As such, if I were to write down every single one of the websites, articles, videos, or conversations that have allowed me to learn about all these things, I would probably have to write another book. Ain't nobody got time for that!

The Internet is sort of like a democratic socialist government in that people pay for it and benefit from services all the other people who use it provide. So consider this a book written in part by the Internet, by the humans that pay for it, by the humans who use it for free, and most of all by the humans who invented it, written for the Earthlings that don't use it. I tried to do my best to make references as directly as possible, and if you do have an Internet connection, I sincerely hope you will do your own research to verify any factual claims I have made that may sound, for lack of a better term, out of this world. As I said in the preface, the guiding light of philosophy is a desire to know the truth. If you can find evidence and provide a coherent argument against a claim I have made, that's great! Being wrong isn't a bad thing, and if you can justify your believes without authority, then that will only bring you closer to the truth. Just try not to act all snobby if I made a mistake, I'm only a Homo Sapien.

But I completely understand if you find it suspicious that a person could have knowledge about all of existence and not cite their sources. So in order to appease my more discerning viewers, allow me to list many of the YouTube Channels that I used as resources, all of which are free to the public, but can also be supported through subscription, and in some cases, donations that can help spread education throughout the world. I will also include the primary textbooks I used whilst studying astrophysics at The University of Colorado in Boulder.

YouTube Channels

AnimalBytesTV, a channel dedicated to educating people about the wonders of nature and inspiring a love of all creatures that live on our planet, reminiscent of the wonderful work of Steve Erwin.

AsapSCIENCE, which makes videos about fun and quirky science using imaginative marker drawings, and applies scientific concepts to everyday anomalies.

BrainCraft, a channel that delves into the complex world of psychology and explores the enigma of how our brains interact with the larger world.

BuzzFeedVideo, which provides social commentary and insight into the human condition from all walks of life and varied socioeconomic backgrounds, with a fun and entertaining outlook. One of the few outlets that covers random fun and deep social issues at once.

Fermilab, my source for all things particle physics. A fun and well-choreographed look into the world of particle accelerators and the modern wonders and insight science has wrought.

It's Okay To Be Smart, a channel that openly demonstrates the benefits of a nerdy lifestyle, encouraging all of its viewers to embrace the wonders of the natural world and stay curious.

Kurzgesagt – In a Nutshell, one of the best animated channels on the Internet, made by individuals who love exploring nature and the deep existential questions of human behavior. Also, the narrator has an exquisite accent.

MinutePhysics, the go-to source for all things physics, presenting in a plain yet elegant fashion the terrifyingly complex world of the physicist, demonstrating the power of stick figures and how a graduate student can use the faculty around them as resources to share complex topics in a comprehensive manner.

NASA, the channel run by the most criminally underfunded branch of the American government, and the best source for space pictures.

Neuro Transmissions, an up-and-coming channel that will enrich anyone's neural synapses, dedicated to explaining concepts in neurology, psychology, and all the intricacies of that bundle of nerves between our ears in an educational setting.

PBS Space Time, a channel that tackles the complexities of relativity, quantum mechanics, and the daunting challenges of space exploration. By far one of the most advanced physics channels made directly for the hardcore physics enthusiast.

Physics Girl, a channel loaded with merriment that exemplifies a passion for science that should be shared with not only youth, but also those who need proof that science indeed works. It provides this by featuring homemade science experiments and explanations of complex physics that are fun for the whole family!

SciShow, one of the premiere channels for updates in scientific research, science history, and scientific insight into everyday questions about modern life. Along with its derivative *SciShow Space*, this league of science has built one of the most important tools for the public to understand the natural world.

TED-Ed, the channel for all types of information about everything from literature, to science, mathematics, history, sociology, and everything in between.

AnimalWonders Montana, a personal introduction to an eclectic cast of nature's creatures and a beautiful insight into the personalities of the animals that we share our home with.

sexplanations, an open minded, safe, professional, and most importantly educational insight into all the complexities of the human animal's sexuality, and a channel that demolishes the age-old stigmas that have made our natural instincts taboo, instead making them casual and smart.

Wisecrack, which provides intellectual and humorous satire about a variety of topics, with programs like Thug Notes that provide literary analysis with a street vernacular, 8-Bit Philosophy about all the metaphysical questions faced by humanity through the medium of old video games, and Earthling Cinema which provides film analysis through the lens of advanced alien civilizations.

thebrainscoop, an exposé of all the strange things inside of The Field Museum in Chicago, proving that animal corpses are not only fascinating, but in fact crucial for our understanding of biology and ecology.

Finally, I'd like to express a special gratitude to the *CrashCourse* channel, which was by far the most useful reference tool for researching the content within this book. Not only does the content match beautifully with the material I already learned in high school and college, it also presents it in a way that makes it bar none the most professional and entertaining educational channel on the Internet. You can learn about a variety of topics on this channel with their series on Games, Physics, Chemistry, Biology, Economics, Government, Astronomy, World History, U.S. History, Philosophy, Anatomy and Physiology, Literature, Psychology, and Ecology. These videos are suited for audiences of all ages, and if you have a student who is just beginning their academic journey, they also have a show for the youth called *Crash Course Kids*.

Textbooks

Carroll, Bradley W., and Dale A. Ostlie. An Introduction to Modern Astrophysics. San Francisco: Pearson Addison-Wesley, 2007. Print.

Griffiths, David J. Introduction to Electrodynamics. Upper Saddle River, NJ: Prentice Hall, 1999. Print.

McIntyre, David H., Corinne A. Manogue, and Janet Tate. Quantum Mechanics: A Paradigms Approach. Boston: Pearson, 2012. Print.

Taylor, John R. Classical Mechanics. Sausalito, CA: U Science, 2005. Print.

Stewart, James. Essential Calculus. Belmont, CA: Brooks/Cole, 2013. Print.

Hartle, J. B. Gravity: An Introduction to Einstein's General Relativity. San Francisco: Addison-Wesley, 2003. Print.

Ryden, Barbara Sue. Introduction to Cosmology. San Francisco: Addison-Wesley, 2003. Print.

Made in the USA
Middletown, DE
16 June 2019